AF361742

Water Management for Climate Smart Agriculture

Water Management for Climate Smart Agriculture

Editors

Michel Riksen
Coen Ritsema
Karrar Mahdi

Basel • Beijing • Wuhan • Barcelona • Belgrade • Novi Sad • Cluj • Manchester

Editors

Michel Riksen
Environmental Science Group
Wageningen University &
Research
Wageningen
Netherlands

Coen Ritsema
Environmental Science Group
Wageningen University &
Research
Wageningen
Netherlands

Karrar Mahdi
Environmental Science Group
Wageningen University &
Research
Wageningen
Netherlands

Editorial Office
MDPI
St. Alban-Anlage 66
4052 Basel, Switzerland

This is a reprint of articles from the Special Issue published online in the open access journal *Water* (ISSN 2073-4441) (available at: www.mdpi.com/journal/water/special_issues/Climate_Smart_Agriculture).

For citation purposes, cite each article independently as indicated on the article page online and as indicated below:

Lastname, A.A.; Lastname, B.B. Article Title. *Journal Name* **Year**, *Volume Number*, Page Range.

ISBN 978-3-7258-0620-1 (Hbk)
ISBN 978-3-7258-0619-5 (PDF)
doi.org/10.3390/books978-3-7258-0619-5

Cover image courtesy of Karrar Mahdi

Contents

About the Editors

Michel Riksen

Michel Riksen has worked in the Soil Physics and Land Management (SLM) group at Wageningen University since 1998. In 1990, he obtained his BSc in Tropical Agriculture with a specialization in Tropical Soil and Water Management at Larenstein Deventer, Netherlands. In 1994, he finished his MSc at Wageningen University in Tropical Land Management, with a specialization in Erosion and Soil and Water Conservation. In 1998, he began his first research position in an EU-funded wind erosion project. During this time, he researched the socioeconomic and policy aspects of wind erosion in the northwest European region. In 2001, he received his PhD fellowship from the Cornelis Lely Foundation for a research project titled: "Making use of Wind and Water Erosion in Landscape Development". For this project, he studied the role of soil erosion processes in nature development. He obtained his PhD in 2006 and continued to work on inland drift sands in the Netherlands as a postdoc researcher. In 2009, he became an Assistant Professor in the SLM group, teaching the planning and design of soil and water conservation measures and land management strategies. His main research activities are in the fields of (1) wind erosion and (coastal) dune formation in the Netherlands and (2) soil and water conservation/land management. His publication list covers a wide range of subjects, including socioeconomic aspects of erosion, conservation policies, soil conservation, geomorphology in landscape ecology, and water harvesting.

Coen Ritsema

Coen Ritsema is a Professor at Wageningen University and Research and Chair of the Soil Physics and Land Management group. He has over 30 years of experience in fundamental and applied research in soil and water-related sciences and coordinating large (inter)national multidisciplinary research projects and programs. During the past decade, he coordinated numerous EU-funded projects involving partners from around the world. His interest has been focused on land–hydrology interactions at different spatial and temporal scales, with special attention to soil physical and chemical processes, and soil degradation and conservation issues. He has published about 150 research papers in many international scientific journals and has been the (co-)editor of eight Special Issues. Furthermore, he has been granted honorary/guest professorships at Deakin University, Australia (2004); the Chinese Academy of Sciences at the Institute of Soil and Water Conservation, Yangling, China (2008); and Moscow State University of Environmental Engineering, Russia (2009). He received the Distinguished Researcher Award of the World Association of Soil and Water Conservation in 2013. Additionally, he is a member of the scoping group Land Degradation and Restoration of the Intergovernmental Science-Policy Platform on Biodiversity and Ecosystem Services (IPBES) (UNEP-UNESCO-FAO-UNDP) and is involved in several other commissions inside and outside the Netherlands.

Karrar Mahdi

Dr. Karrar Mahdi has over 15 years of experience as a researcher, teacher, project coordinator, and engineer. In 2004, he completed his BSc in Civil Engineering, and in 2009, he completed his MSc in Environmental Engineering. In 2018, he obtained his PhD from Wageningen University in the Soil Physics and Land Management group (SLM). Since 2018, he has been working at the Soil Physics and Land Management group in the Environmental Science Department at Wageningen University & Research in the Netherlands. His primary research and teaching interests include soil and water processes, pollution, rainwater harvesting, drought, and climate-smart agriculture.

Preface

Global climate change has resulted in significant changes in local weather conditions in many countries. Changes in rainfall and temperature threaten agricultural production and increase the vulnerability of individuals who are dependent on agriculture. In particular, water shortages in arid and semi-arid regions will become more prominent. In such regions, irregularly distributed rainfall will result in increased droughts as well as extreme rainfall events. Climate-smart agriculture (CSA) is an approach that aims to transform and reorient agricultural systems to adapt to the effects of climate change.

This Special Issue, titled "Water Management for Climate-Smart Agriculture," was initiated as part of the institutional collaboration project (OKP-IRA-104278) titled "Efficient Water Management in Iraq By Switching to Climate-Smart Agriculture: Capacity Building and Knowledge Development" between Wageningen University in the Netherlands and the following six universities in Iraq: the University of Anbar, the University of Basra, the University of Kerbala, the University of Kufa, the University of Mosul, and Salahaddin University-Erbil. The project was led by the Soil Physics and Land Management group (SLM) at Wageningen University & Research, funded by the Dutch Ministry of Foreign Affairs, and co-funded by the aforementioned six Iraqi universities.

The goal of this Special Issue was to gather the latest research, resulting from the mentioned project as well as from other researchers, on topics related to climate-smart agricultural water management to transfer knowledge and guide and support related stakeholders, researchers, and students in achieving more sustainable and resilient agricultural systems, particularly in Iraq and other arid/semi-arid regions. This Special Issue covers a wide range of topics, including rainwater harvesting, drought, efficient farming practices, irrigation, and groundwater management and recharge. As such, the knowledge presented in this Special Issue is relevant for other regions around the world that are also coping with similar conditions of climate change and increasing water shortages. Therefore, we trust that the knowledge provided in this Special Issue will eventually contribute to the widespread and targeted adoption and implementation of climate-smart agricultural water management practices where needed and where achievable.

We express our sincere appreciation to all the authors who made valuable contributions to this Special Issue. Without their input, this Special Issue would never have been released. We also express our gratitude to the Nuffic 'Orange Knowledge Programme' (OKP) for managing and funding the OKP-IRA-104278 project.

Michel Riksen, Coen Ritsema, and Karrar Mahdi
Editors

Article

Water Quality, Availability, and Uses in Rural Communities in the Kurdistan Region, Iraq

Shwan Seeyan [1,*], Ammar Adham [2], Karrar Mahdi [3] and Coen Ritsema [3]

1   Soil and Water Department, Agricultural Engineering Science College, Salahaddin University, Erbil 44002, Iraq
2   Dams and Water Resources Engineering Department, Engineering College, University of Anbar, Ramadi 31001, Iraq; engammar2000@uoanbar.edu.iq
3   Soil Physics and Land Management Group, Wageningen University & Research, 6700 AA Wageningen, The Netherlands; Karrar.mahdi@wur.nl (K.M.); coen.ritsema@wur.nl (C.R.)
*   Correspondence: shwan.seeyan@su.edu.krd; Tel.: +964-750-4487884

Citation: Seeyan, S.; Adham, A.; Mahdi, K.; Ritsema, C. Water Quality, Availability, and Uses in Rural Communities in the Kurdistan Region, Iraq. *Water* **2021**, *13*, 2927. https://doi.org/10.3390/w13202927

Academic Editor: Antonio Lo Porto

Received: 8 September 2021
Accepted: 17 October 2021
Published: 18 October 2021

Abstract: Water resource management and the investigation of the quality and quantity of groundwater and surface water is important in the Kurdistan Region of Iraq. The growing population, as well as agricultural and industrial projects, consume huge amounts of water, especially groundwater. A total of 572 ground and surface water samples were collected for physicochemical analysis to determine the availability and quality of the water in the Kurdistan region. The physicochemical parameters such as pH, electrical conductivity, and total dissolved solids were analyzed to evaluate the suitability of the water for different purposes like livestock, irrigation, and agriculture. GIS-based multi-criteria decision analysis (MCDA) was used to determine the suitability map of water for irrigation purposes. Most of the groundwater samples were suitable for irrigation except for some samples from Erbil City, especially those taken in the Makhmur district, and samples from some small areas in the cities of Sulaymania and Duhok. All groundwater samples were acceptable for all types of agricultural crops, except for 15 well samples that were determined not to be usable for fruit crops. However, this water was acceptable for livestock and poultry. Most of the water wells provided freshwater except for 36 deep wells, which supplied slightly brackish to brackish water. Water samples were found to have low to medium salinity levels except for 26 well samples and one spring sample that had high salinity levels, and 2 well samples with very high salinity levels. Most of the samples had an excellent to good water classification except for 85 samples classified as permissible, 8 classified as doubtful, and 4 classified as unsuitable for irrigation according to the Todd classification. According to the Rhoades classification, all water samples were non-saline to slightly saline except for 11 samples that were moderately saline.

Keywords: water quality; suitability map; water classification and uses; Kurdistan Region

1. Introduction

Water is considered the most important resource to consider when trying to achieve sustainable agricultural development worldwide. Improving the management of water supplies and concentrating on reducing water consumption are both necessary in order to establish sustainable and efficient agricultural systems, especially more efficient irrigation systems. Agricultural activities must concentrate on both the quantity and quality of water to prevent water contamination, unsustainable usage, land loss, and desertification.

The Mediterranean region is one of the most sensitive areas in the world, with significant decreases in rainfall and increases in temperature expected in the future [1,2]. Climate conditions greatly affect crop production. Due to water shortages, especially during the dry season, surface water and groundwater are used more frequently to increase crop production. Improvements in water management are necessary to increase and diversify food production in order to meet the needs of a growing population while simultaneously reducing crop vulnerability to droughts, floods, and climate change.

To ensure food security and sustainable water management for agriculture, more crops need to be irrigated drop by drop to ensure improvements in water use without negative impacts on the quantity and quality of downstream water supplies. Given the present and future food demands, increasing water scarcity will put a range of stresses on agricultural productivity and exacerbate sustainability problems [3]. Water demands in the urban, industrial, and commercial sectors increasingly exceed acceptable water supply limits, resulting in local depletion of surface and groundwater resources. An accurate assessment of water demand and supply outside the agricultural sector is a prerequisite to effective water management [4]. The study of chemical characteristics of groundwater is very important for municipal, commercial, industrial, agriculture, and drinking water supplies. Development can contribute to the pollution of groundwater, and consideration must be given to the protection of water quality. Physicochemical parameters were analyzed in this study, including the temperature of water well samples (T °C), salinity in terms of total dissolved solids (TDS), electrical conductivity (EC), and reactivity in terms of (pH). One of the most conservative properties of groundwater is temperature. It is a standard physical characteristic that is important in the concentration of the chemical properties of water. Temperature is an important factor for geochemical reactions and organism life [5]. TDS is defined by the content of all dissolved solids in water, ionized or non-ionized, but does not include colloidal materials, suspended sediment, and dissolved gasses [6]. EC, or the conductance of groundwater, is a function of temperature, the type of ions present, and the concentration of various ions' specific conductance. Readings are usually adjusted to 25 °C so that variations in conductance are a function only of the concentration and type of dissolved constituents present [7]. pH is the negative logarithm of hydrogen ion activity, and its value expresses the intensity of activity or alkalinity of water under normal temperature (T °C) and pressure conditions [8]. Water quality is typically calculated by comparing the measurements of physicochemical parameters to standard measurements, which provides an estimation of possible pollutants without providing any precise data on the quality of groundwater [9,10]. Improper management of groundwater resources results not only in a scarcity of water, but also in a change in water quality [11]. The multi-criteria decision approach (MCDA) is a decision-making technique that integrates qualitative and quantitative data by decomposing problems into systematic orders based on a set of criteria [12]. Irrigation of cropland has become a widely used practice and has greatly increased the productivity of farmland. Irrigation has made it possible to farm in regions that otherwise would not be farmable. There were several objectives of this research: to give some first information of the groundwater quality; to assess the quality of water for different purposes; to classify groundwater and irrigation water using different methods; to determine water needs for drinking, livestock, irrigation, and agriculture; and to create a suitability map for irrigation using MCDA for the Kurdistan Region in northern Iraq.

2. Materials and Methods

2.1. Study Area Description

The Kurdistan Region is located in northern Iraq. It covers an area of 40,643 km^2 and has a population of about 5.1 million people. Kurdistan is bordered by Turkey in the north, the Republic of Iran in the east, the Mosul province in the west, and the Kirkuk province in the south (Figure 1). The area includes two main rivers: the Greater Zab and the Lesser Zab, which flows from the Tigris river.

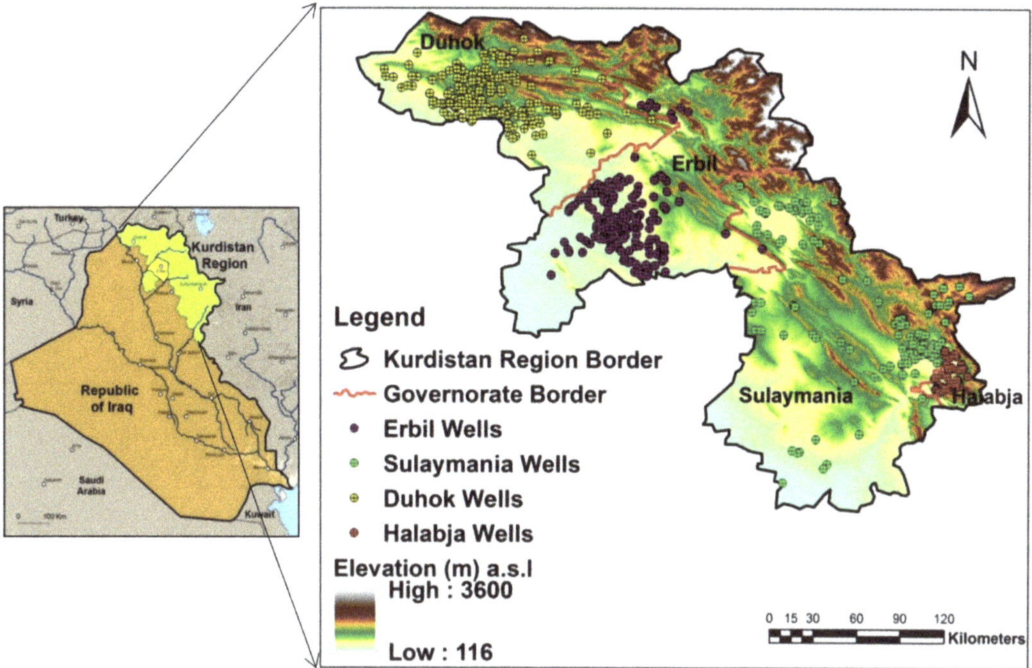

Figure 1. Map showing the origin of the groundwater samples taken across the Kurdistan Region.

This area is known for its semi-arid Mediterranean-type climate. Most places in this region experience cold, rainy winters and long, hot, dry summers. Meteorological data were collected from different meteorological ground stations in the Kurdistan region between 2005 and 2019.

Meteorological data obtained from different meteorological stations in the cities of Erbil, Sulaymania, Duhok, and Halabja show that the annual precipitation was about 400.3 mm in Erbil, 685.3 mm in Sulaymania, 569 mm in Duhok, and 497.7 mm in Halabja. The maximum and minimum mean monthly relative humidity were 69.9% in January and 27.7% in July in Erbil, 70.3% in January and 24.01% in August in Sulaymania, 67.4% in January and 26.7% in August in Duhok, and 67.7% in January and 13.7% in July in Halabja. The maximum monthly temperature in Erbil was 35.3 °C in July, and the minimum monthly temperature was 8.9 °C in January. In Sulaymania, the maximum monthly temperature was 33.7 °C in July, and the minimum was 6.8 °C in January. In Duhok, the maximum monthly temperature was 33.1 °C in July, and the minimum was 7.7 °C in January. In Halabja, the maximum monthly temperature was 35.2 °C in July, and the minimum was 6.2 °C in January (Figure 2). In terms of evaporation, in Erbil City, the maximum mean monthly evaporation was 13.4 mm in July, and the minimum monthly evaporation was 1.8 mm in January. In Sulaymania City, the maximum monthly mean evaporation was 11.8 mm in July, and the minimum was 2.3 mm in January. For Duhok City, the maximum was 11.1 mm in July, and the minimum was 1.4 mm in December. In Halabja City, the maximum mean evaporation was 11.9 mm in July, and the minimum was 2.3 mm in December. The mean annual sunshine duration was 8.5 h/day in Erbil City, 7.4 h/day in Sulaymania City, 7.6 h/day in Duhok City, and 7.5 h/day in Halabja City. The annual mean wind speed was 1.7 m/s in Erbil City, 1.3 m/s in Sulaymania City, 1.12 m/s in Duhok City, and 0.81 m/s in Halabja City (Appendix A).

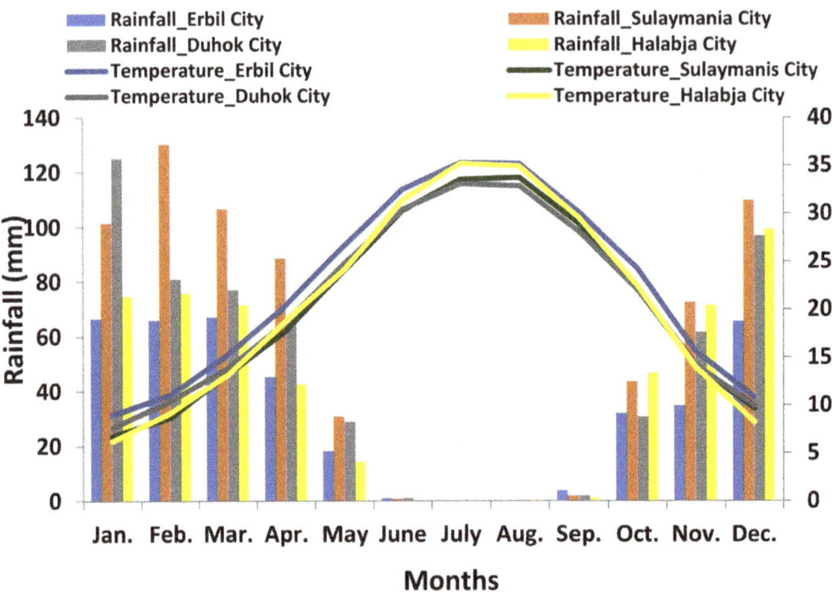

Figure 2. Mean monthly rainfall and temperature in the study area for the period 2005–2019.

2.2. Geological and Tectonic Setting

The exposed geological units in the Kurdistan Region are represented by formations that date from the Ordovician to the Tertiary period (Table 1 and Figure 3).

Table 1. Age and lithological description of the geological units.

Age	Geological Unit	Lithological Description
Holocene	Recent alluvial deposits	Different sized clastics, mixture of clay, sand, and pebbles
Pleistocene	River Terraces	Mixture of clay, sand, and pebbles
Pliocene	Bai Hassan and Muqdadiya formation	Thick sandstone, siltstone, and conglomerate
Late Miocene	Injana formation	Red sandstone, siltstone, and intercalations of red clay and pebbly sandstone
Middle Miocene	Fatha formation	Layers of red claystone, limestone, marl, and lenses of gypsum with some thin layers of siltstone
Middle-Late Eocene	Pila Spi formation	Well-bedded, recrystallized limestone, dolomite, and marly limestone
Early Eocene	Gercus formation	Red mudstone, sandstone, and shale, with rare conglomerates
Paleocene	Kolosh and Khurmala formations	Mainly clastics: shale, limestone, marl, and mudstone with tongues of white limestone
Late Cretaceous	Dokan, Gulneri, Komitan, Aqra, Bekhme, Shiranish, and Tanjero formations	Limestone, grey dolomite-containing bituminous limestone, blue-grey marl, and beds of marly limestone
Early Cretaceous	Chiagara, Balambo, Sarmord, Garagu, and Qamchuqa formations	Dolostone, dolomitic limestone, some calcareous marl, and limited shale
Late Jurassic	Naokelekan and Barsrin formations	Limestone, dolomitic limestone, shaley limestone, carboniferous shale, and bituminous dolomitic shales

Table 1. *Cont.*

Age	Geological Unit	Lithological Description
Middle Jurassic	Sargelu formation	Thin bedded shaley black limestone and shale with black chert and brown dolomitic marl, highly fossiliferous
Early Jurassic	Sarki and Sehkanian formations	Dolomitic limestone with splintery fractures, which are generally bituminous and fossiliferous
Late Triassic	Baluti, Kurrachina, Beduh, and Avroman formations	Alternations of shales, limestone, dolomites, and dolomitic limestone
Late Permian	Chaizairi formation	Beds of shale, limestone, and some evaporates
Ordovician	Khabourr formation	Thick sandstone-shale cyclic alternations

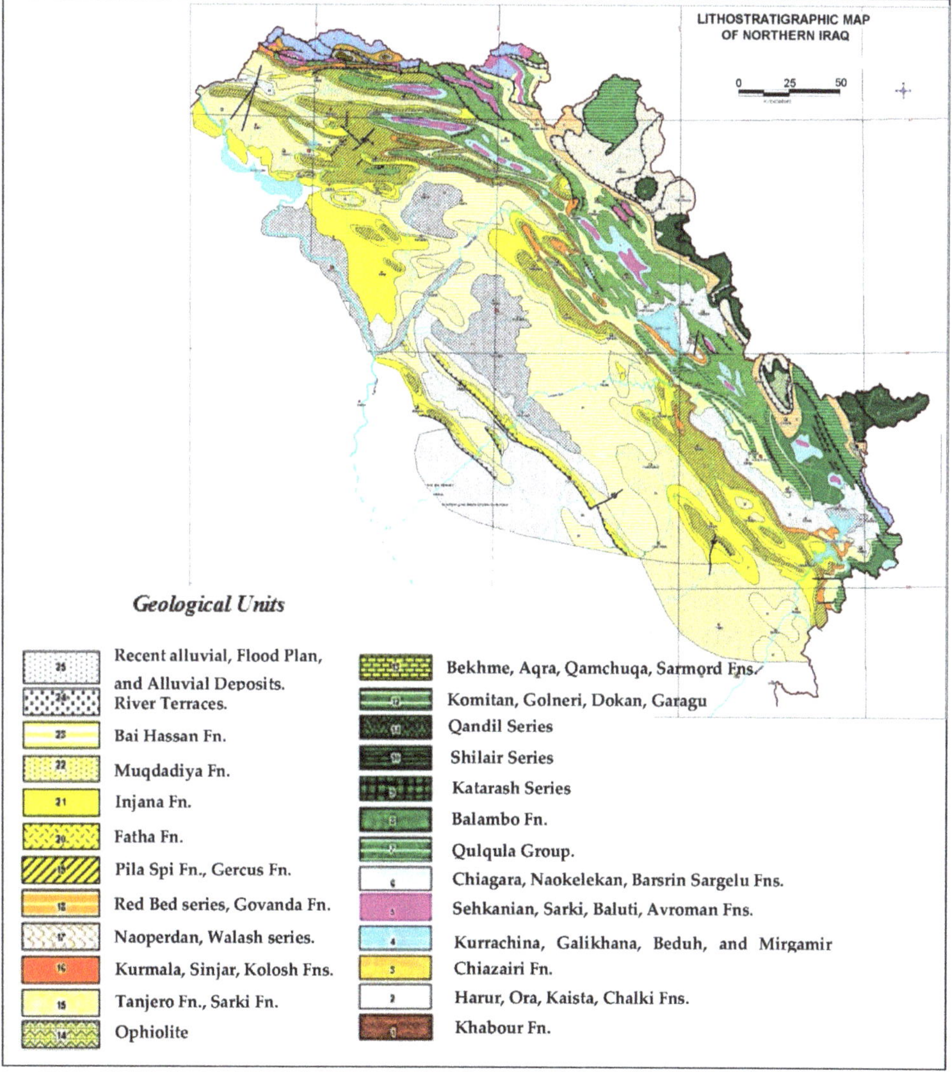

Figure 3. Geological map showing the lithological units of the study area (After Stevanovic and Marcovich, 2003).

The oldest unit from the Ordovician is the Khabour formation which is comprised of thick sandstone-shale cyclic alternations. The Late Permian period is represented by the Chaizairi formation, which includes shale, limestone, and some evaporates. The Late Triassic period is represented by Baluti, Kurrachina, Beduh, and Avroman formations which are generally composed of alternations of shales, limestone, dolomites, and dolomitic limestone. The Early Jurassic period is represented by Sarki and Sehkanian formations which consist of dolomitized limestone with splintery fractures, which are generally bituminous and fossiliferous. The Middle Jurassic period is represented by the Sargelu formation, which consists of thinly bedded shaley black limestone and shale with black chert and brown dolomitic marl, and is highly fossiliferous. The Late Jurassic period is represented by Naokelekan and Barsrin formations which consist of limestone, dolomitic limestone, shaley limestone, carboniferous shale, and bituminous dolomitic shales. The Early Cretaceous period is represented by Chiagara, Balambo, Sarmord, Garagu, and Qamchuqa formations which include dolostone, dolomitic limestone, some calcareous marl, and limited shale. The Late Cretaceous period is represented by Dokan, Gulneri, Komitan, Aqra, Bekhme, Shiranish, and Tanjero formations which include limestone, grey dolomite-containing bituminous limestone, blue-grey marl, and beds of marly limestone. The Paleogene period is represented by Khurmala, Kolosh, Pila Spi, and Gercus formations which include interchanging layers of grey claystone, shale, silt, and sandstone with conglomerate lenses. Tongues of white limestone can be found in the Kolosh and Khurmala formations. Red clay, siltstone, and sandstone, as well as a tongue of limestone, can be found in the Gercus formation. Well-bedded, recrystallized limestone, dolomite, and marly limestone can be found in the Pila Spi formation. The Neogene period is represented by Fatha, Injana, Muqdadeya, and Bai Hassan formations. The Fatha formation includes layers of red claystone, limestone, marl, and lenses of gypsum with some thin layers of siltstone. The Injana formation includes red sandstone, siltstone, and intercalations of red clay and pebbly sandstone. The Muqdadeya and Bai Hassan formations include sandstone, siltstone, and conglomerate. The youngest units are represented by Quaternary deposits, which include river terraces, the flood plain, and recent alluvial deposits, which include poorly cemented conglomerate, muddy sandstone, and a cover of pebbly clay [13,14].

2.3. Data Collection and Water Sample Analysis

The physicochemical properties for the analysis of 572 water samples, including 535 deep well samples (169 wells in Erbil, 119 wells in Sulaymania, 209 wells in Duhok, and 18 wells in Halabja), 33 spring samples, and 4 river water samples in the study area were collected from the database of Ministry of Agriculture and Water Resources—Groundwater Directorate in the Kurdistan Region and field work carried out during 2020–2021. The parameters for the samples included temperature, pH, electrical conductivity (EC), salinity, and total dissolved solids (TDS); all these parameters were measured in situ in the field by the portable device (HANNA instrument model Hi8314).

2.4. Interpolation and Statistical Analysis

The interpolation was done in ArcGIS 10.1 using the Kriging method to plot the parameter distribution for the well samples. Kriging spatial interpolation assumes that the distance or direction between sample points reflects a spatial correlation that can be used to explain variations in the surface. This approach is an efficient geostatistical interpolation technique focused on the special correlation of sampled points [15]. Statistical analysis of the analyzed parameters was carried out using the SPSS program.

2.5. Geo-Information Technique

GIS-based multi-criteria decision analysis (MCDA) was used to create a suitability map for using groundwater and surface water for irrigation purposes based on water availability and quality. The criteria layers were assessed using the multi-criteria decision approach combined with the weighted overlay function in ArcGIS 10.1. This process was used to evaluate the suitability of a specific area for a specific purpose.

2.6. Water Use and Suitability for Different Purposes

The suitability of the water for any particular use is determined by comparing the calculated and measured physical, chemical, and biological parameters with set standards for a particular use. In this study, the physicochemical parameters were used to compare the calculated and measured analysis. Train classification [16] was used to determine the suitability of the water for irrigation purposes based on the total dissolved solid (TDS).

2.7. Water Type Classification

2.7.1. Classification According to TDS

Water samples classified according to Hillel [17], Drever [18], Altoviski [19], and Gorrell [20], depending on TDS (Table 2).

Table 2. Classifications of water according to TDS in (mg/L).

Water Class	Gorrell (1958)	Altoviski (1962)	Drever (1997)	Hillel (2000)
Fresh water	0–1000	0–1000	<1000	<500
Slightly brackish water (Marginal)	—	1000–3000	—	500–1000
Brackish water	1000–10,000	3000–10,000	1000–20,000	1000–2000
Salty water	10,000–100,000	10,000–100,000	—	—
Saline water	—	—	35,000	5000–10,000
Highly Saline Water	—	—	—	10,000–35,000
Brine water	100,000	>100,000	>35,000	>35,000

2.7.2. Classification According to EC

Water samples classified according to USDA [21] and Mayer et al. [22], depending on the EC parameter (Table 3).

Table 3. Classifications of water according to EC in (µS/cm).

Water Class	USDA (1954)	Mayer et al. (2005)
Low salinity water	100 < EC < 250	550–1200
Medium salinity water	250 < EC < 750	1200–2200
High salinity water	750 < EC < 2250	2200–5000
Very high salinity water	2250 < EC < 5000	—

2.8. Classification of the Irrigation Water

2.8.1. Todd Classification (1980)

This classification depends on electrical conductivity (Table 4).

Table 4. Water classification according to Todd [23].

EC (µS/cm)	Water Class
<250	Excellent
250–750	Good
750–2000	Permissible
2000–3000	Doubtful
>3000	Unsuitable

2.8.2. Rhoades Classification (1992)

Rhoades classified irrigation water into six types based on TDS and EC (Table 5).

Table 5. Water classification according to Rhoades [24].

Water Class	EC (µS/cm)	TDS (mg/L)
Non-saline	<700	<500
Slightly-saline	700–2000	500–1500
Moderately saline	2000–10,000	1500–7000
Highly-saline	10,000–25,000	7000–15,000
Very highly saline	25,000–45,000	1500–35,000
Brine	>45,000	>35,000

2.8.3. Don Classification (1995)

This classification depends on electrical conductivity and total dissolved solid. Don classified irrigation water into five types (Table 6).

Table 6. Irrigation water classification according to Don [25].

Water Quality	EC (µS/cm)	TDS (mg/L)
Excellent	250	175
Good	250–750	175–525
Permissible	750–2000	525–1400
Doubtful	2000–3000	1400–2100
Unsuitable	>3000	>2100

3. Results and Discussion

3.1. Physicochemical Parameters

The physicochemical characteristics are shown in Table 7 and Appendices B–F and H and Section G. The electrical conductivity for the water well samples ranged between 134 and 5090 µS/cm, the spring samples ranged between 196.6 and 796.5 µS/cm, and the river samples ranged between 297 and 480 µS/cm. The highest concentration was measured in Erbil City in the Said-Ubaid village well, while the lowest concentration was measured in Sulaymania City in the Qalaga village well (Figure 4).

Table 7. Basic statistics of the physicochemical parameters of water samples in the study area.

Sample	Parameters	EC (µS/cm)	pH	TDS (mg/L)	Temperature °C
Wells Erbil City	Maximum	5090	9.1	3309	31
	Minimum	286	6.5	186	17
	Mean	643.9	7.7	419	23
	SD *	560	0.5	365	1.9
Wells Sulaymani City	Maximum	3290	9.5	2139	31
	Minimum	134	6.8	87	13
	Mean	563.2	8.2	366.1	20
	SD *	447.5	0.6	290.8	3.4
Wells Duhok City	Maximum	2400	8.6	1560	29
	Minimum	220	6.3	143	14
	Mean	687.5	7.5	446.9	20.4
	SD *	256.6	0.3	166.8	2.2
Wells Halabja City	Maximum	2540	9.6	1651	29
	Minimum	304	6.5	198	18
	Mean	530.3	8.3	344.7	21.9
	SD *	373.4	0.7	242.7	2.1

Table 7. *Cont.*

Sample	Parameters	EC (µS/cm)	pH	TDS (mg/L)	Temperature °C
Spring Samples	Maximum	796.5	8.2	509.8	26.5
	Minimum	196.6	7.4	128	16.5
	Mean	432.4	7.7	280.3	20.4
	SD *	138.3	0.2	88.9	3.1
River Samples	Maximum	388.5	8.2	252.5	22.9
	Minimum	361	8	234.7	22.2
	Mean	371.5	8.2	241.5	22.5
	SD *	12.7	0.1	8.3	0.3

* SD Standard Deviation.

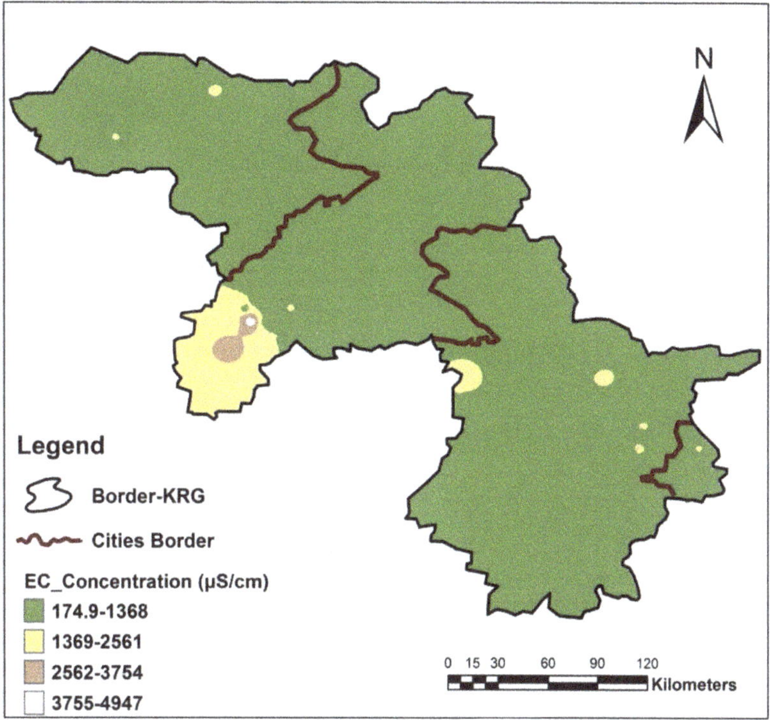

Figure 4. Distribution of the electrical conductivity of groundwater across the Kurdistan Region.

The pH for the deep wells ranged between 6.5 and 9.6, the samples from the springs ranged between 7.4 and 8.2, and the samples from the river ranged between 7.9 and 8.4. The highest concentration was measured in a deep well in Amura located in Halabja City, and the minimum concentration was measured in a deep well in Chrostana in Halabja City (Figure 5).

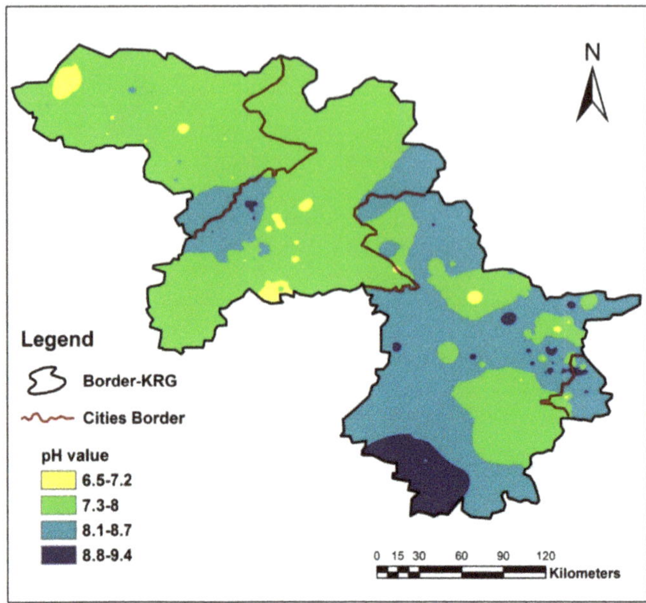

Figure 5. Distribution of the pH of groundwater across the Kurdistan Region.

Total dissolved solids in the deep wells ranged between 87 and 3309 mg/L, the spring samples ranged between 128 and 509.8 mg/L, and the river samples ranged between 193 and 312 mg/L. The highest concentration of TDS was found in a well in the village of Said-Ubaid in Erbil City, while the lowest concentration was measured in a well in Sulaymania City in the village of Qalaga (Figure 6).

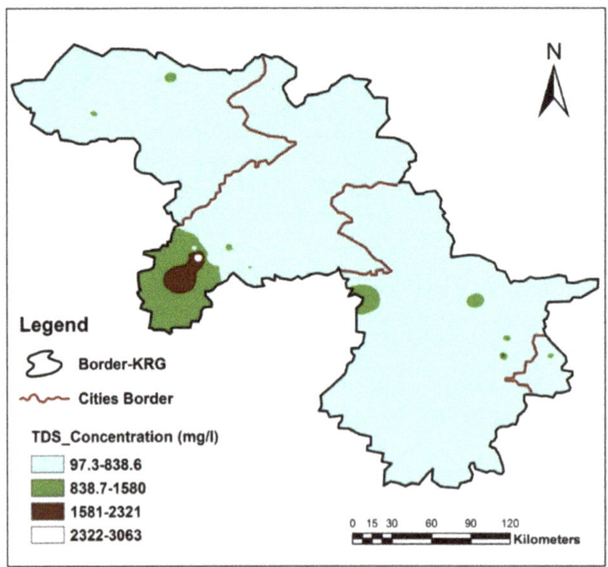

Figure 6. Distribution of the TDS of groundwater across the Kurdistan Region.

The temperature in the deep wells ranged between 10 and 31 °C, the spring samples ranged between 16.5 and 26.5 °C, and the river samples ranged between 21.4 and 23.3 °C. The highest value was measured in Erbil City in the Chamadubz village well, and the lowest value was also measured in Erbil City in the Hasarok village well (Figure 7).

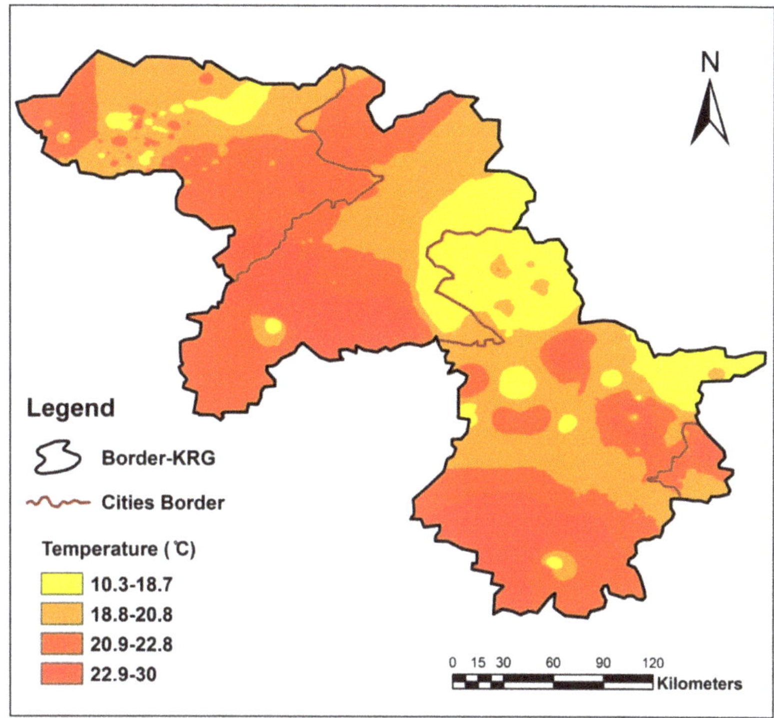

Figure 7. Distribution of the temperature of groundwater across the Kurdistan Region.

3.2. Water Uses and Suitability Analysis

3.2.1. Water Use for Livestock Purposes

In order to determine water quality for livestock purposes, the water samples were compared with the Ayers and Westcot classification of groundwater suitability for livestock and poultry according to electrical conductivity concentration (Table 8) [26]. All the water samples were acceptable for livestock and poultry purposes because the electrical conductivity fell within acceptable ranges except for the Said-Ubaid water well sample taken from Erbil city. This water was acceptable for livestock but unacceptable for poultry because the EC concentration was more than 5000 µS/cm which has been shown to reduce growth and increase mortality in poultry (Figure 8).

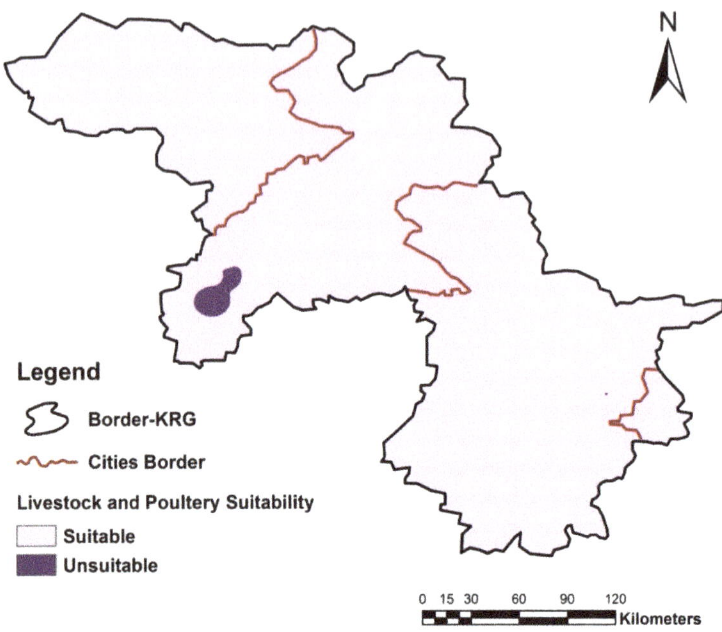

Figure 8. Suitability map for livestock and poultry according to Ayers and Westcot (1989).

Table 8. Water quality for livestock and poultry compared with Ayers and Westcot (1989) standards.

EC (μS/cm)	Specifications	Remarks	Water Samples
<1500	Excellent	This water has a relatively low level of salinity and should present no serious burden to any livestock or poultry	Erbil City 286–5090 (μS/cm)
1500–5000	Acceptable	This water should be satisfactory for all classes of livestock and poultry. It may cause temporary and mild diarrhea in livestock not accustomed to it or watery droppings in poultry (especially at the higher levels) but should not affect health or performance	Sulaymania City 134–3290 (μS/cm)
5000–8000	Acceptable for livestock, unacceptable for poultry	Causes temporary diarrhea in livestock and reduced growth and death in poultry	
8000–11,000	Limited for livestock, unacceptable for poultry	Avoid use for pregnant and lactating animals as levels increase Not acceptable water for poultry	Duhok City 220–2400 (μS/cm)
11,000–16,000	Limited	Not acceptable for animals	
>16,000	Not used	The risks posed by highly saline waters are so great that they cannot be recommended for use under any circumstances	Halabja City 304–2540 (μS/cm)

3.2.2. Water Use for Agricultural Purposes

The properties of Todd's classification [23] for Agricultural crops depending on total dissolved solids were applied for assessing water use purposes. This assessment showed that nearly all water samples were acceptable for all types of agricultural crops barring a few exceptions. Specifically, 16 well samples from Erbil City, 8 well samples from Sulaymania

City, 14 well samples from Duhok City, and 2 wells from Halabja City were not suitable for fruit crops (Table 9, Figure 9 and Appendices B–F and H and Section G).

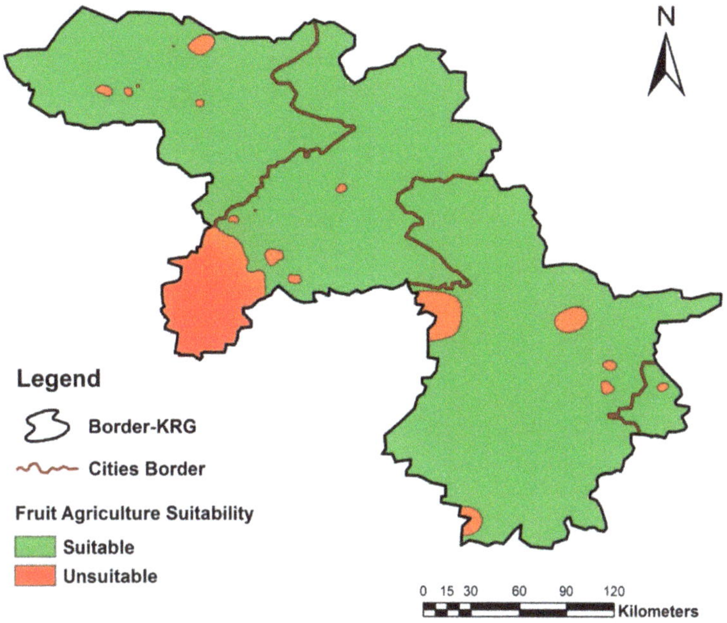

Figure 9. Suitability map for livestock and poultry according to [19].

Table 9. Todd classification (1980) for agricultural crops compared with water samples from the study area.

Crop Divisions	Low TDS Endurance	Medium TDS Endurance	High TDS Endurance	Water Samples
Fruit	<300 µS/cm Avocado, lemon, orange, apple strawberry, picot prune, plum	300–400 µS/cm Olive, date, fig cantaloupe, pomegranate	400–1000 µS/cm Palm	Erbil City 286–5090 (µS/cm) Sulaymania City 134–3290 (µS/cm)
Vegetable	300–400 µS/cm Green bean, celery, radish	400–1000 µS/cm Cucumber, onion, peas carrot, potato, cauliflower lettuce, squash	1000–12,000 µS/cm Spinach, kale, asparagus	Duhok City 220–2400 (µS/cm)
Field crops	400–600 µS/cm Field bean	600–1000 µS/cm Sunflower, corn, rice, flax, castor bean, wheat	1000–10,000 µS/cm Cotton, sugar beet, barley	Halabja City 304–2540 (µS/cm)

3.2.3. Water Use for Irrigation Purposes

One problem caused by irrigating cropland is the possibility of groundwater contamination. Fertilizer and pesticide use need to be more carefully restricted in order to reduce the risk of contamination [27]. The suitability of irrigation water is dependent on the effects of its mineral content on both plants and soil, as well as the effect of salts which could cause changes in soil structure. Infiltration increases with increasing TDS, which is used for evaluating soil permeability [28]. Most of the water samples were acceptable for irrigation and would not have detrimental effects on crops. Several samples of well water proved that the water was not suitable for irrigation, including 17 well samples from Erbil City, 7 well samples from Sulaymania City, 52 well samples from Duhok City, and 2 well samples from Halabja City that could potentially have harmful effects on crops that are sensitive to salinity. Additionally, 4 well samples from Erbil City, 4 well samples from Sulaymania City, 2 well samples from Duhok City, and 1 well sample from Halabja City could be harmful to sensitive crops. Only 4 well samples (3 wells in Erbil City and 1 well in Sulaymania City) could be used for highly tolerant crops (Table 10 and Appendices B–F and H and Section G).

Table 10. Train classification (1979) for irrigation water and compared with water samples from the study area.

TDS (mg/L)	Specifications	Water Samples TDS Range
500	Used for irrigation; does not cause harmful effects	Erbil City 186–3309 mg/L
500–1000	Used for irrigation but causes harmful effects on crops sensitive to salinity	Sulaymania City 87–2139 mg/L
1000–2000	Causes harmful effects on crops, so use carefully	Duhok City 143–1560 mg/L
2000–5000	Used only for irrigating highly tolerant crops	Halabja City 198–1651 mg/L

3.3. Suitability Analysis

A suitability map was created by combining the derived layers to define suitable groundwater locations for irrigation purposes. The steps for this procedure started with reclassifying datasets using a model in ArcGIS to reclassify the interpolation maps of the physicochemical parameters into relative classes. In this approach, for every criterion input, each cell in the study area has a different value for each layer. To determine irrigation suitability, the suitability map was created by integrating the derived layers. Because combining these layers in this format is not possible, the next step was to reclassify the previous maps into a relative four classes with a common value. In the resulted maps, the suitable locations are referred to as number one, while number four indicates unsuitable locations (Figure 10). After reclassification, the weighted overlay analysis was used to create an integrated study of common values for a variety of dissimilar and miscellaneous inputs and to produce a final suitability map for groundwater irrigation.

According to the results, the region was divided into three classes: high suitability, low suitability, and unsuitable with respect to the input factors using the weight overly method. In the resulting maps, the suitable locations are referred to as number one, while the number four indicates unsuitable locations. Figure 8 shows the reclassified map of the four criteria used in this study. High suitability defines water samples that have parameters and concentrations within the acceptable limit, and unsuitable defines the water samples that have concentrations over the standard or acceptable limit. Most of the groundwater samples were suitable for irrigation except for some samples from the Makhmur district in Erbil City, the Chamchamal and Kfri districts in Sulaymania City, and the Zawita district in Duhok City (Figure 11).

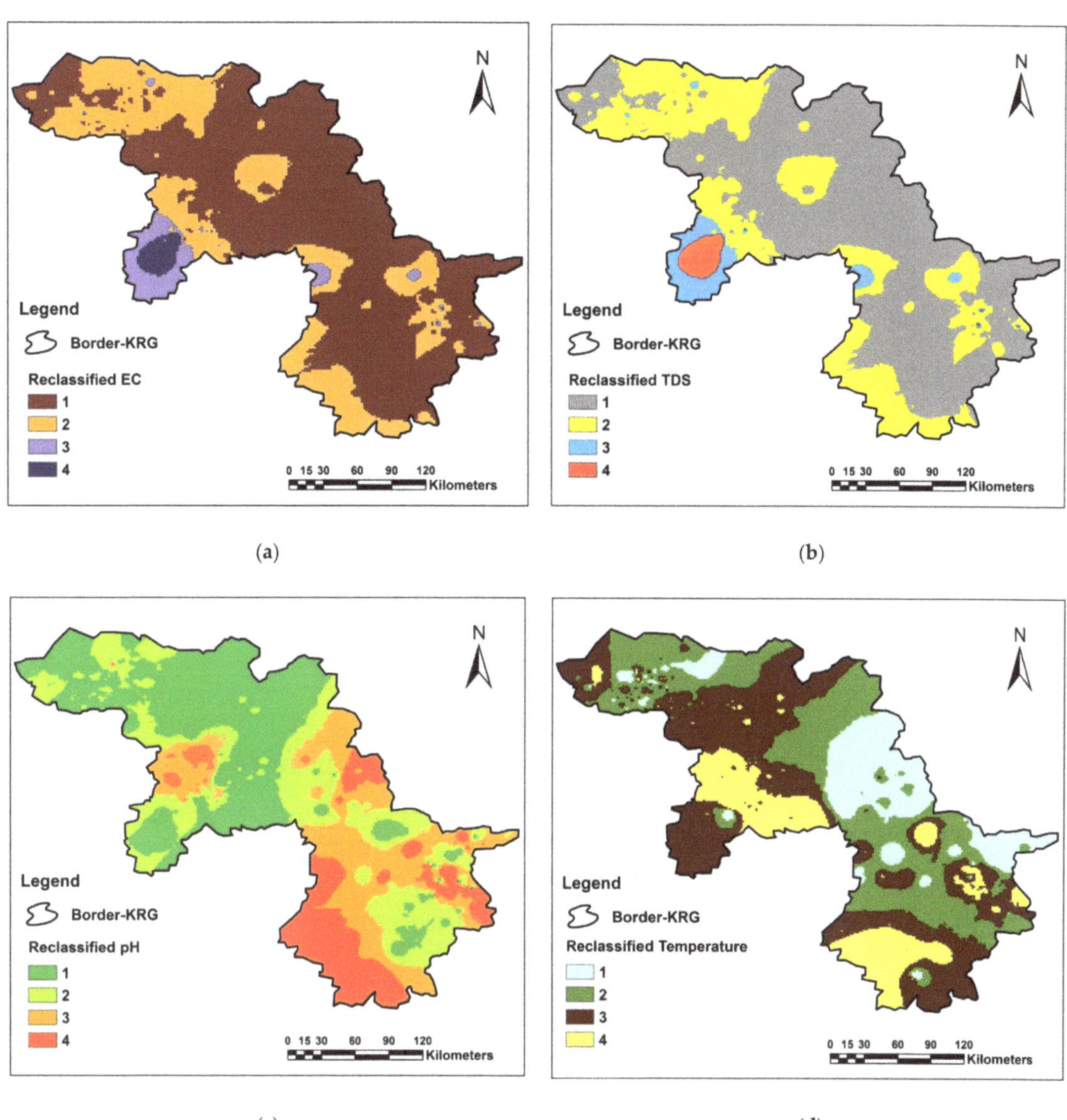

Figure 10. Reclassified map of the studied criteria; (**a**) Reclassified electrical conductivity, (**b**) Reclassified total dissolved solid, (**c**) Reclassified pH value, and (**d**) Reclassified temperature value.

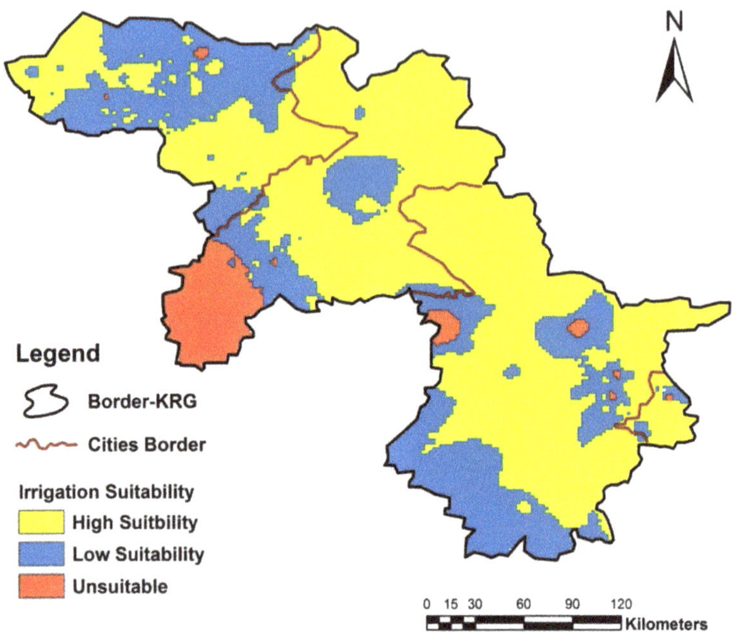

Figure 11. Irrigation water suitability map of the Kurdistan Region.

3.4. Water Type and Classification of the Irrigation Water

Most of the well water samples were freshwater except for 36 deep well samples that ranged from slightly brackish to brackish water. Considering TDS results, all the spring samples were considered freshwater according to [17–20] (Appendices B–F and H and Section G).

According to the [21,22], based on the EC, most of the water samples had low to medium salinity except for 26 well samples and one spring sample that had high salinity, and 2 well samples that had very high salinity (Appendices B–F and H and Section G).

Most of the samples had an excellent to good water classification except for 85 samples that were classified as permissible, 8 samples that were classified as doubtful, and 4 samples that were classified as unsuitable for irrigation according to the Todd classification based on EC. According to the Rhoades classification, all water samples were non-saline to slightly saline except for 11 samples that were moderately saline. According to the Don Classification, most of the samples were excellent to good except for 85 samples that were permissible, 8 samples that were doubtful, and 4 samples that were unsuitable for irrigation (Appendices B–F and H and Section G).

4. Conclusions

This study examined water quality and availability as well as water use for different purposes and the suitability of water for irrigation in the Kurdistan Region, Iraq. The water samples were acceptable for all types of agricultural crops with the exception of 16 well samples from Erbil City, 8 well samples from Sulaymania City, 14 well samples from Duhok City, and 2 well samples from Halabja City, which should not be used for fruit crops. Most water samples were acceptable for livestock and poultry purposes except for a well water sample from the Said Ubaid area of Erbil city that was acceptable for livestock but unacceptable for poultry because of its high electrical conductivity which causes reduced growth and increased mortality in poultry. Most of the water samples were acceptable for irrigation and would not cause detrimental effects on crops except for

17 well samples from Erbil City, 7 well samples from Sulaymania City, 52 well samples from Duhok City, and 2 well samples from Halabja City that could be harmful to crops that are sensitive to salinity. Additionally, 4 well samples from Erbil City, 4 well samples from Sulaymania City, 2 well samples from Duhok City, and 1 well sample from Halabja City could be harmful to crops. A total of 4 well samples (3 wells in Erbil city and 1 well in Sulaymania city) could be used for irrigating highly tolerant crops.

Most of the deep well and spring samples were considered freshwater except for some deep well samples that contained slightly brackish to brackish water according to the total dissolved solids. Most samples contained water with low to medium salinity except for some wells and one spring sample that contained water with high salinity, and two well samples that had water with very high salinity. Suitability analysis shows that most of the groundwater samples were suitable for irrigation except for the samples taken from the Makhmur District in Erbil City, the Chamchamal and Kfri districts in Sulaymania City, and the Zawita district in Duhok City.

Agricultural activities may have adverse effects on water quality due to the release of nutrients (as a result of soil management and fertilizer application) and other chemicals like pesticides into aquatic environments. Biological contamination (e.g., from microbiological organisms in manure), soil erosion, and sediment burdens may increase due to poor farming practices. As a result, farmers and other water users should try to reduce negative effects on water quality.

Author Contributions: Conceptualization, S.S. and C.R.; methodology, S.S.; software, S.S.; validation, S.S., C.R. and K.M.; formal analysis, S.S.; investigation, S.S., C.R., A.A.; resources, S.S.; data curation, S.S.; writing—original draft preparation, S.S.; writing—review and editing, C.R. and K.M.; visualization, S.S., A.A. and K.M.; supervision, C.R.; project administration, C.R. and K.M.; funding acquisition, C.R. All authors have read and agreed to the published version of the manuscript.

Funding: This Research is fully funded by NUFFIC Orange Knowledge Programme OKP-IRA-104278.

Institutional Review Board Statement: Not applicable.

Informed Consent Statement: Not applicable.

Data Availability Statement: Some data in this manuscript obtained from Ministry of Agriculture and Water Resources-Groundwater Directorate, Kurdistan Region, Iraq and the other data from the field work during 2020–2021 and analyzing the parameters.

Acknowledgments: This study was funded by the NUFFIC Orange Knowledge Programme (OKP-IRA-104278) and coordinated by Wageningen University & Research, The Netherlands. The authors thank the Ministry of Agriculture and Water Resources and Groundwater Directorate in the Kurdistan Region for making data available for this study.

Conflicts of Interest: The authors declare no conflict of interest.

Appendix A

Table A1. Mean Monthly Climatic Parameters of the study area for the period 2005–2019 in Erbil, Sulaymania, Duhok, and Halabja.

Meteorological Station	Parameter	Months											
		Jan.	Feb.	Mar.	Apr.	May	June	July	Aug.	Sep.	Oct.	Nov.	Dec.
Erbil City	Rainfall (mm)	66.6	65.9	67.1	45.4	18.3	1	0	0.1	3.7	32	34.7	65.6
	R.H%	69.9	67.3	61.2	54.4	40.7	30. 9	27.7	29	34	45.3	59	66.4
	Temp. (°C)	8.9	11	15.3	20.4	26.6	32.5	35.3	35.2	30.2	24.2	15.6	10.8
	Sunshine Duration (h/day)	5.4	6.1	6.6	7. 9	9.2	11.9	12	11.6	10.5	8.1	7.2	5.9
	Wind Speed (m/s)	1.7	1.8	2	2.1	1.9	1.9	1.6	1.6	1.4	1.7	1.4	1.5
	Evaporation (mm)	1.8	2.5	4.4	6	9.3	12.7	13.4	12.6	9.5	6.3	3.2	2
Sulaymania City	Rainfall (mm)	101.3	130	106.5	88.4	30.8	0.7	0	0.01	1.8	43.6	72.5	109.7
	R.H%	70.3	66.8	58.9	55.7	41.7	27	24.9	24	30.1	44.1	59.7	65.5
	Temp. (°C)	6.8	8.7	13.2	17.8	24	30.3	33.5	33.7	29	22.3	14.1	9.6
	Sunshine Duration (h/day)	4.8	5.2	5.6	7.1	7.9	9.9	10.5	10.4	9.3	7.3	6	5.1
	Wind Speed (m/s)	1.1	1.1	1.4	1.2	1.2	1.7	1.6	1.4	1.3	1.2	1.1	0.8
	Evaporation (mm)	2.3	2.6	3.3	4.7	7	10.8	11.8	11	7.9	4.9	2.9	2.3
Duhok City	Rainfall (mm)	125	80.9	77	65	28.8	1.1	0.1	0.1	2	30.6	61.7	96.8
	R.H%	67.4	66.5	59.8	54.3	43.1	30.9	26.8	26.7	31.8	43	58.3	64.8
	Temp. (°C)	7.7	10.4	13.9	18.6	24.7	30.4	33.1	32.8	28.1	22	14.2	9.9
	Sunshine Duration (h/day)	4.2	5.2	5.7	7	9.1	11.2	11.5	11.1	9.2	7	5.5	4.4
	Wind Speed (m/s)	1.3	1.1	1.3	1.2	1.3	1.2	1.1	1	1	1	0.9	1
	Evaporation (mm)	1.4	1.9	3.2	4.9	7.5	10.3	11.1	10.5	7.82	4.6	2.1	1.4

Table A1. *Cont.*

Meteorological Station	Parameter	Months											
		Jan.	Feb.	Mar.	Apr.	May	June	July	Aug.	Sep.	Oct.	Nov.	Dec.
Halabja City	Rainfall (mm)	74.6	75.7	71.5	42.7	14.3	0.3	0	0.4	1.1	46.7	71.3	99.2
	R.H%	67.7	57.6	56.1	48.9	32	15.9	13.7	13.9	18.6	32.7	57.9	64.3
	Temp. (°C)	6.2	9.1	13.2	18.7	24.1	31.3	35.2	34.9	29.6	22.4	13.9	8.2
	Sunshine Duration (h/day)	4.7	5.4	5.5	7.1	7.6	10	10.7	10.5	9.3	7.3	6.2	5.2
	Wind Speed (m/s)	0.7	0.7	0.8	0.8	1	0.9	0.9	0.8	0.75	0.7	0.6	1.1
	Evaporation (mm)	2.3	2.6	3.5	4.8	7	10.8	11.9	11.4	7.97	5	2.9	2.5

Appendix B

Table A2. Physicochemical Parameter Analysis for the Deep Wells in Erbil City, UTM-WGS84 Coordination System.

SN	Well Name	X	Y	Z	Well Depth (m)	EC (µS/cm)	Temp. °C	pH	TDS (mg/L)
1	Tendura	395413	3993026	335	180	709	24	7.3	461
2	Siaw	387823	3986646	294	135	522	25	7.8	339
3	Mastawa	395425	3989611	313	110	2380	22	7.1	1547
4	Chiman	398291	3992781	332	140	701	25	7.6	456
5	Sirawa	390621	3991566	288	170	684	25	7.2	445
6	Surbash kakalla	399116	3986546	328	150	1440	22	7.3	936
7	Surbash Hawez	399119	3990618	342	50	1090	24	7.8	709
8	Sorizha	399758	3984480	327	160	685	24	7.6	445
9	Shekh Sherwan	385648	3988509	297	112	492	25	8	320
10	Delogoli Khwaru	396086	3995879	339	162	706	23	7.8	459
11	Daldaghan	399945	3992666	339	140	771	24	7.4	501
12	Dustapa	399176	3995794	325	124	576	25	7.9	374

Table A2. *Cont.*

SN	Well Name	X	Y	Z	Well Depth (m)	EC (µS/cm)	Temp. °C	pH	TDS (mg/L)
13	Shekh Sherwan	385318	3988765	296	200	508	25	7.9	330
14	Yadiqizlar	398385	3982621	322	200	960	25	7.4	624
15	Mastawa	393502	3989935	314	132	1160	24	7.3	754
16	ArabKand	401207	4000412	348	135	686	23	7.6	446
17	Binberzi Gichka	396837	4001782	342	200	534	25	7.9	347
18	Binberz	395933	4001268	352	180	452	24	7.9	294
19	Binberz	396224	4000855	345	153	455	25	7.6	296
20	Jimka	395878	4000194	326	100	608	24	7.4	395
21	Yarmja	394854	3999461	331	120	362	24	8.2	235
22	SwarayGawra	396326	3998153	374	110	670	23	7.9	436
23	Khazna	391529	3997369	311	142	713	23	8	463
24	Lajan Harki	391734	3999080	335	175	1200	25	8.5	780
25	Dusara Jabar	400708	3990293	341	170	970	24	8.3	631
26	Qoritan Chukil	405209	3992905	357	240	442	24	8.4	287
27	SwarayGichka	401598	4004542	401	250	646	24	8.1	420
28	Beryat	404024	3996473	342	173	467	23	7.8	304
29	ArabKand	400867	4000423	352	128	681	22	7.6	443
30	Awena	382629	3991199	290	101	940	26	8.4	611
31	Delogoli Khwaru	394152	3997089	325	120	356	23	8.3	231
32	Turaq	404792	4002899	376	200	578	24	7.8	376
33	SwarayGichka	402800	4004166	370	240	714	23	8.1	464
34	Tobzawa	410115	3997736	398	170	365	23	8.1	237
35	Trpaspian	403751	3983092	336	174	660	22	8.2	429
36	Pirdaud	403042	3986933	348	150	613	23	7.8	398

Table A2. *Cont.*

SN	Well Name	X	Y	Z	Well Depth (m)	EC (µS/cm)	Temp. °C	pH	TDS (mg/L)
37	Serkarez	406857	3998106	377	245	386	25	8.2	251
38	Gird Muhammad	407322	3997855	383	249	438	24	8.5	285
39	Serkarez	406998	3997742	376	240	430	25	7.9	280
40	TimaryGawra	408089	3996528	374	204	501	24	8.4	326
41	Mizahmed	405693	3996774	370	164	450	23	7.8	293
42	qatawy	406891	3997283	376	208	412	25	8	268
43	Serkeez	407100	3997486	388	200	496	24	8.4	322
44	Baghlominara	405721	4001433	378	160	499	21	8.3	324
45	Dugirdkan	405002	3981465	357	170	740	24	7.3	481
46	Elinjagh	420484	3979467	429	180	410	24	7.5	267
47	BerAraban	410425	3974764	360	125	670	25	7.3	436
48	Qushtapa	412908	3984905	398	60	470	24	7.4	306
49	SebiranyAdo	409447	3981910	378	185	530	25	7.5	345
50	Doshewan	426219	3982081	496	175	480	25	7.4	312
51	Dolazay Nawand	412989	3972710	334	180	620	25	7.2	403
52	QaziKhana	408707	3973989	343	205	1600	23	7	1040
53	SurbashKhidr	406064	3975330	346	50	1300	25	7.5	845
54	OmerMamka	416194	3969285	317	160	710	24	7.4	462
55	GirdMala	415049	3985162	421	100	470	23	7.5	306
56	Aliawa	412033	3967864	331	160	600	24	7	390
57	Omaraway Gawra	419638	3968151	319	140	630	23	7.1	410
58	Rolka	428855	3980903	517	200	460	26	7.3	299
59	KanyBizra	426232	3985419	538	108	440	24	7.3	286
60	Rolka	426608	3977937	458	200	530	26	7.3	345

Table A2. *Cont.*

SN	Well Name	X	Y	Z	Well Depth (m)	EC (µS/cm)	Temp. °C	pH	TDS (mg/L)
61	Aziana	427596	3974853	443	180	380	25	7.3	247
62	Pongena Mantik	428654	3976444	485	126	430	21	7.5	280
63	Dokala	406733	3977235	357	150	660	26	7.4	429
64	Qashqa	429168	3967811	273	82	440	25	7.5	286
65	Seequchan	419075	3964982	299	186	680	21	7	442
66	Sinala	418483	3972394	349	198	570	24	7.1	371
67	Shekhan	416152	3964799	309	153	440	24	7	286
68	Omerawa Bichuk	417486	3965549	303	150	690	24	7.3	449
69	GirdaSor	423034	3971717	361	200	360	23	7.3	234
70	Girdasor	423584	3971885	362	200	360	23	7.4	234
71	Mirakany Afandi	424219	3979606	458	153	540	28	7.2	351
72	Mirakany khedr	428579	3984536	520	254	490	24	7.2	319
73	GirdLanka	414267	3966889	301	200	560	27	7.2	364
74	Qarajnagha	413443	3992743	398	190	330	24	7.6	215
75	Aliawa mardan	407215	3991827	380	50	510	23	7.8	332
76	Aliawa mardan	406343	3990223	354	220	560	22	7.4	364
77	Mortka shahab	412765	3988467	408	194	330	24	7.7	215
78	Kardiz	420011	3983698	452	50	400	23	7.6	260
79	Majidawa	373717	3988067	244	206	928	21.8	7.8	603
80	Garasor	366198	3966290	236	123	3100	21	7.3	2015
81	Hasarok	383687	3980931	320	111	400	20	7.9	260
82	Milhurt	368641	3980649	259	143	2220	23	7.4	1443
83	Said Ubaid	376313	3981264	265	143	5090	20	7.6	3309
84	shorazartka	375605	3983239	259	157	3770	20	7.7	2451

Table A2. *Cont.*

SN	Well Name	X	Y	Z	Well Depth (m)	EC (µS/cm)	Temp. °C	pH	TDS (mg/L)
85	Gomaspan	437276	4014456	817	160	697	21	7.7	453
86	Almawani khwaru	430030	4028563	773	120	1850	21	7.2	1203
87	SaryBlind	430375	4020969	829	103	640	21	7.2	416
88	Dawdawa	430149	4028189	797	145	723	18	6.5	470
89	Shekhan Harki	423363	4028611	589	100	577	20	7.5	375
90	Azhga	426637	4031996	761	121	676	20	7.4	439
91	Zagros	427847	4030919	783	100	585	20	7.6	380
92	Qalasinji Saru	439237	4024471	1195	130	456	20	7.8	296
93	Tobzawa	414660	4024692	484	50	405	20	7.8	263
94	Harbo	434247	4069154	650	175	331	20	7.8	215
95	Khardan	442530	4074013	812	255	442	20	7.4	287
96	Kalak	382135	4015828	287	112	330	24	8.4	215
97	Konakalak	379603	4012701	320	105	512	22	7.9	333
98	Malaomer	386802	4015630	242	102.8	1150	25	8	748
99	Khabat	376462	4008393	246	170	950	25	8.4	618
100	Chamadubz	376782	4010057	252	180	1270	31	8.8	826
101	Bastam	378033	4011366	261	180	1150	25	8.5	748
102	KonaSekhora	397099	4005687	409	160	526	24	8.6	342
103	Kany Qirzhala	397768	4007730	450	220	545	25	8.5	354
104	Sebirani Gaura	387589	4011243	402	143	542	24	8	352
105	Kawraban	389545	4009579	420	160	452	24	8.3	294
106	Qariatagh	399065	4002583	355	117	678	24	8.3	441
107	Qalatga	398014	4012250	381	200	448	25	8.3	291
108	Satoor	392155	4004737	446	160	391	25	8.4	254

Table A2. *Cont.*

SN	Well Name	X	Y	Z	Well Depth (m)	EC (µS/cm)	Temp. °C	pH	TDS (mg/L)
109	Girdarasha zab	383730	4018266	277	120	434	24	8.3	282
110	Kawr Gosk	389147	4023536	297	100	1210	24	8.5	787
111	Gainji Gaura	393451	4020664	314	100	447	24	8.1	291
112	Agholan Assad	390878	4020421	298	155	386	24	8.5	251
113	Agholan bichuk	388452	4020472	321	160	474	21	8.2	308
114	Girdasor	396053	4017280	325	150	511	23	7.9	332
115	Shewarash Kon	391955	4025303	282	150	930	23	8.7	605
116	Shewarash	394680	4025042	365	136	419	24	8.5	272
117	Shewarash Diwan	392407	4026303	291	99	605	23	8.6	393
118	mamalok	395288	4023005	355	88	286	24	8.3	186
119	Kharaba Draw	398390	4018460	369	100	411	22	8.2	267
120	Smailawa	395643	4014878	345	140	583	23	8.4	379
121	Halajay gaura	423849	3989468	534	120	360	23	7	234
122	Palany	419985	3986363	461	114	380	23	7.5	247
123	Sablagh	420927	3989485	506	125	340	23	7.4	221
124	Baghmera shahab	420519	3996696	476	136	330	23	7.3	215
125	Daratoo	414839	3998493	430	154	289	23	7.3	188
126	Girdarashay Mufti	411909	3997991	413	100	340	22	7.4	221
127	Kasnazan	422851	4007170	581	182	390	22	7.5	254
128	Mam choghan	429660	4009737	867	130	360	21	7.7	234
129	Sharaboty Gichka	428527	4011232	850	75	390	21	7.6	254
130	Mala Omar	422978	4017757	635	122	360	23	7.7	234
131	Tobzawa	462649	3993215	723	115	507	17	7.9	330
132	Shiwashan	482780	3982821	702	281	504	21	7	328

Table A2. *Cont.*

SN	Well Name	X	Y	Z	Well Depth (m)	EC (μS/cm)	Temp. °C	pH	TDS (mg/L)
133	Sarkarezy zrary	404494	4030763	378	185	420	22	8	273
134	Sarkawr Harky	402843	4029145	389	98	378	20	9.1	246
135	Qafar	407471	4028044	438	111	430	21	8	280
136	Rashkin	406162	4007467	385	200	505	23	7.5	328
137	Przin	416741	4013421	493	160	412	22	7	268
138	Bahrka	413063	4019111	464	168	370	21	7	241
139	Bahirka	414097	4020895	494	300	375	23	7	244
140	Bark Bichuk	412902	4012230	469	240	381	24	7.3	248
141	ShekhaShil	411314	4017350	436	145	436	23	7.4	283
142	Grdachal	404224	4024290	389	134	433	23	8	281
143	Qalanchoghan	400623	4016125	357	240	398	24	8.3	259
144	Shakholan	398649	4027829	359	138	420	20	9	273
145	Barhushter	400202	4024210	357	150	568	23	9	369
146	Saidan	398445	4021755	347	56	468	22	8	304
147	Daraban	401047	4018612	348	240	440	24	9	286
148	Darashakran	409834	4029370	409	180	590	20	7.5	384
149	Hababan	412796	4043905	597	145	360	21	8	234
150	Binaslawa	420902	4001510	525	136	370	21	7	241
151	Binaslaway Bchuk	424875	4001969	581	120	360	22	7	234
152	Binaslawa	421933	4001655	533	147	370	22	8	241
153	Ankawa	409437	4010729	421	132	379	21	7	246
154	Serwaran Qtr	415487	4005967	454	200	350	22	7.7	228
155	Nawand	413828	4003996	444	170	423	21	7.7	275
156	Polisan	413679	4004401	435	300	370	20	7.5	241

Table A2. *Cont.*

SN	Well Name	X	Y	Z	Well Depth (m)	EC (μS/cm)	Temp. °C	pH	TDS (mg/L)
157	Badawa	413828	4003167	432	151	310	21	7.6	202
158	Zanko	413428	4001349	429	150	330	21	7.7	215
159	Park	409071	4006018	401	171	490	22	7.5	319
160	Nawand	406202	4005323	383	250	475	21	7	309
161	Taajil	410466	4005273	410	166	585	21	7.6	380
162	nawroz	408067	4003647	397	156	661	21	7.3	430
163	Sarkavr Well	416765	4078678	637	139	489	21	7.3	318
164	Fakiran village well	421428	4080281	478	123	436	23	7.5	283
165	Shuri Village Well	420125	4076941	619	104	457	22.4	7.4	297
166	Pirasal Village Well	424872	4077161	492	100	400	21	7.8	260
167	Havendika Village	433114	4072683	452	100	414	21.2	7.2	269
168	Harbo Village Well	434247	4069154	650	175	331	20	7.8	215
169	Mergasor well 2	438635	4076840	1101	113	850	21.5	7.6	553

Appendix C

Table A3. Physicochemical Parameter Analysis for the Deep Wells in Sulaymani City, UTM-WGS84 Coordination System.

SN	Well Name	X	Y	Z	Well Depth (m)	EC (μS/cm)	Temp. °C	pH	TDS (mg/L)
1	Bosken	492819	4011595	501	94	380	21	7.6	247
2	saidawa	491491	4011704	563	130	620	21	7.4	403
3	Hartal	470555	4024921	1255	102	310	13	8.5	202
4	Sarwchawa	478114	4014519	586	100	580	17	7.5	377
5	awazhe	514478	3986624	1179	95	520	16	7.9	338
6	Nolichka	511480	3989031	937	63	370	18	8.1	241

Table A3. *Cont.*

SN	Well Name	X	Y	Z	Well Depth (m)	EC (µS/cm)	Temp. °C	pH	TDS (mg/L)
7	Yoliana	483853	4007714	549	85	350	16	7.8	228
8	Sarbasti Quarter	483804	4007117	557	92	350	15	7.8	228
9	Rizgari Quarter	484297	4006723	550	88	350	15	7.8	228
10	For Collective	484834	4007381	556	90	360	17	7.7	234
11	Girdaspian	510816	4004613	672	155	279	19	8.4	181
12	khirajo	500642	4008443	524	116	378	18	8.5	246
13	sultanadei taza	503697	4008416	545	103	279	18	8.6	181
14	Qalaway New	503690	4013512	607	125	225	19	8.8	146
15	Bastaseny Khwaroo	503424	4011839	578	105	248	17	8.6	161
16	Kanjaray New village	502202	4010943	564	130	237	16	8.4	154
17	Banwaqal	497550	4010438	549	85	284	20	8.5	185
18	zorkani khwaroo	522819	4014921	637	111	269	17	8.6	175
19	Qadirawa	501068	4014722	617	130	299	17	8.5	194
20	haji awa	481268	4008715	566	120	440	17	7.7	286
21	Hanarok	514012	4003962	720	120	465	19	8.3	302
22	binawshan	509246	4001922	602	164	367	19	8.3	239
23	Sedallan	507569	4014372	628	150	416	20	8.1	270
24	karsonan	505069	4007123	543	119	260	17	8.6	169
25	Dolla Bfra	504867	4009490	553	120	312	17	8.4	203
26	Zharawa collective	506231	4008932	559	45	265	17	8.5	172
27	Kawibabasan	508161	4009222	575	145	285	19	8.5	185
28	Tagaran	547775	3948290	894	129	1960	17	9	1274
29	Kele	541286	3962029	833	180	230	19	7.5	150
30	Chokhmakh	527503	3962832	1054	154	599	30	7	389

Table A3. *Cont.*

SN	Well Name	X	Y	Z	Well Depth (m)	EC (µS/cm)	Temp. °C	pH	TDS (mg/L)
31	Shewashan	485615	3992677	630	117	350	22	7.7	228
32	kani watman	479968	3993488	871	153	361	16	8.1	235
33	kwna mare	505221	3986108	586	99	283	18	7.6	184
34	Merzarostami Gawra	493836	3989437	531	162	334	20	8.6	217
35	Kani bnaw	492351	3986961	685	151	537	18	8.28	349
36	Nuraden	514021	3998667	553	107	454	19	8.4	295
37	Palkarash	478870	3947771	674	150	2000	21	8.3	1300
38	Gazalan	482058	3947557	636	90	2150	23	8.2	1398
39	sadun awa	494714	3925980	621	38	234	23	8.5	152
40	Kani Shaitan	500018	3945710	901	96	590	22	7.9	384
41	Chalaw	477100	3941102	697	100	1420	22	8.4	923
42	Sofi Hassan	511080	3926397	848	92	450	22	7.8	293
43	Zhallay Darband	512901	3924145	750	82	840	22	7.7	546
44	Kani shaitan	500819	3944918	884	80	410	17	8.3	267
45	Kani shaitan	501257	3944696	873	80	490	17	8.4	319
46	Bani maqan	501125	3944744	877	200	320	17	8.4	208
47	Banimaqan	479218	3928952	870	110	470	15	8.6	306
48	Qalaga	482100	3928926	875	65	134	20	8.8	87
49	Sewsenan	534547	3895245	992	300	177	20	7.4	115
50	Garazil	545864	3898366	1000	28	590	19	7.5	384
51	Tangisar	531698	3916913	938	108	380	20	7.9	247
52	Tatan	526853	3921515	912	80	305	17	8.8	198
53	Masydar	583659	3957601	1206	87	329	16	9	214
54	Kanisef	592707	3958965	1248	40	458	16	7.5	298

Table A3. Cont.

SN	Well Name	X	Y	Z	Well Depth (m)	EC (μS/cm)	Temp. °C	pH	TDS (mg/L)
55	Kanimasian	579488	3946254	1322	133	400	19	7.5	260
56	Sarkan	582093	3946353	1260	150	444	17	7	289
57	Kura mewy saroo	575665	3941050	1238	23	447	13	7	291
58	Kani merani Komary	585024	3935339	1249	54	485	16	7.8	315
59	Keloo	578390	3953483	1290	65	463	16	8.4	301
60	Nizara	586038	3946372	1290	60	511	14	8.2	332
61	Gokhlan	587028	3949089	1279	63	530	19	8.4	345
62	Uch tapan	598197	3952425	1280	70	472	20	8.4	307
63	Nawgirdan	580542	3915229	542	125	448	19	8.6	291
64	Said sadiq center	581926	3909386	518	110	602	21	8.3	391
65	Sara Quarter/Said sadiq	577757	3912741	515	100	555	20	8.4	361
66	Hassar Project	577930	3912496	524	40	496	20	8.5	322
67	Haji Qadr	579461	3912596	528	68	531	21	8.6	345
68	Hassar Water Project	580321	3912476	520	80	552	21	8.6	359
69	Moryas	579245	3912385	524	88	445	20	9.3	289
70	Mayawa	565778	3925360	1071	107	790	24	8.1	514
71	Geldara	559294	3926455	1236	92	473	18	8.8	307
72	Kazhaw	560580	3932330	1226	84	600	27	8.4	390
73	Tapi karam	557295	3933568	1128	120	429	24	8.7	279
74	Qalijo	568567	3910902	507	152	518	21	8.7	337
75	Bard Bard	564676	3914456	550	300	1140	31	8.3	741
76	Sarawy Khwaroo	572066	3927602	836	167	544	22	8.4	354
77	Greza village	574837	3914244	512	90	350	21	8.1	228
78	Sherabara Village	566765	3917246	613	120	601	22	8.6	391

Table A3. *Cont.*

SN	Well Name	X	Y	Z	Well Depth (m)	EC (μS/cm)	Temp. °C	pH	TDS (mg/L)
79	Qawela Village/3	567313	3920629	742	265	2730	24	6.8	1775
80	Qawela	571488	3925050	837	183	613	23	8.8	398
81	Mirmam	571620	3926245	774	199	454	21	9.4	295
82	Hozy Khwaja	570526	3927690	799	260	445	23	8.8	289
83	Mizgawta	574270	3926586	750	88	483	22	8.7	314
84	Kani Pankai Khwaroo	578970	3920647	617	95	465	19	8.9	302
85	Qumashy Saroo	564686	3915211	552	83	739	20	8.6	480
86	Barda Rash	571627	3913618	498	132	436	24	8.9	283
87	Wandarena	573438	3920197	645	110	708	21	8	460
88	Shoke	561237	3937364	1320	55	678	17	8.2	441
89	Barzinja	559097	3936440	1202	151	650	22	7.3	423
90	Barzinja	562508	3934493	1307	115	650	21	7.3	423
91	Gelara	563283	3933352	1308	100	500	22	7.3	325
92	Kani Panka	560518	3932550	1279	90.5	480	24	8.7	312
93	Kani Spika	565734	3914671	542	133	515	22	8.5	335
94	Kani Spekae	572643	3917416	563	75	556	22	9.3	361
95	Kani Speka	571821	3916995	562	55	500	22	7.3	325
96	Sarawy saro	571962	3917302	570	63	673	23	8.5	437
97	Be rashka	575575	3915205	513	52.6	485	19	8.5	315
98	Auch quba village	580698	3908617	502	115	382	23	8.1	248
99	Shanadary Kon	583202	3922138	616	127	630	21	7.7	410
100	Warmawa	581302	3919741	583	120	413	21	7	268
101	Warmawa	561452	3907065	587	133	448	21	7.2	291
102	Jollana	561420	3907764	560	116.7	424	19	7.5	276

Table A3. *Cont.*

SN	Well Name	X	Y	Z	Well Depth (m)	EC (µS/cm)	Temp. °C	pH	TDS (mg/L)
103	Daq	554720	3905994	674	146	510	20	7.1	332
104	Cham w Zhala	568733	3901292	667	60	322	21	7.9	209
105	Awakala	561888	3900517	705	68	560	19	7.7	364
106	Yakhshee khwaro	550792	3904199	890	31	562	19	7.8	365
107	Ashtokan	564619	3907908	545	72	3290	24	8.8	2139
108	Jabara	494074	3827861	149	100	1270	25	8.9	826
109	Khidran	479803	3998683	552	127	387	19	8.4	252
110	Warmin	572771	3884546	415	42	550	19	7.4	358
111	Saedawa	519335	3840306	305	86	500	15	7.6	325
112	Saeeda	517150	3837429	293	120	620	23	8.8	403
113	Fatah homar	503774	3848391	348	120	570	23	9.4	371
114	Bakrashal	502589	3847738	345	197	630	25	9.1	410
115	Homar bli gawra	499503	3849244	336	100	820	27	8.6	533
116	Chanakhchian	561631	3917366	645	129	351	22	8.9	228
117	Kullajoi Hama jan	512916	3856680	513	108	517	25	8	336
118	Zangi Gawra	537252	3860126	561	147	537	24	7.4	349
119	Zerinjo Khwaro	562798	3913032	549	111.3	787	29	7.5	512

Appendix D

Table A4. Physicochemical Parameter Analysis for the Deep Wells in Duhok City, UTM-WGS84 Coordination System.

SN	Well Name	X	Y	Z	Well Depth (m)	EC (μS/cm)	Temp. °C	pH	TDS (mg/L)
1	Avrik	332933	4078301	862	180	862	21	7.5	560
2	Ekmale	326002	4085892	730	184	864	21	7.7	562
3	Gre Qesrok	323283	4084514	673	107	857	19	7.8	557
4	Etot	327324	4080718	580	191	785	18	7.7	510
5	Bade	329095	4086427	8190	302	835	23	8	543
6	Banye	341132	4087887	850	210	783	21	7.6	509
7	Duhok	325430	4086000	591	131	1200	22	7.3	780
8	Bagerat	336632	4090235	8310	150	745	23	7.7	484
9	Botya	323524	4089547	763	160	1300	21	7.5	845
10	Berebhar	330449	4083264	742	153	755	20	7.4	491
11	Malta Saro	316261	4081602	512	153	680	20	7.8	442
12	Malta Khwaro	316028	4080686	492	200	1100	23	7.5	715
13	Zawite	335164	4088211	790	134	845	21	7.2	549
14	Sindor	326803	4085922	7280	203	768	20	7.6	499
15	Shakhke	319249	4083645	671	220	700	21	7.8	455
16	linava	319127	4089788	6720	194	687	21	7.5	447
17	Duhok	317032	4080858	498	186	600	21	7.2	390
18	Shakhke	319426	4083407	6750	200	625	21	7.5	406
19	Zirka	315728	4084302	599	180	620	21	7.3	403
20	Warmele	336886	4115352	1169	67	860	21	7.9	559
21	Gjabara	322976	4081776	656	232	451	21	7.9	293
22	Nizarke/10	327177	4078190	700	150	545	20	7.5	354

Table A4. *Cont.*

SN	Well Name	X	Y	Z	Well Depth (m)	EC (µS/cm)	Temp. °C	pH	TDS (mg/L)
23	Bakoze	313436	4089081	757	141	2400	22	7.2	1560
24	Baroshke	324248	4080204	584	140	780	17	7.9	507
25	Gaverke	316818	4080277	492	200	550	18	7.5	358
26	Qarqarava	324757	4086611	705	155	1500	23	8.1	975
27	Segirka	322038	4079194	625	240	682	18	7.3	443
28	Nezarke	323178	4078768	665	152	700	19	7.6	455
29	Bagera Khwaro	336388	4090200	802	106	600	21	7.6	390
30	Eminke	329887	4081165	859	174	750	21	7.3	488
31	Koret Gavana	335110	4088056	818	165	585	18	7.6	380
32	Berebuhar	330356	4083178	712	172	790	18	7.8	514
33	Khrabiya	340713	4090371	921	112	500	18	7.8	325
34	Ronahi	323086	4078936	674	220	520	14	7.9	338
35	Serhildan	325542	4081164	757	260	562	16	7.4	365
36	Zari land	317834	4082918	569	200	1410	18	7.9	917
37	Shakhki	319424	4083645	686	190	490	15	7.7	319
38	Shindokha	318237	4092506	571	90	650	19	7.4	423
39	Mezringan	408599	4073901	839	71	500	22	7.4	325
40	Nihawe	321642	4081097	547	124	672	23	7.5	437
41	Gondik	392620	4072732	736	89	785	23	7.4	510
42	Jem Sine	387615	4077588	594	180	783	23	7.4	509
43	Tobzawe	388907	4047021	457	146	585	21	7.2	380
44	Drin Khaje	398058	4059939	557	126	524	23	7.3	341
45	Shoshe	388592	4072203	759	195	496	23	7.3	322
46	Meroke	422094	4068676	798	115	569	23	7.3	370

Table A4. *Cont.*

SN	Well Name	X	Y	Z	Well Depth (m)	EC (µS/cm)	Temp. °C	pH	TDS (mg/L)
47	Serderava	348867	4098992	1099	232	738	19	7.4	480
48	Miska	348098	4115084	979	188	650	19	8	423
49	Syretika	342060	4098101	1141	225	580	20	7.6	377
50	Barashe	348006	4094969	1228	136	700	20	8	455
51	Bibava	347129	4100414	811	150	550	19	7.4	358
52	Dihe	338122	4111800	980	130	580	19	7.4	377
53	Dokare	332058	4112710	720	200	980	20	7.5	637
54	Bamerne	345738	4109899	1139	130	540	18	7.4	351
55	Shrty	343173	4108183	1009	200	965	19	7.6	627
56	Teni	343807	4106965	1017	166	970	19	8	631
57	Dokary	332058	4112710	720	200	980	18	7.9	637
58	Zewa shikh pirmos	340849	4110891	1196	128	520	19	7.6	338
59	Hloora	381168	4101451	625	158	850	19	7.6	553
60	Kanya mala	366482	4107925	1205	100	437	15	7.7	284
61	Khlbish	340747	4107188	853	135	500	16	8	325
62	Qadish	357073	4108234	1233	100	485	16	7.2	315
63	Kerbraski	340985	4100222	828	170	600	19	7.9	390
64	Hdene	353749	4123526	1514	112	460	22	7.7	299
65	Pase	332028	4113791	7175	174	510	19	7.5	332
66	Ekmale	361241	4114833	1161	102	2015	20	7.9	1310
67	Bilminde	381566	4079678	512	145	549	21	7.6	357
68	Jimbilke	348070	4115396	954	120	1300	20	7.2	845
69	Baretin	346145	4076550	640	157	732	22	7.5	476
70	Dize	343544	4079892	637	195	784	24	7.5	510

Table A4. *Cont.*

SN	Well Name	X	Y	Z	Well Depth (m)	EC (µS/cm)	Temp. °C	pH	TDS (mg/L)
71	Mersida	366533	4076234	656	144	748	23	7.2	486
72	Shkeft hindiyan	349332	4072117	750	130	618	22	7.6	402
73	Shlya	356012	4080442	873	150	930	22	7.7	605
74	Shehiya	353080	4075300	528	180	645	21	7.6	419
75	Der khidre	354326	4075470	564	205	655	21	7.4	426
76	Mkirs	351824	4075092	544	170	930	22	7.7	605
77	Baratin	345076	4076563	624	90	658	21	7.2	428
78	Avriva	355110	4068857	570	167	543	22	7.9	353
79	Khinis	358591	4069166	461	91	548	23	7.5	356
80	Geli roman	343896	4078297	772	162	540	18	7.6	351
81	Migara	352123	4071416	778	134	657	20	7.9	427
82	Badinava	356927	4075417	568	230	552	20	7.8	359
83	Mam yezdin	343619	4072331	760	146	745	23	7.8	484
84	Ba'adre	344241	4067121	549	94	758	21	7.7	493
85	Basewa	334650	407379	650	177	650	24	7.7	423
86	Jeman	337429	4076432	783	70	689	22	7.5	448
87	Brifka	341464	4075308	1010	86	765	22	7.7	497
88	Shekh Hesen	339922	4073189	755	178	602	24	7.1	391
89	Ba'adre	344234	4064329	465	200	520	20	7.2	338
90	Esyan	347180	4065072	548	86	397	21	7.5	258
91	Beroshka Sa'adon	328933	4103109	999	194	600	22	7.4	390
92	Beshinke	324021	4094686	747	176	925	18	7.5	601
93	Mangesh	330420	4099214	1002	180	600	21	7.8	390
94	Majilmakht	337016	4097365	1035	95	700	23	7.5	455

Table A4. *Cont.*

SN	Well Name	X	Y	Z	Well Depth (m)	EC (µS/cm)	Temp. °C	pH	TDS (mg/L)
95	Besifke alsufla	331784	4095148	885	161	510	18	7.3	332
96	Besifke	331016	4095108	880	115	548	18	7.3	356
97	Kamaka	329540	4092271	834	198	1300	21	7.3	845
98	Ekmala khabor	319632	4104534	824	175	600	29	7.5	390
99	Alkish	339549	4097669	1040	86	879	23	7.9	571
100	Zeka abu	321969	4100997	825	150	898	22	7.9	584
101	Dilya	315739	40959446	895	148	550	18	7.4	358
102	Shawreke	320859	4097417	682	169	380	26	7.8	247
103	Gre pete	327983	4096002	794	116	410	21	7.8	267
104	Gond kose	317281	4107436	558	124	786	21	7.3	511
105	Kovle	334505	4101971	897	184	500	20	7.4	325
106	Rostinke	344994	4098613	990	182	765	22	7.3	497
107	Navishke	319951	4103261	899	196	480	16	7.4	312
108	Ozmana	317377	4103967	772	180	472	16	7.6	307
109	Koreme	332693	4104079	1032	125	450	17	7.9	293
110	Milhimban	325857	4099653	904	198	448	15	7.4	291
111	Alindke	317491	4098277	704	157	590	17	7.8	384
112	Derke	334586	4093466	942	150	573	17	7.6	372
113	Grepte	328113	4095997	824	182	420	15	7.7	273
114	Zinava	309400	4099865	826	130	832	16	7.8	541
115	Ashanke	313466	4102931	633	201	520	18	7.4	338
116	Dergijnik	328190	4096028	834	198	555	19	7.4	361
117	Kerble	325584	4095738	719	105	620	21	7.5	403
118	Qesrok	364315	4352009	432	200	539	22	7.8	350

Table A4. *Cont.*

SN	Well Name	X	Y	Z	Well Depth (m)	EC (μS/cm)	Temp. °C	pH	TDS (mg/L)
119	Mitka Seri	370774	4072489	489	140	465	23	7.6	302
120	Selke	361728	4067073	443	162	560	22	7.4	364
121	Baviyan	357519	4066265	432	173	529	21	7.6	344
122	Mitka alsufli	370535	4072636	485	73	612	20	7.3	398
123	Piran	363081	4059678	404	169	520	21	7.5	338
124	Hinjirok	356781	4046063	484	182	366	21	8	238
125	Mam Reshan	359131	4058926	398	180	459	23	8.1	298
126	Shekhan	352873	4063659	554	120	452	21	7.9	294
127	Doshivan	348219	4058619	408	190	570	21	7.5	371
128	Almeman	344915	4058922	412	180	598	22	7.5	389
129	Shiv shrin	346445	4057335	395	190	570	22	7.6	371
130	Said Zari	303956	4081666	439	200	822	21	7.5	534
131	Sertank	311095	4081010	482	155	700	19	7.3	455
132	Sumail	308642	4081337	464	200	1020	20	7.3	663
133	Sershor	303855	4080492	483	220	650	17	7.8	423
134	Sertank	310616	4081046	475	180	600	19	7.3	390
135	Domize	311857	4072838	420	200	670	20	7.7	436
136	Khorshinia	317072	4072797	479	130	450	19	7.4	293
137	Qasreen	322750	4088132	520	190	750	19	7.9	488
138	Sharya	319509	4071272	469	190	650	20	7.2	423
139	Sharia	319509	4071272	469	190	620	20	7.2	403
140	Sharya complex	324681	4068938	588	151	710	19	7.2	462
141	Upper Deleb	313919	4080894	507	192	600	18	7.6	390
142	Bakhtme	311877	4073095	426	204	935	20	7.7	608

Table A4. *Cont.*

SN	Well Name	X	Y	Z	Well Depth (m)	EC (µS/cm)	Temp. °C	pH	TDS (mg/L)
143	Domiz	311857	4072838	420	200	1000	20	7.7	650
144	Bakhetme	311877	4073095	426	204	935	20	7.7	608
145	Dostka	328516	4073408	560	70	746	20	7.6	485
146	Rezgari complex	312107	4072823	422	134	746	16	7.6	485
147	Qsreen	321431	4065377	540	232	487	18	7.4	317
148	Sharya	319846	4071251	485	170	1085	21	7.9	705
149	Shekh Khedre	323702	4075661	605	200	518	18	7.5	337
150	Uppe deleb	313919	4080894	507	192	220	18	7.8	143
151	Meserik	305001	4082573	452	162	800	19	7.9	520
152	Ivzorok Shane	278649	4100051	446	170	455	20	7.6	296
153	Batel	293169	4093222	502	195	750	22	8	488
154	Khrabdem complex	287050	4092307	391	180	650	17	7.6	423
155	Kilke	282263	44099441	451	178	490	21	7.6	319
156	Sershor	302376	4088065	482	158	766	18	7.8	498
157	Ave zerik miri	290289	4093611	500	180	450	23	7.6	293
158	ALasy	295729	4099390	655	126	600	21	7.4	390
159	Pebzne	278050	4104759	561	203	780	22	7.4	507
160	Aloka	315966	4078631	543	153	620	19	7.6	403
161	Sumail	309339	4081170	463	190	755	21	7.4	491
162	Sitke	325368	4070241	464	184	635	23	7.7	413
163	Meserik	306104	4081504	454	200	740	22	7.3	481
164	Avzorok Mere	290287	4093611	500	180	450	24	7.6	293
165	Bakhetme	309607	4075254	357	200	745	23	7.5	484
166	Kilke	282813	4099445	467	178	430	22	8.3	280

Table A4. *Cont.*

SN	Well Name	X	Y	Z	Well Depth (m)	EC (µS/cm)	Temp. °C	pH	TDS (mg/L)
167	Hajeya	293067	4099617	592	175	430	28	7.8	280
168	Gre gawre	304000	4082966	452	200	998	26	7.4	649
169	Tanahi	312500	4081614	483	200	530	24	7.5	345
170	Qsara	316054	4078975	543	200	802	24	7.4	521
171	Kwashi	303958	4096995	756	210	571	20	7.6	371
172	Tenahi	313130	4081411	501	198	414	22	7.5	269
173	Qeshefre	312110	4087446	670	143	586	22	7.5	381
174	Tobzawe	302558	4088235	486	180	822	23	7.2	534
175	Selan Mamik	384752	4074279	835	164	585	20	7.2	380
176	Sercaf	370214	4076013	604	210	483	21	7.3	314
177	Basifre	365416	4076433	647	194	595	21	7.7	387
178	Shekhka	360730	4069702	522	116	390	21	7.3	254
179	Rkava	340688	4073959	455	190	570	21	7.8	371
180	Mam Yezdin	342976	4072779	738	140	470	21	7.9	306
181	Ba'adre	344273	4066695	532	180	470	21	7.6	306
182	Bakhirnif	320421	4089908	676	135	900	21	6.7	585
183	Zawite	335164	4088211	785	165	840	21	7.2	546
184	Memane	329270	4089449	963	201	780	20	7.7	507
185	Gelbok	380339633	47090271	892	237	575	21	7.3	374
186	Betase	334477	4075080	659	84	545	20	7.3	354
187	Jeman	337061	4075691	746	130	1460	21	7.5	949
188	Pishta Gre	323350	4085115	683	129	510	21	6.3	332
189	Bagera	335314	4092774	545	200	480	21	7.3	312
190	Rkava	340592	4073943	766	163	540	20	6.8	351

Table A4. *Cont.*

SN	Well Name	X	Y	Z	Well Depth (m)	EC (µS/cm)	Temp. °C	pH	TDS (mg/L)
191	Mangesh	330406	409923	993	210	675	21	7.5	439
192	Tehlava	332960	4109484	719	196	490	20	8.6	319
193	Banka	340243	4113227	1509	245	854	19	7.6	555
194	Shrty	341723	4108183	1009	200	770	20	7.8	501
195	Berashe	346381	4096204	1327	170	750	20	6.8	488
196	Spindare	349166	4094425	1177	86	790	20	7.7	514
197	Siare	357226	4092036	1043	72	545	20	8	354
198	Tazika	347020	4098568	999	92	320	20	7.1	208
199	Rostinke	343510	4098339	1138	290	648	20	7.8	421
200	Kani golan	367404	4108752	1382	223	582	19	7.5	378
201	Metin	341655	4112259	1595	242	893	19	7.6	580
202	Sarke	395247	4096108	598	110	759	19	7.5	493
203	Migara	347800	412885	1464	193	913	20	7.5	593
204	Bircat	360549	4081089	1045	250	1250	20	6.5	813
205	Rkava	340494	4073953	767	220	540	20	6.8	351
206	Baadre	317465	4079326	570	191	610	21	6.8	397
207	Qesrok	3643167	4352016	446	170	735	22	7.3	478
208	Hasan Iva	291494	4107300	703	168	660	23	6.5	429
209	Hezel	296303	4115226	465	160	400	22	6.9	260

Appendix E

Table A5. Physicochemical Parameter Analysis for the Deep Wells in Halabja City, UTM-WGS84 Coordination System.

SN	Well Name	X	Y	Z	Well Depth (m)	EC (µS/cm)	Temp. °C	pH	TDS (mg/L)
1	Bakhtiary	592071	3892595	825	163	390	20	8.5	254
2	Bawakochak	588873	3890000	859	134	385	19	8.3	250
3	Zamaqi	588203	3894673	660	120	443	18	8.1	288
4	Near to ababaile	593610	3892759	943	109	519	20	8.5	337
5	Jalila	592722	3895712	756	100	348	22	8.7	226
6	Anab—Jalila	592572	3895984	765	97	365	22	8.3	237
7	Kishadary	586062	3907468	503	115	441	20	8.4	287
8	Kani Too	581276	3894130	652	130	304	21	6.5	198
9	Belanga	583882	3892091	695	61	527	23	8.3	343
10	Miraelly	580016	3890642	633	121	623	21	8	405
11	Chrostana	581313	3890799	644	185	1151	22	7.9	748
12	Gunda Village	582370	3890069	684	126	408	20	8.2	265
13	Hana Zhalla Village	582179	3888734	675	100	950	23	7.5	618
14	Saraw Village	584382	3889504	786	105	516	19	8	335
15	Presy Saroo	585853	3891312	809	95	432	19	8.1	281
16	Byawella	592426	3897181	734	100	382	22	9.1	248
17	Anab-Byawella	592117	3897069	729	89	384	22	8.5	250
18	Khakukholl	586980	3904073	519	129	484	21	7.6	315
19	Basharaty Khwaroo	586841	3902459	533	105	387	23	8.4	252
20	Kagrdal	583261	3900444	513	110	452	22	8.1	294
21	Ghwlami khwaroo	579248	3897246	510	99	596	24	8.1	387
22	Imam zamin	579938	3898233	511	120	439	24	8.2	285
23	Sharazor Project-6	586483	3899015	562	128	420	22	8.3	273

Table A5. *Cont.*

SN	Well Name	X	Y	Z	Well Depth (m)	EC (µS/cm)	Temp. °C	pH	TDS (mg/L)
24	Khurmal	594307	3907412	563	98	2540	29	8.4	1651
25	Amwra	591361	3912840	660	112	425	25	9.6	276
26	Mirt soor	590598	3914380	702	95	346	21	6.6	225
27	Qulkhurd	587237	3913052	526	74	742	22	9.3	482
28	Mala waisa	587657	3911858	531	63	570	22	9.3	371
29	Aliawa	585467	3913493	528	160	437	23	9.5	284
30	Shashki khwaroo	589662	3904454	558	72	360	21	8.3	234
31	Rostum bag	595651	3904943	632	84	351	24	8.2	228
32	Gomalar	592687	3903819	609	101	365	24	7.8	237
33	De kon	594328	3904138	627	100	362	21	7.5	235
34	Zardahal	598089	3899419	906	181	448	21	7.9	291
35	Qainaja	585183	3916320	543	165	520	20	7.3	338
36	Tapy safa	589053	3905505	550	129	514	24	8.3	334
37	Dalamar	591410	3892298	798	156	356	20	8.5	231
38	Shakrally	587155	3904786	521	164	468	22	8.7	304

Appendix F

Table A6. Physicochemical Parameter Analysis for the Spring and River Samples in the Study Area, UTM-WGS84 Coordination System.

SN	Spring Name	X	Y	Z	EC (μS/cm)	Temp (°C)	TDS Mg/L	pH
1	Hiran Spring	454550	4014850	925	536	24.5	348.4	7.5
2	Kani Hanjeer Spring	445441	4036667	744	525	24.3	341.3	7.6
3	Sisawa Spring	447906	4038217	861	536	24.2	348.4	7.6
4	Amokan Spring	435472	4053369	634	507.5	24.9	329.9	7.8
5	Kani Chirgan Spring	432855	4054601	556	644.5	24.6	418.9	7.8
6	Kani Qura Bag Spring	425821	4052478	407	469.5	24.7	305.2	7.9
7	Graw Spring Spring	426026	4043286	627	656	24.6	426.4	7.6
8	Kani Khazal Spring	429244	4049134	503	571	24.8	371.2	7.8
9	Aspendara Spring	450966	4023173	974	540	24.8	351.0	7.5
10	Gomashin Spring	513551	3940143	889	309	17.8	200.8	8.1
11	Kanisarwchawa Spring	501402	3950379	939	343	19.4	223.1	7.6
12	Cholmak Spring	503981	3939024	925	349	18	227.1	7.6
13	Mortka Spring	505940	3936511	935	335	16.7	218.0	7.7
14	Zekan Spring	510020	3934824	811	346	17.2	225.1	7.6
15	Khaldan Spring	510846	3936146	823	253	18.7	164.3	7.9
16	Alibzaw Spring	513125	3933774	826	293	18.5	190.6	7.8
17	Qushqaya Spring	519069	3935353	827	197	18.8	127.8	8.2
18	Kani shaya Spring	519303	3933194	783	250	19	162.2	7.7
19	Warmziar Spring	522429	3930660	805	406	19.6	263.6	7.6
20	Barowi gawra Spring	526007	3926479	937	463	17.6	301.1	7.7
21	Darikali Spring	523300	3924898	957	371	17.6	241.3	7.8
22	Halai sarwchawa Spring	518219	3932409	793	367	18.3	238.3	7.7
23	Shekhmand Spring	514267	3929614	815	351	26.5	228.2	8.1

Table A6. *Cont.*

SN	Spring Name	X	Y	Z	EC (µS/cm)	Temp (°C)	TDS Mg/L	pH
24	Gomatagach Spring	514365	3933015	810	353	19	229.2	7.8
25	Hanjeera Spring	509826	3932541	890	365	16.6	237.3	7.8
26	Aligoran Spring	510422	3931134	989	510	20.5	331.6	8.1
27	Delezha Spring	517743	3923677	791	362	16.5	235.2	7.7
28	Gurbaz Spring	520783	3921370	803	413	21	268.7	7.8
29	Azaban Spring	571219	3898928	870	345.5	18	221.1	7.6
30	Siyara Spring	582344	3898235	602	459	20	293.8	7.4
31	Birke Spring	574532	3896998	835	704	17.6	450.6	7.5
32	Qashti Spring	565342	3893567	699	796.5	19	509.8	7.4
33	Ahmed Brenda Spring	575461	3881721	899	342.5	18.5	219.2	7.7
34	Greater Zab River	423363	4051506	368	388.5	22.2	252.5	8.1
35	Lesser Zab River	430190.4	3966505	302	362.5	22.7	235.6	8.2
36	Sirwan River	563919.9	3884792	367	374	22.9	243.1	8.2
37	Tanjero River	574854.6	3895831	457	361	22	235	8

G.

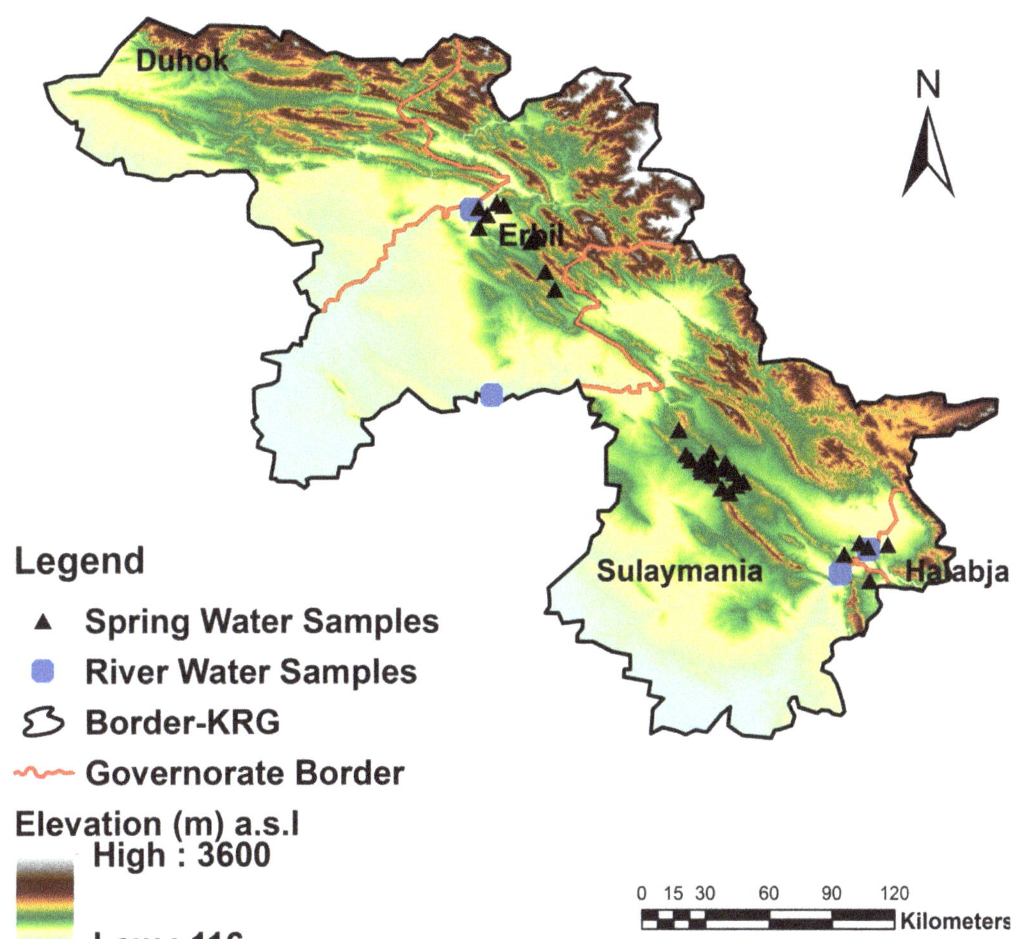

Figure A1. Location Map Showing the Spring and River Samples of the Study Area.

Appendix H

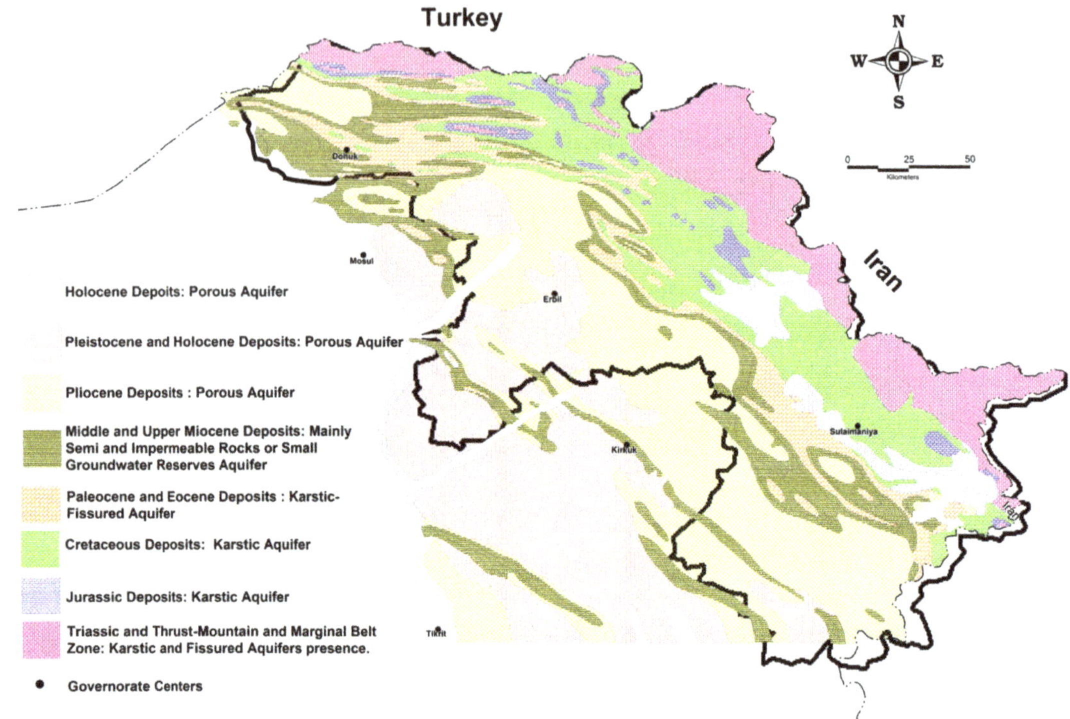

Figure A2. Hydrogeological Map (Aquifer System) across the Kurdistan Region (after Stevanovic and Marcovich, 2004).

References

1.   Erol, A.; Randhir, T.O. Climatic change impacts on the Ecohydrology of Mediterranean watersheds. *Clim. Chang.* **2012**, *114*, 319–341. [CrossRef]
2.   Christensen, J.H.; Kumar, K.K.; Aldria, E.; An, S.-I.; Cavalcanti, I.F.A.; De Castro, M.; Dong, W.; Goswami, P.; Hall, A.; Kanyanga, J.K.; et al. Climate Phenomena and Their Relevance for Future Regional Climate Change Supplementary Material. Contribution of Working Group I to the Fifth Assessment Report of the Intergovernmental Panel on Climate Change, 2013, Climate Change 2013: The Physical Science Basis 62. Available online: http://www.climatechange2013.org (accessed on 29 January 2013).
3.   Booker, J.F.; Trees, W.S. Implications of Water Scarcity for Water Productivity and Farm Labor. *Water* **2020**, *12*, 308. [CrossRef]
4.   Habib, Z.; Wahaj, R. *Water Availability, Use and Challenges in Pakistan—Water Sector Challenges in the Indus Basin and Impact of Climate Change*; FAO: Islamabad, Pakistan, 2021. [CrossRef]
5.   Seather, O.M.; Caritat, P.D. *Geochemical Processes, Weathering and Groundwater Recharge in Catchments*; A. A. Balkama: Rotterdam, The Netherlands, 1997; 400p.
6.   Davis, S.N.; Dewiest, R.J. *Hydrology*; John Wiley & Sons. Inc.: New York, NY, USA, 1966; 463p.
7.   Walton, C.W. *Groundwater Resource Evaluation*; McGraw Hill Co.: New York, NY, USA, 1970; 664p.
8.   Langmuir, D. *Aqueous Environmental Geochemistry*; Prentice Hall: Hoboken, NJ, USA, 1997; 600p.
9.   Houatmia, F.; Azouzi, R.; Charef, A.; Bédir, M. Assessment of groundwater quality for irrigation and drinking purposes and identification of hydrogeochemical mechanisms evolution in northeastern, Tunisia. *Environ. Earth Sci.* **2016**, *75*, 746. [CrossRef]
10. Chakraborty, S.; Kumar, R.N. Assessment of groundwater quality at a MSW landfill site using standard and AHP based water quality index: A case study from Ranchi, Jharkhand, India. *Environ. Monit. Assess.* **2016**, *188*, 335. [CrossRef] [PubMed]
11. Xu, P.; Zhang, Q.; Qian, H.; Zheng, L. Spatial distribution characteristics of irrigation water quality assessment in the Central-Western Guanzhong Basin, China. *IOP Conf. Ser. Earth Environ. Sci.* **2021**, *647*, 012143. [CrossRef]
12. Chan, F.; Kumar, N.; Tiwar, M.; Lau, H.; Choy, K. Global supplier selection: A fuzzy-AHP approach. *Int. J. Prod. Res.* **2008**, *46*, 3825–3857. [CrossRef]
13. Stevanovic, Z.; Markovic, M. *Hydrogeology of Northern Iraq, Climate, Hydrology, Geomorphology & Geology*; FAO: Rome, Italy, 2003; Volume 1, pp. 1–122.

14. Stevanovic, Z.; Markovic, M. *Hydrogeology of Northern Iraq, Hydrogeology and Aquifer System*; Specific edition; FAO UN: Rome, Italy, 2004; Volume 2, pp. 1–175.
15. Vajsáblová, M. Variograms and regression methods of their creation. *Acad. J.* **2013**, *8*, 1998–2007.
16. Train, R.E. *Quality Criteria for Water*; Castle House Publication, Ltd.: London, UK, 1979; 256p.
17. Hillel, D. *Salinity Management for Sustainable Irrigation Integrating Science, Environment, and Economics, Environmentally and Socially Sustainable Development: Rural Development*; The World Bank: Washington, DC, USA, 2000.
18. Drever, J.I. *The Geochemistry of Natural Water, Surface and Groundwater Environment*, 3rd ed.; Prentice Hall: Hoboken, NJ, USA, 1997; 436p.
19. Altoviski, M.E. *Hand Book of Hydrogeology*; Gosgeolitzda Moscow, USSR: Moscow, Russia, 1962; 614p. (In Russian)
20. Gorrell, H.A. *Classification of Formation Waters Based on Sodium Chloride Content*; American Association of Petroleum Geologists Bulletin: Tulsa, OK, USA, 1958; Volume 42, No. 10.
21. U.S.D.A. *USDA Census of Agriculture*; United State Department of Agriculture: Washington, DC, USA, 1954.
22. Mayer, X.M.; Ruprecht, J.K.; Bari, M.A. *Stream Salinity Status and Trends in South-West Western Australia*; Report No. SLUI 38; Department of Environment, Salinity and Land Use Impacts Series; Department of Environment, Government of Western Australia: Perth, WA, Australia, 2005.
23. Todd, D.K. *Groundwater Hydrology*, 2nd ed.; John Wiley & Sons, Inc.: New York, NY, USA, 1980; 535p.
24. Rhoades, J.D.; Kandiash, A.; Mashali, A.M. *The Use of Saline Water for Crop Production*; Irrigation and Drainage paper No. 48; FAO: Rome, Italy, 1992.
25. Don, C.M. *A Grows Guide to Water Quality*; University College Station: College Station, TX, USA, 1995.
26. Ayers, R.S.; Westcot, D.W. *Water Quality for Agriculture. Irrigation and Drainage Paper 29*; Rev.1; FAO: Rome, Italy, 1989; 174p.
27. Grossman, Z. International Environmental Problems & Policy, a Class Website on Water Privatization and Commodification. Produced by Students of Geography at the University of Wisconsin-Eau Claire, USA. 2004. Available online: https://view.officeapps.live.com/op/view.aspx?src=https%3A%2F%2Fkuliah.rizaldi.web.id%2FESL%2520223_Ekonomi%2520Sumber%2520Daya%2520Air%2FWater%2520is%2520Life.doc (accessed on 8 September 2021).
28. Leavy, D.B.; Kearney, W.F. Irrigation of Native Rangeland Using Treated Waste Water From Institute Uralian processing. *J. Environ. Qual.* **1999**, *28*, 208–217. [CrossRef]

Article

Groundwater Quality Evaluation and the Validity for Agriculture Exploitation in the Erbil Plain in the Kurdistan Region of Iraq

Shwan Seeyan [1,2,*], Haifa Akrawi [1], Mohammad Alobaidi [3], Karrar Mahdi [4], Michel Riksen [4] and Coen Ritsema [4]

[1] Soil and Water Department, Agriculture Engineering Sciences College, Salahaddin University-Erbil, Erbil 44002, Iraq

[2] Department of Petroleum and Mining Engineering, Engineering Faculty, Tishk International University, Erbil 44002, Iraq

[3] Soil and Water Department, Agriculture and Forestry College, Mosul University, Mosul 41001, Iraq

[4] Soil Physics and Land Management Group, Wageningen University & Research, 6700 AA Wageningen, The Netherlands

[*] Correspondence: shwan.seeyan@su.edu.krd; Tel.: +964-750-4487884

Citation: Seeyan, S.; Akrawi, H.; Alobaidi, M.; Mahdi, K.; Riksen, M.; Ritsema, C. Groundwater Quality Evaluation and the Validity for Agriculture Exploitation in the Erbil Plain in the Kurdistan Region of Iraq. *Water* **2022**, *14*, 2783. https://doi.org/10.3390/w14182783

Academic Editor: Yuanzheng Zhai

Received: 13 August 2022
Accepted: 3 September 2022
Published: 7 September 2022

Publisher's Note: MDPI stays neutral with regard to jurisdictional claims in published maps and institutional affiliations.

Abstract: Climate change and the fast growth of industrial and agricultural enterprises can have a negative impact on groundwater quality. The evaluation of groundwater quality is an important issue to determine the suitability of water for agriculture and other purposes in the Kurdistan Region of Iraq. The quality of water is an important indicator for selecting the best Climate Smart Agriculture practices that can be applied in the region. Industrial and agricultural enterprises use massive amounts of groundwater pollutants such as fertilizers and pesticides, especially in the agriculture sectors. Groundwater samples were collected from varying depths of 110 to 200 m for chemical and physical analysis to determine water availability and quality as well as the effect of water use and of drought on groundwater level fluctuation in Erbil City. The analysis includes pH, electrical conductivity, temperature, total dissolved solids, major cations (Ca^{2+}, Mg^{2+}, Na^+, K^+) and major anions (SO_4^{2-}, HCO_3^-, Cl^-, CO_3^-). The high TDS value is founded in the central part of the study area according to groundwater flow which originates from the mountain area toward the center of the plain. The results of the sodium adsorption ratio (SAR) shows that all water well samples are suitable for irrigation which have a low sodium hazard and use on sodium sensitive crops must be cautioned against, and the sodium hazard shows that there is no toxic effect on the plants because all the groundwater samples fall in the standard limits of sodium percent, which is less than 60%. The sodium hazard is low, based on RSC results, because it falls below the standard limit which is less than 1.5 meq/L. All groundwater samples are classified as having excellent-to-good permeability. The classification of the potential salinity of groundwater samples shows that nine water samples are in the class excellent-to-good, three water samples are good-to-injurious, and four samples are injurious-to-unsatisfactory. The water type in the area is mostly sulfate except for three samples, two of which are of the chloride type and the third is bicarbonate.

Keywords: groundwater quality; water classification; agriculture purposes; hydrochemical indictors; plain area

1. Introduction

Groundwater is the primary source of drinking water in the Erbil region in Kurdistan, Iraq. As the population continues to rise, more water is required for industrial, domestic, environmental, recreational, and agricultural purposes. When water resources are limited, rising demand for water necessitates efficient water resource management and assessment, particularly when the water is to be used for human consumption [1] and crop production. Water management improvements are required to enhance and diversify food production to fulfil the needs of a growing population, while minimizing crop vulnerability to droughts, floods, and climate change [2]. Water management in climate smart agriculture includes techniques such as drip irrigation and hydroponics, which are more dependent on good water quality. Water quality can have a negative influence on the performance of an irrigation system due to the plugging of emitters and sprinklers. These problems can be caused by inorganic solids (silt and sand), organic solids (algae, bacteria and slime) and dissolved solids (calcium, iron and manganese) [3].

Therefore, groundwater chemistry based on hydro-chemical data is necessary for obtaining basic information on water types, categorizing water for various applications, identifying distinct groundwater aquifers, and studying various chemical processes.

The physical and chemical characteristics that impact groundwater quality in a given area are substantially influenced by geological formations and anthropogenic activity [4]. Electrical conductivity levels reflected by salinity damage to plants are highly important considerations in evaluating the quality of water used for irrigation because of its impact on the osmotic pressure of the soil solution and the capacity of plants to absorb water via their roots [5].

Groundwater chemical characteristics play an important role in identifying and assessing water quality, and chemical classification shown by the concentration of various predominant cations, anions and their interrelationships. Ion dissolution in groundwater occurs more frequently as a result of interactions between groundwater and rock or soil, and the evaporation process, than as a result of precipitation or other sources. The composition of rainwater, mineralogy of the watershed and aquifers, topography, and climate controls the chemical composition of surface- and groundwater [6].

Groundwater fluctuation analysis estimated the variations in stored water, renewable storage water quantity, and investment of groundwater uses [7]. Fluctuation is affected by many factors such as rainfall intensity and quantity, Infiltration capacity of the soil and bed rocks, groundwater depth above sea level, topography, evapotranspiration, and water well discharge [8].

Climate change, in the form of longer and more severe droughts or more intense rainfall events leading to flooding, can affect both the quality and quantity of water, necessitating planning and management to mitigate its negative effects on drinking water supplies.

The main objectives of this study were to investigate the possible sources of ions in the groundwater, and to understand the hydrogeological processes and the hydro-chemical characteristics of the groundwater by analyzing irrigation water parameters such as major cations and anions. This will allow for a discussion of the possibility of using groundwater for different purposes. The spatial distribution of hydro-chemical constituents of groundwater related to its suitability for different purposes, groundwater classification, water (quality) type, hypothetical salts and the groundwater level fluctuation were identified for the selected monitoring wells in the study area.

2. Materials and Methods

2.1. Study Area Description

The area of interest is located in the southwestern part of Erbil City and north of Gwer district, which situated in Shamamek district, extending between ($43°39'17''$–$44°0'11''$ E) and ($35°55'10''$–$36°12'24''$ N). The area covers about 663 km^2, the elevation ranging from 300 m to 500 m above sea level within the foothill zone (Figure 1). The area is bordered by the Zurga Zraw Dagh anticline in the south and southwest and Erbil City in the northeast. The crops in the area are mainly wheat and barley, and the irrigation system is surface irrigation.

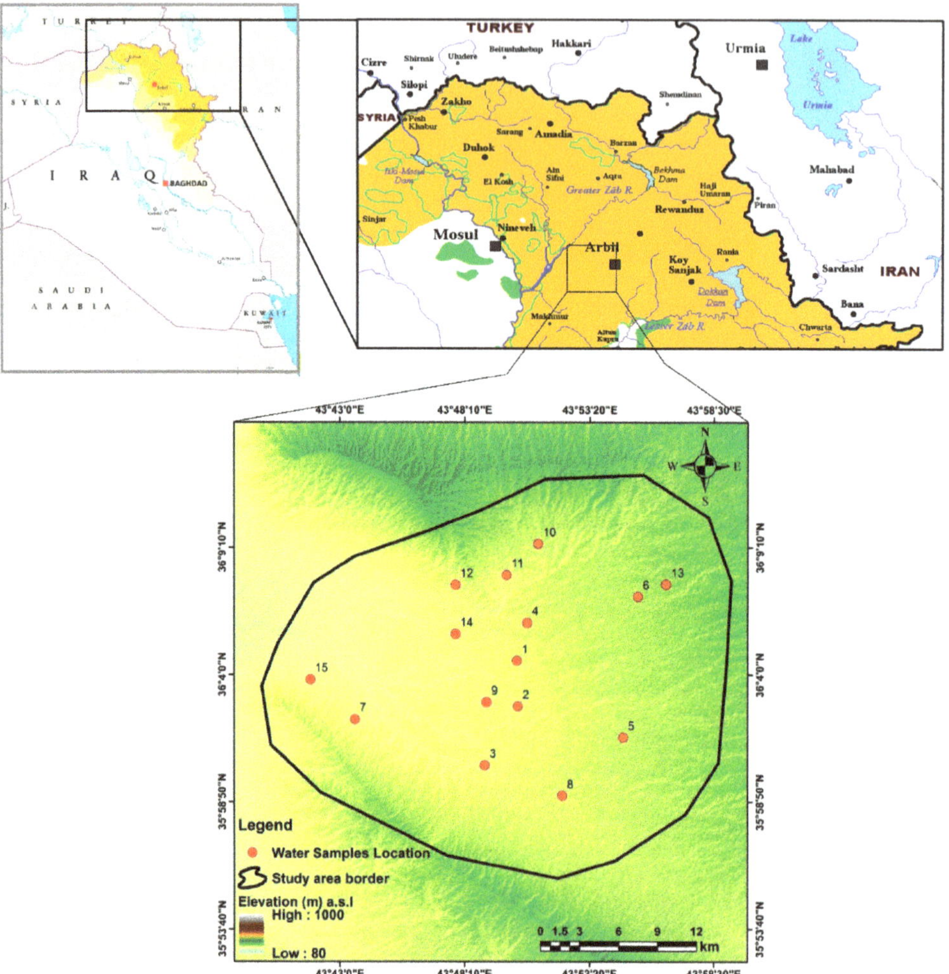

Figure 1. Location map of the study area with water well sample locations.

2.2. Climate

The climate in the study area belongs to the semi-arid Mediterranean type. It is characterized by cold and rainy winters, and long, hot, and dry summers. Meteorological data obtained from the Erbil meteorological station for the period from 2003 to 2020 (Figure 2) shows that annual precipitation is about 456.2 mm, maximum and minimum mean monthly relative humidity is 70.9% in January and 27.3% in July, respectively. Maximum monthly temperature is about 39.9 °C in August and the minimum is about 11.6 °C in January. Maximum evaporation is 136 mm in July and the minimum is 18.7 mm in January. The mean annual sunshine duration is 8.2 h/day, and wind speed is between 1.4 and 2.1 m/s with an annual mean wind speed of 1.8 m/s.

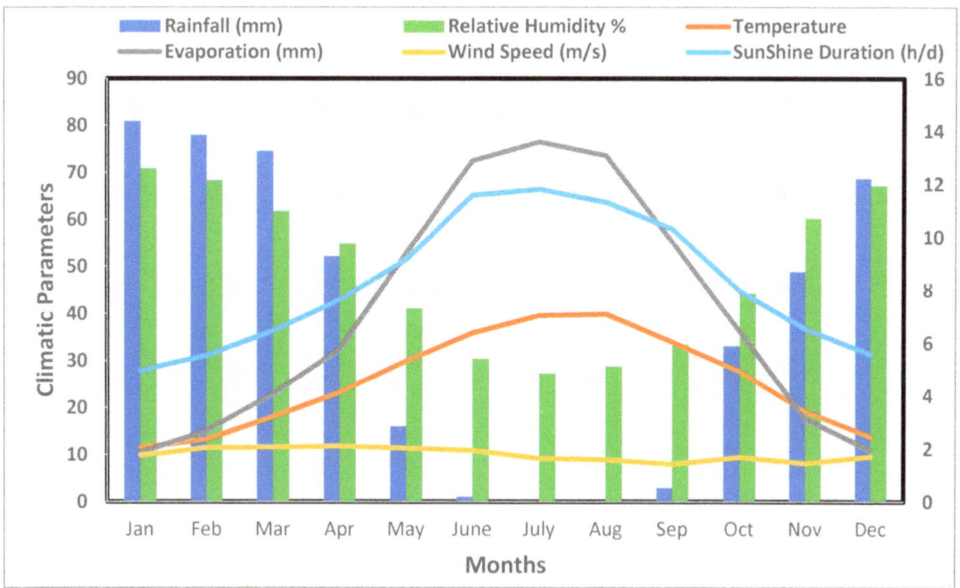

Figure 2. Mean monthly climatic parameters in the study area for the period 2003–2020.

2.3. Lithological and Tectonic Framework

The main outcrops in the study area are Pleistocene and Holocene deposits represented by residual and older terraces consists of conglomerate, gravel, sand, clay, silt; Pliocene deposits represented by the Bai Hassan formation consist of molasses sediment represented by alternating claystone and conglomerate with some sandstone and siltstone, and the Muqdadiya formation, which was laid down in a fluvial environment in a strongly sinking fore deep, and might be considered as typical fresh water molasses with mostly Pliocene age. The aquifer in the study area is porous aquifer.

Tectonically, the study area is a part of the Unstable Shelf Zone that was affected by the Alpine orogeny in the Mesozoic in the Chamchamal-Butma sub-zone of the Foothill Zone. The unstable shelf is characterized by structural trends and faces changes that are parallel to the Zagros-Taurus suture belts [9].

2.4. Water Sampling and Analysis

Sixteen water well samples were collected in the study area in February 2022 to investigate quality, suitability, uses and classification of the water in the study area. Garmin eTrex 20 GPS device was used for field data collection and determining the coordination of the deep well locations, which generally showed a spatial accuracy of ± 4 m (Table 1). Physical parameters for the samples such as temperature, pH and electrical conductivity (EC) were measured in the field using portable EC, T and pH meters. Chemical parameters such as major cations (Ca^{2+}, Mg^{2+}, Na^+, K^+) and major anions (SO_4^{2-}, HCO_3^-, Cl^-, CO_3^-) were analyzed in the laboratory of the University of Mosul using ion chromatography instruments. Total hardness results from the presence of divalent metallic cations of calcium and magnesium, which are very abundant in water. The total hardness (TH) was calculated using the equation given by Hem, 1985 [10]:

$$TH \text{ (as } CaCO_3) \text{ mg/L} = (Ca^{2+} + Mg^{2+}) \times 50 \tag{1}$$

Table 1. Locations and depths of the wells in the Erbil plain area, UTM Coordination system.

#	Name of Wells	Easting	Northing	Elevation	Well Depth (m)
1	Tandura Village Well	395,413	3,993,026	335	180
2	Mastawa Village Well	395,425	3,989,611	313	110
3	Aliawa Shekh Village Well	393,348	3,985,207	993	150
4	Dil uguleKhwaru Village Well	396,086	3,995,879	339	162
5	Doosarafatah Village Well	401,961	3,987,191	337	195
6	Haza Village Well	403,004	3,997,753	354	200
7	Shekh Sherwan Village Well-1	385,318	3,988,765	296	200
8	YadiQizlar Village Well	398,120	3,982,848	321	150
9	Dheivan Village Well-1	393,502	3,989,935	314	132
10	Binberze Gichka Village Well	396,837	4,001,782	342	200
11	Yarmja Village Well	394,854	3,999,461	331	120
12	Lajan Harki Village Well-1	391,684	3,998,770	331	170
13	Sardar Village Well	404,773	3,998,649	367	171
14	Dhemat Village Well-1	391,635	3,995,103	304	150
15	Awena Village Well-1	382,574	3,991,776	285	180
16	Bryat Village Well	404,024	3,996,473	342	173

2.5. Groundwater Quality Assessment

The concentration of cations and anions was interrelated, and the irrigation indexes were calculated including the sodium adsorption ratio (SAR) [11], sodium percentage (Na%) [12], residual sodium carbonate (RSC) [11], magnesium hazard (MH) [13], potential salinity (Ps) [14], permeability index (PI) [15], and monovalent cation adsorption ration (MCAR) [16] were used to assess groundwater quality.

The indexes were calculated using the equations below:

$$\text{SAR} = Na^+ \ (epm)/[Ca^{+2} + Mg^{+2} \ (epm)/2]^{0.5} \tag{2}$$

$$Na\% = [Na^+ + K^+ \ (epm)/Ca^{+2} + Mg^{+2} + Na^+ + K^+ \ (epm)] \times 100 \tag{3}$$

$$\text{RSC (in epm)} = (CO_3^{-2} + HCO_3^{-}) - (Ca^{+2} + Mg^{+2}) \tag{4}$$

$$\text{MH} = Mg^{+2}/(Ca^{+2} + Mg^{+2}) \times 100 \tag{5}$$

$$\text{Ps} = Cl^- + \sqrt{SO_4} \tag{6}$$

$$\text{PI} = [[Na^+ + \sqrt{HCO_3^{-}}]/[Ca^{+2} + Mg^{+2} + Na^+]] \times 100 \tag{7}$$

$$\text{MCAR} = Na^+ + K^+/(Ca^{+2} + Mg^{+2}/2)^{0.5} \tag{8}$$

The interpolation for the parameter's concentration was carried out in ArcGIS 10.1 using the Kriging method to plot the parameter distribution for the well samples in the study area.

2.6. Cation Ratio of Structural Stability (CROSS)

The Cation ratio of structural stability CROSS was used to assess the soil permeability hazard.

$$CROSS = C_{Na} + 0.56C_K / [(C_{Ca} + 0.60C_{Mg})/2]^{0.5} \qquad (9)$$

The major cations commonly occur in irrigation water in soil solutions, and on soil cations exchange sites, with concentrations and relative distributions influenced by both natural and anthropogenic factors [17].

Rengasamy and Marchuk [18] proposed that CROSS should be more predictive than SAR in assessing irrigation water quality for soil permeability hazard because it includes the dispersive effect of K in addition to that of Na and differentiates the flocculating effect of Mg from that of Ca.

2.7. Hydrochemical Formula and Water Type

Water type is always represented by account of major cations and anions in (epm%) it exceeds than (15%) in the hydro-chemical formula, and the formula are determined according to Ivanov (1968) formula [19]:

$$TDS\,(mg/L) \quad \frac{Anion\,(epm\%)\,in\,decreasing\,order}{Cation\,(epm\%)\,in\,decreasing\,order} \quad pH \qquad (10)$$

2.8. Groundwater Uses for Irrigation Purposes

The irrigation of cropland has become a widely used practice and has greatly increased the productivity of farmland. It has made it possible to farm in regions that would not be farmable without irrigation. A problem with irrigated cropland is the possibility of groundwater contamination and the stricter restrictions that are going to have to be implemented on the quantity of fertilizers and pesticides used to reduce the risk of contamination [20].

The classification of irrigation water depends on variables such as: Total Dissolved Solids (TDS); Sodium Adsorption Ratio; Residual Sodium Carbonate; and Chloride.

2.8.1. Total Dissolved Solids

The suitability of irrigation water is dependent on the effect of the mineral constituent of water on both the plant and soil, and the effect of salts on soil causing changes in soil structure. Infiltration is increased with increase in (TDS), and is then used for evaluating soil permeability [21].

Train classification (1979) [22] was used to assess the suitability of the water for irrigation, comparing this classification with water samples in the study area (Table 2).

Table 2. Train classification for suitability of irrigation water.

TDS	Specifications
<500	Use for irrigation does not have a harmful effect
500–1000	Use for irrigation has a harmful effect on sensitive crops for salinity
1000–2000	Has a harmful effect on crops so needs experience to use
2000–5000	Use for high tolerance crop irrigation and needs experience to use

2.8.2. Sodium Adsorption Ratio

General classification of water sodium hazard based on SAR according to Bauder et al. (2004) [23] were used to determine the suitable water uses for irrigation (Table 3).

Table 3. General classification of water sodium hazard based on SAR values [22].

SAR	Sodium Hazard	Specification
1–9	Low	Use on sodium sensitive crops must be cautioned
10–17	Medium	Amendments (such as gypsum) and leaching needed
18–25	High	Generally unsuitable for continuous use
>26	Very high	Generally unsuitable for use

2.8.3. Residual Sodium Carbonate

Higher RSC values suggest that a significant amount of calcium and some magnesium ions precipitate from the solution, increasing the percentage of sodium in water and soil particles and thus increasing the risk of a sodium hazard [24].

The relation between RSC and suitability of water for irrigation purposes is as in the table below (Table 4):

Table 4. Suitability of water for irrigation purposes according to RSC.

RSC	Suitability of Water for irrigation
RSC > 2.5	Unsuitable for irrigation
1.5 < RSC < 2.5	Range between suitable and unsuitable water for irrigation
RSC < 1.5	Water suitable for irrigation purposes

2.8.4. Chloride

Chloride is not adsorbed by soils but readily moves with the soil water; it is taken up by plant roots and moves upward to accumulate in the leaves [25]. Chloride is essential to plants in very small amounts; it can cause toxicity to sensitive crops at high concentration. The Bauder [23] classification was used to determine the suitability of water uses for irrigation (Table 5).

Table 5. Classification based on chloride and its effect on the crops [23].

Chloride (ppm)	Effect on Crops
Below 70	Generally safe for all plants
70–140	Sensitive plants show injury
141–350	Moderately tolerant
Above 350	Plants show injury

2.9. Groundwater Classification

Classification of groundwater according to chemical indicators depends on hydrochemical parameters. Different types of classification were applied in this research to classify the water such as: Piper Diagram Classification; Sholler Classification; Chadha Classification; and Gibbs diagram classification.

2.9.1. Piper Diagram Classification (1944) [26]

This classification can be combined with the classification based on the dominant ions present in the water. Most classifications of this type use a percentage of anion and cation equivalents per million [27].

2.9.2. Sholler Classification (1972) [28]

In this classification, the ion concentration in (epm) units is plotted on semi-logarithm paper. This type of diagram facilitates a visual comparison of the composition of different water types in descending order, shown in Table 6 [29].

Table 6. Water type according to Schoeller classification.

	Cations		Anions
A	$r(Na+K) > rMg > rCa$	1	$rCl > rSO_4 > rHCO_3$
B	$r(Na+K) > rCa > rMg$	2	$rCl > rHCO_3 > rSO_4$
C	$rMg > r(Na+K) > rCa$	3	$rSO_4 > rCl > rHCO_3$
D	$rMg > rCa > r(Na+K)$	4	$rSO_4 > rHCO_3 > rCl$
E	$rCa > r(Na+K) > rMg$	5	$rHCO_3 > rCl > rSO_4$
F	$rCa > rMg > r(Na+K)$	6	$rHCO_3 > rSO_4 > rCl$

According to this classification, parallel relationships in the hydro-chemical composition for the water reflect the effect of dissolution processes or weathering of rocks by the water, otherwise the water composition is from another source [30].

2.9.3. Chadha Classification (1999) [31]

Chadha (1999) created a new schematic dividing the origins of ions into eight categories. The square or rectangular field in a Chadha diagram represents the overall ion distribution and character of groundwater and is used to demonstrate geochemical composition and hydro-chemical processes. The rectangular field is divided into eight sub-fields, each of which symbolizes a different water type, in order to determine the basic character of groundwater (Figure 3).

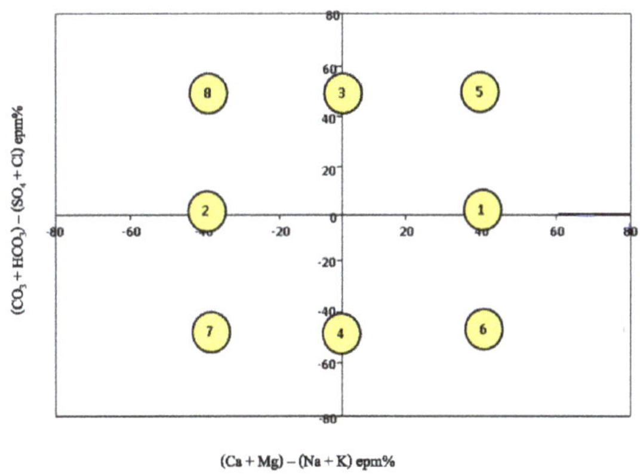

Field   Descriptions

1 Alkaline earths exceed alkali metals.

2 Alkali metals exceed alkaline earths.

3 Weak acidic anions exceed strong acidic anions.

4 Strong acidic anions exceed weak strong acidic anions.

5 Alkaline earths and weak acidic anions exceed both alkali metals and strong anions, respectively.

6 Alkaline earths exceed alkali metals and strong acidic anions exceed weak acidic anions.

7 Alkali metals exceed alkaline earths and strong acidic anions exceed weak acidic anions.

8 Alkali metals exceed alkaline earths and weak acidic anions exceed strong acidic anions.

Figure 3. Water type according to Chadha classification in the study area.

2.9.4. Gibbs Diagram Classification (1970) [32]

The Gibbs diagram is a method for estimating the origin of ions in groundwater by focusing on the correlation between the concentration of cations (Na^+, Ca^{2+}) and anions (Cl^-, HCO_3^-), and total dissolved solids (Figure 4).

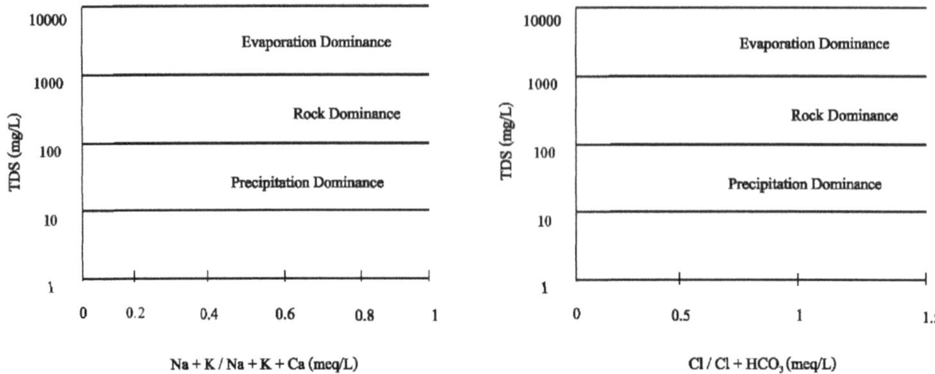

Figure 4. Gibbs Diagram for cations and anions in the study area.

2.10. Groundwater Level Fluctuation

The data of seven monitored wells were obtained to determine groundwater level fluctuation during the period between 2010 until 2020, and the effect of drought periods caused by climate change on the groundwater in the study area. The data was obtained from the Groundwater Directorate-Erbil, Kurdistan Region.

3. Results and Discussion

3.1. Physical and Chemical Analysis

The physico-chemical analysis is shown in Table 7, and the special distribution of all parameters analyzed are shown in Appendices A–C. The map of total dissolved solids shows that the high value is founded in the central part of the study area according to the groundwater flow which originates from the mountain area toward the center of the Erbil plain (Appendices A–E).

Table 7. Physico-chemical Parameter Analysis for the Deep Wells in the study area, cations and anions in mole/m^3 unit.

SN	Well Name	EC μs/cm	pH	TDS ppm	SO_4^{2-}	Cl^-	HCO_3	CO_3^{-2}	Ca^{+2}	Mg^{+2}	Na^+	K^+	TH
1	Tandura Village Well	1045	7.89	679	5.5	1.6	2.5	0	5.7	1.3	2.4	0.3	350
2	Mastawa Village Well	2740	7.52	1781	8.5	1.5	4.6	0	5.3	1.2	1.8	0.3	325
3	Aliawa Shekh Village Well	1618	7.82	1052	5.3	0.6	3.2	0	5.9	1.6	1.6	0.1	375
4	Dil uguleKhwaru Village Well	1189	7.65	773	4.1	3.4	2.3	0	4.6	3.4	1.6	0.2	400
5	Doosarafatah VillageWell	921	7.8	599	5.7	0.5	3.3	0	4	2.5	2.9	0.1	325
6	Haza Village Well	1434	7.85	932	5.4	1.9	2.4	0	4.8	2.2	2.4	0.3	350
7	Shekh Sherwan Village Well-1	917	7.86	596	6.5	1.4	2.3	0	6.3	1.24	2.5	0.3	375
8	YadiQizlar Village Well	888	7.82	577	3.1	0.9	3.1	0	3.6	1.9	1.5	0.1	275
9	Dheivan Village Well-1	2060	7.6	1339	24.1	2.6	4.2	0	14.5	13	3.4	0.1	1375
10	Binberze Gichka Village Well	965	7.76	627	6.5	32	4.5	0	15	12	3.5	0.1	1350
11	Yarmja Village Well	1144	7.63	744	1.66	5.9	2.6	0	5.3	1.2	3.5	0.2	325
12	Lajan Harki Village Well-1	1406	7.64	914	9.0	4.5	2.7	0	7	5.5	3.5	0.2	625
13	Sardar Village Well	912	7.86	593	4.5	1.6	2.4	0	5.4	1.1	1.8	0.3	325
14	Dhemat Village Well-1	2075	7.6	1349	18.2	4.3	4.3	0	14.7	8.3	3.8	0.1	1150
15	Awena Village Well-1	1455	7.86	946	6.3	1.5	2.4	0	4.5	2.5	2.9	0.3	350
16	Bryat Village Well	2350	7.88	1528	24.2	2.8	3.7	0	16.2	11.8	3.1	0.1	1400

3.2. Groundwater Quality Assessment Parameters

The sodium adsorption ratio (SAR) ranges from 1.1 to 2.7, with an average of 1.7 meq/L; the sodium percentage (Na%) ranges from 5.4% to 22%, with an average of 12.8%; the residual sodium carbonate (RSC) ranges from -24.3 to -1.9 meq/L, with an average of -8.9 meq/L; the monovalent cation adsorption ration (MCAR) ranges from 0.2 to 0.9, with an average of 0.5 meq/L; the cation ratio of structural stability (CROSS) ranges from 0.9 to 2.1, with an average of 1.3 meq/L; the Ps ranges from 2.5 to 35.3, with an average of 8.4 meq/L; the PI percent ranges from 2.6% to 9.4%, with an average of 6.2%, and the magnesium hazard (MH) ranges from 16.5 to 47.3, with an average of 31.7 meq/L (Table 8 & Appendix D). The special distribution of quality assessment parameters is shown in Appendix E.

Table 8. Basic statistics of the chemical analysis and field parameters of groundwater samples.

Parameter	Maximum	Minimum	Mean	SD
EC (μs/cm)	2740	888	1444.9	577.2
pH	7.9	7.5	7.8	0.1
TH (ppm)	1400.0	275.0	604.7	435.3
TDS (ppm)	1781.0	577.0	939.3	375.2
SO_4^{2-} (mole/m^3)	24.2	1.7	8.7	7.0
Cl^- (mole/m^3)	32.0	0.5	4.2	7.6
HCO_3^- (mole/m^3)	4.6	2.3	3.2	0.8
Ca^{2+} (mole/m^3)	16.2	3.6	7.7	4.5
Mg^{2+} (mole/m^3)	13.0	1.1	4.4	4.3
Na^+ (mole/m^3)	3.8	1.5	2.6	0.8
K^+ (mole/m^3)	0.3	0.1	0.2	0.1
Na%	22	5.4	12.8	4.9
SAR	2.74	1.14	1.71	0.55
Ps	35.3	2.5	8.43	8.27
RSC	-1.9	-24.3	-8.93	8.16
MCAR	0.9	0.2	0.5	0.25
CROSS	2.1	0.9	1.3	0.37
MH	47.3	16.5	31.7	11.4
PI%	9.4	2.6	6.3	2.23

Increased sodium levels in irrigation water cause the breakdown of well-structured soils, which limits aeration and water permeability, resulting in lower crop development [33]. A sodium percentage of more than 60% is considered toxic to plants; in the samples studied there is no toxic effect on the plants because all the water samples are less than 60% Na. General classification of the irrigation water is according to Bauder et al., 2004 [23] and based on SAR, the water well samples which are suitable for irrigation have a low sodium hazard and use on sodium sensitive crops must be cautioned against. According to Eaton (1995) [34] and based on RSC, all water samples are suitable for irrigation because they fall below the standard limit, which less than 1.5 meq/L.

The sodium, calcium, magnesium, and bicarbonate concentrations in the soil influence soil permeability, which also affects the quality of irrigation water over time. Nagaraju et al. (2014) [35] classified water quality on the basis of PI into Classes I, II, and III. Classes I and II indicate good water quality for irrigation purposes, while Class III water is unsuitable for irrigation (Figure 5). A high permeability index is linked to underlying structural elements that allow for widespread groundwater contamination. The groundwater samples of the study area fall into Class I (29.05–72.75%) and were described as having excellent-to-good permeability.

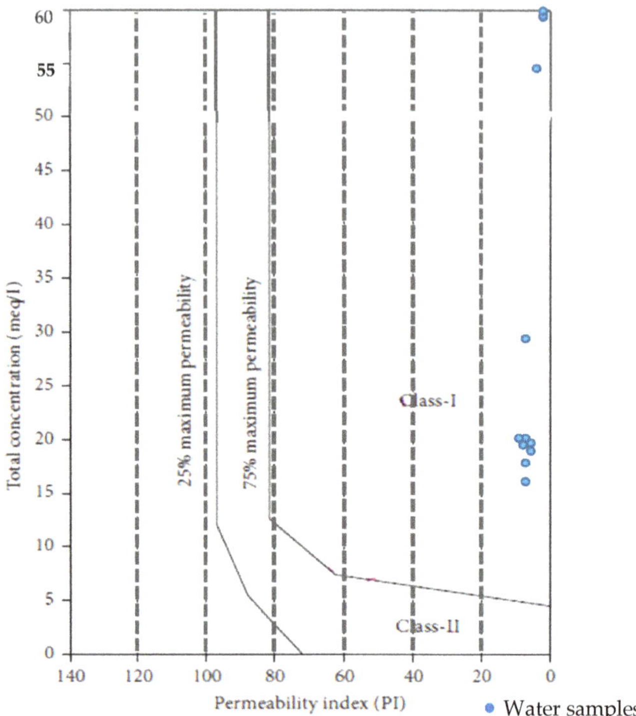

Figure 5. Permeability index diagram classification of groundwater quality.

An increased proportion of Mg^{2+} relative to Ca^{2+} increases sodication in soils, which causes the dispersion of clay particles, which destroys soil structure and lowers relative hydraulic conductivity [36]. Magnesium ratios of more than 50 are deemed hazardous and unsuitable for irrigation. As soils grow more alkaline, this will have a negative impact on crop yield. All the water samples in the study area are acceptable for irrigation purposes.

The suitability of water for irrigation is not dependent on soluble salts. Because low-solubility salts precipitate in the soil and accumulate with each irrigation treatment, the soil salinity rises as the concentration of highly soluble salts rises [14].

The classification of the potential salinity of groundwater samples includes three classes; 1. Excellent to Good (<5), 2. Good to Injurious (5–10), and 3. Injurious to Unsatisfactory (>10). Nine water samples are classified as excellent-to-good, three water samples are good-to-injurious, and four samples are in the injurious-to-unsatisfactory water class.

The MCAR ratio may predict the adsorption of monovalent ions by soil colloids on the basis of cation exchange isotherms, but it fails to weigh the relative efficacies of Na^+ and K^+ in the numerator and of Ca^{2+} and Mg^{2+} in the denominator and treats members of each pair as identical [16].

When the soil's K^+ and Mg^{+2} levels are low, CROSS will be similar to SAR in predicting soil behavior. However, when these cations are present in higher amounts, CROSS will be more effective than either SAR or MCAR [18].

The water quality types in the study area according to Ivanov, 1968 [19], are of the Sulfate water type, except that two samples were of the chloride type and one sample of the Bicarbonate water type (Table 9).

Table 9. Water type of the water well samples.

Well Number	Water Type
1	Na-Ca-Cl-HCO$_3$ Sulfate
2	Na-Ca-HCO$_3$ Sulfate
3	Na-Mg-Ca-HCO$_3$ Sulfates
4	Na-Mg-Ca-HCO$_3$-Cl Sulfate
5	Mg-Na-Ca-HCO$_3$ Sulfate
6	Mg-Na-Ca-Cl-HCO$_3$ Sulfate
7	Na-Ca-HCO$_3$ Sulfate
8	Na-Mg-Ca-SO$_4$ Bicarbonate
9	Mg-Ca Sulfate
10	Mg-Ca-SO$_4$ Chloride
11	Na-Ca-SO$_4$-HCO$_3$ Chloride
12	Na-Mg-Ca-HCO$_3$-Cl Sulfate
13	Na-Ca-Cl-HCO$_3$ Sulfate
14	Mg-Ca-Cl-HCO$_3$ Sulfate
15	Mg-Na-Ca-Cl-HCO$_3$ Sulfate
16	Mg-Ca Sulfate

3.3. Groundwater Uses for Irrigation

Train classification of the water samples based on Total Dissolved Salts (TDS) shows that eleven water samples, when used for irrigation, had a harmful effect on sensitive crops for salinity, while five water samples had harmful effects on crops, and so experience is needed before using them. According to Bauder et al., 2004 [23] and based on SAR, all water samples have a low sodium hazard and use on sodium sensitive crops must be carried out with caution. According to RSC, all water samples are suitable for irrigation purposes. According to Bauder et al., 2004 [23] and based on chloride, all water samples are suitable for irrigation and generally safe for all plants.

3.4. Groundwater Classification

The Piper Diagram Classification represents one possible system of nomenclature in which water represented by point (A) would be called calcium bicarbonate water, and point (B) would represent calcium, sodium, chloride water. Point (C) would represent sodium, calcium, magnesium, chloride, and sulfate water. The plotting parameters of water samples on this diagram indicate that most of the samples are of class (C) "sulfate water type" except for two samples in class (B) "Chloride water type", and one sample in class (A) "Bicarbonate water type" (Figure 6).

According to the Shoeller Classification, all the water samples are in class E and F, meaning that calcium is a dominant cation in the water and the anions vary, but with a dominance of sulfate anions in most of the water samples (Table 10).

According to the Chadha classification, the water type ranges from class 4 and 6: strong acid anion with prevailing weak acid anion and earthy alkaline prevailing felsic alkaline and strong anion acid with prevailing weak anion acid (Figure 7).

The Gibbs classification is a method for estimating the origin of ions in groundwater by focusing on the correlation between the concentrations of cations, anions, and TDS (Figure 8).

The Gibbs diagram shows that the origin of the concentration of ions in the groundwater is evaporation and rock dominance. This characteristic suggests that ion dissolution in groundwater occurs more frequently as a result of interactions between groundwater and rock or soil and the evaporation process than as a result of precipitation or other sources.

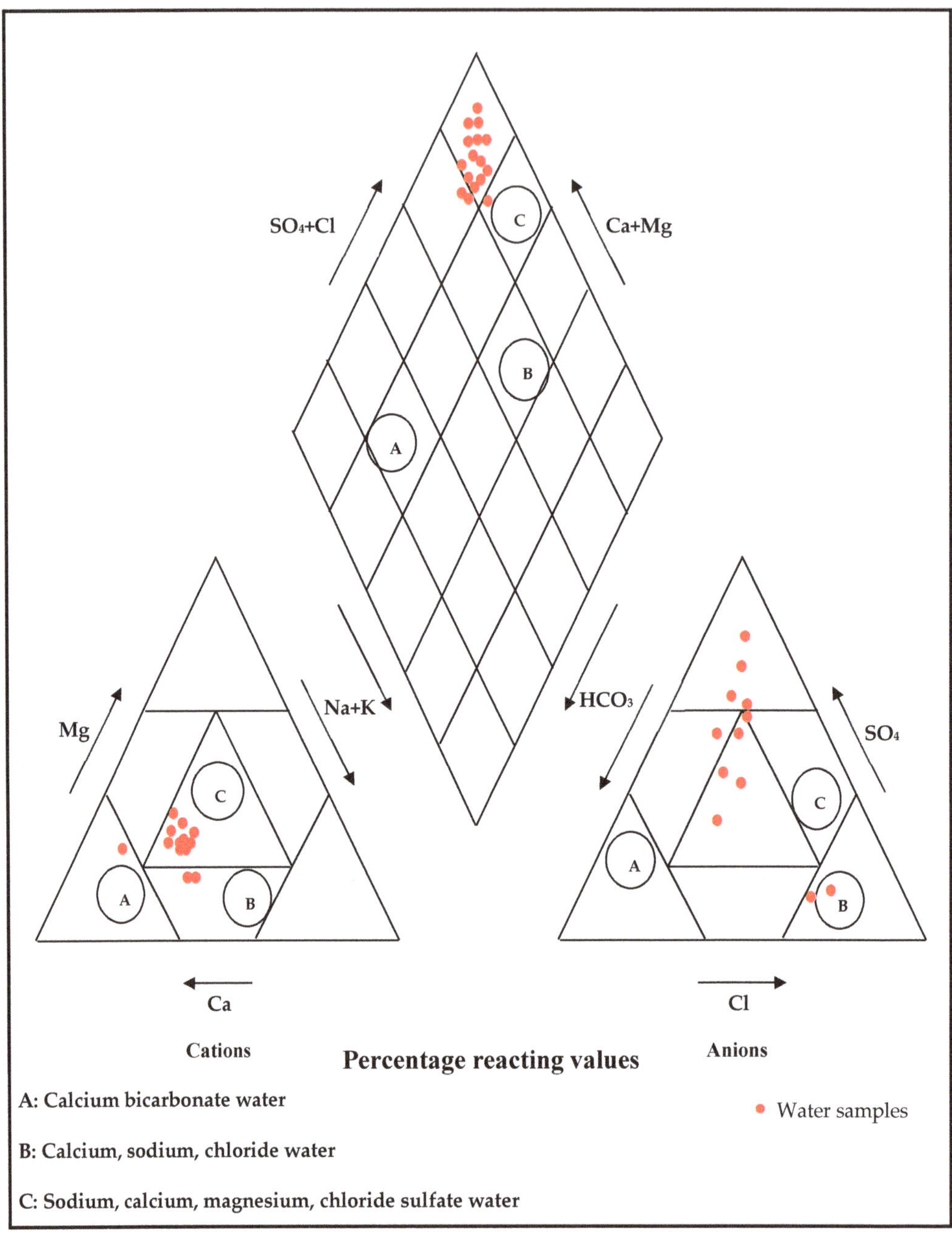

Figure 6. Piper trilinear diagram of major ions of water samples studied ([27]).

Table 10. Water class according to the Shoeller Classification of water samples.

Well Number	CatiIons	Anions	Water Class
1	rCa > r(Na+K) > rMg	rSO_4 > $rHCO_3$ > rCl	E 4
2	rCa > r(Na+K) > rMg	rSO_4 > $rHCO_3$ > rCl	E 4
3	rCa > r(Na+K) > rMg	rSO_4 > $rHCO_3$ > rCl	E 4
4	rCa > rMg > r(Na+K)	rSO_4 > rCl > $rHCO_3$	F 3
5	rCa > r(Na+K) > rMg	rSO_4 > $rHCO_3$ > rCl	E 4
6	rCa > r(Na+K) > rMg	rSO_4 > $rHCO_3$ > rCl	E 4
7	rCa > r(Na+K) > rMg	rSO_4 > $rHCO_3$ > rCl	E 4
8	rCa > rMg > r(Na+K)	$rHCO_3$ > rSO_4 > rCl	F 6
9	rCa > rMg > r(Na+K)	rSO_4 > $rHCO_3$ > rCl	F 4
10	rCa > rMg > r(Na+K)	rCl > rSO_4 > $rHCO_3$	F 1
11	rCa > r(Na+K) > rMg	rCl > $rHCO_3$ > rSO_4	E 2
12	rCa > rMg > r(Na+K)	rSO_4 > rCl > $rHCO_3$	F 3
13	rCa > r(Na+K) > rMg	rSO_4 > $rHCO_3$ > rCl	E 4
14	rCa > rMg > r(Na+K)	rSO_4 > $rHCO_3$ > rCl	F 4
15	rCa > r(Na+K) > rMg	rSO_4 > $rHCO_3$ > rCl	E 4
16	rCa > rMg > r(Na+K)	rSO_4 > $rHCO_3$ > rCl	F 4

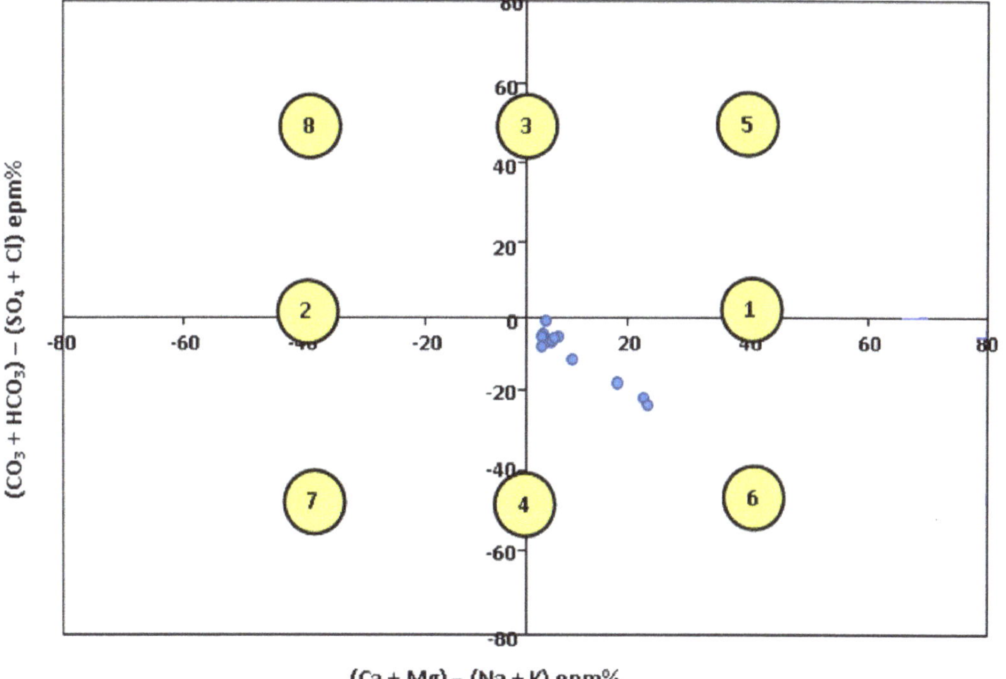

Figure 7. Water type according to the Chadha classification (1999) [30].

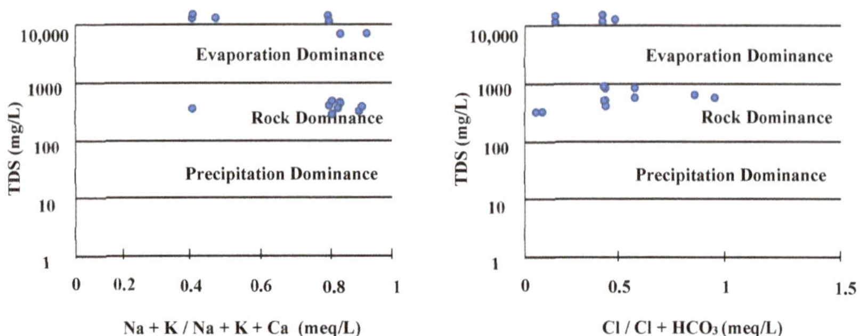

Figure 8. Gibbs diagram for groundwater classification in the study area.

3.5. Groundwater Level Fluctuation

Many factors influence groundwater level fluctuation, including rainfall intensity, rainfall quantity, infiltration capacity of soil and rock beds, groundwater depth above sea level, terrain, evapotranspiration, and water well discharge [8]. Fluctuations in stored water, renewable storage water amount, and groundwater investment were calculated using groundwater fluctuation analysis.

Groundwater recharge in the Erbil plain depends on rainfall quantity. In rainy months, the recharge quantity is greater than the discharge quantity in the wells, which leads to a rise in groundwater levels, while in dry months (the summer season), the discharge quantity is greater than recharge quantity (these may be absence recharges), which leads to a reduction in groundwater levels.

The groundwater level was measured on a weekly basis over ten years by the Groundwater Directorate in Erbil Governorate (from 2010 to 2020) for six monitored wells in the study area (Daldaghan well, Khazna well, Mastawa well, Peerdawood well, Shekh Sherwan, and Tendura-1 well). All the wells penetrate quaternary deposits and Bai Hassan Formation. Figure 9 and Appendix F represent the groundwater fluctuation for these six wells, which shows the depression of the groundwater table (increasing depth to groundwater) during these periods due to the effects of climatic change as well as the effects of drought periods on the study area.

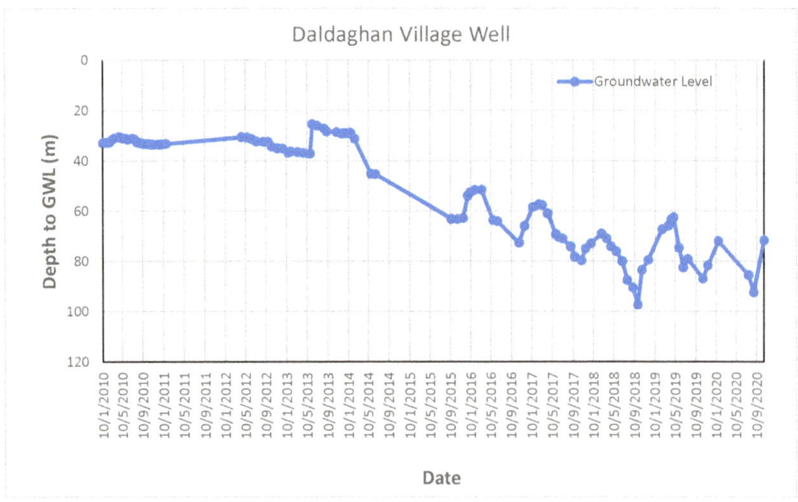

Figure 9. Groundwater level fluctuation curve in Daldaghan Village monitored wells in the study area.

4. Conclusions

The present research assessed the quality evaluation of the groundwater in the Erbil Plain region; the parameters were analyzed using the concentration distribution of ions in the groundwater. The findings demonstrate that using ground water for irrigation will not harm plants because all water samples had a sodium content of less than 60%. Based on RSC, all water samples were shown to be suitable for irrigation because they fall under the standard limit, which is less than 1.5 meq/L. SAR indicates that the water wells are suitable for irrigation because they have a low sodium hazard, though caution should be exercised when using them on sodium-sensitive crops.

According to the classification of water quality of the permeability index, groundwater samples from the study area fall into Class I, and aquifers were found to have excellent-to-good permeability. Nine water samples were classed as excellent-to-good, three water samples were good-to-injurious, and four samples were injurious-to-unsatisfactory, according to the classification of the potential salinity for groundwater samples.

Since CROSS is not based on the exchange isotherm, it cannot determine how much Na^+ and K^+ have been adsorbed. However, it might be used to predict dispersive effects on soil stability and hydraulic properties based on the relative amounts of the four cations present in the equilibrium soil solution. According to the criteria used to evaluate the water quality, the majority of the samples are suitable for irrigation.

All the water samples are in the sulfates water type except for two groundwater samples, which are of the chloride type (Binberze Gichka Village Well and Yarmja Village Well) and one groundwater sample is of the bicarbonate water type (Yadi Qizlar Village Well).

Over the ten years measuring the groundwater table, fluctuation in the area for some wells shows a depression in the groundwater table due to drought periods during this time, as well as to global climate change.

In the future, especially in agricultural areas, we propose the inclusion of contaminants such as pesticides in the monitoring of ground water quality. When pesticides are sprayed on crops, they can penetrate the surface of the ground and reach water-containing aquifers. Groundwater becomes contaminated as a result, making it unusable for both agricultural and human purposes. We also recommend conducting research on the toxic anion and trace elements in the study area to find out the effects of heavy metals and some toxic anions in the area.

Author Contributions: Conceptualization, S.S. and C.R.; methodology, S.S.; software, S.S.; validation, S.S., C.R. and K.M.; formal analysis, S.S. and M.A.; investigation, S.S., H.A. and C.R.; resources, S.S. and H.A.; data curation, S.S.; writing—original draft preparation, S.S.; writing—review and editing, C.R., M.R. and K.M.; visualization, S.S., H.A. and K.M.; supervision, C.R.; project administration, C.R., M.R. and K.M.; funding acquisition, C.R. All authors have read and agreed to the published version of the manuscript.

Funding: This Research is fully funded by the NUFFIC Orange Knowledge Programme (OKP-IRA-104278).

Institutional Review Board Statement: Not applicable.

Informed Consent Statement: Not applicable.

Data Availability Statement: Some data in this manuscript obtained from Ministry of Agriculture and Water Resources-Groundwater Directorate, Kurdistan Region.

Acknowledgments: This study has been funded by NUFFIC, the Orange Knowledge Programme, through the OKP-IRA-104278 project titled "Efficient water management in Iraq by switching to climate smart agriculture: capacity building and knowledge development" coordinated by Wageningen University & Research, The Netherlands.

Conflicts of Interest: The authors declare no conflict of interest.

Appendix A

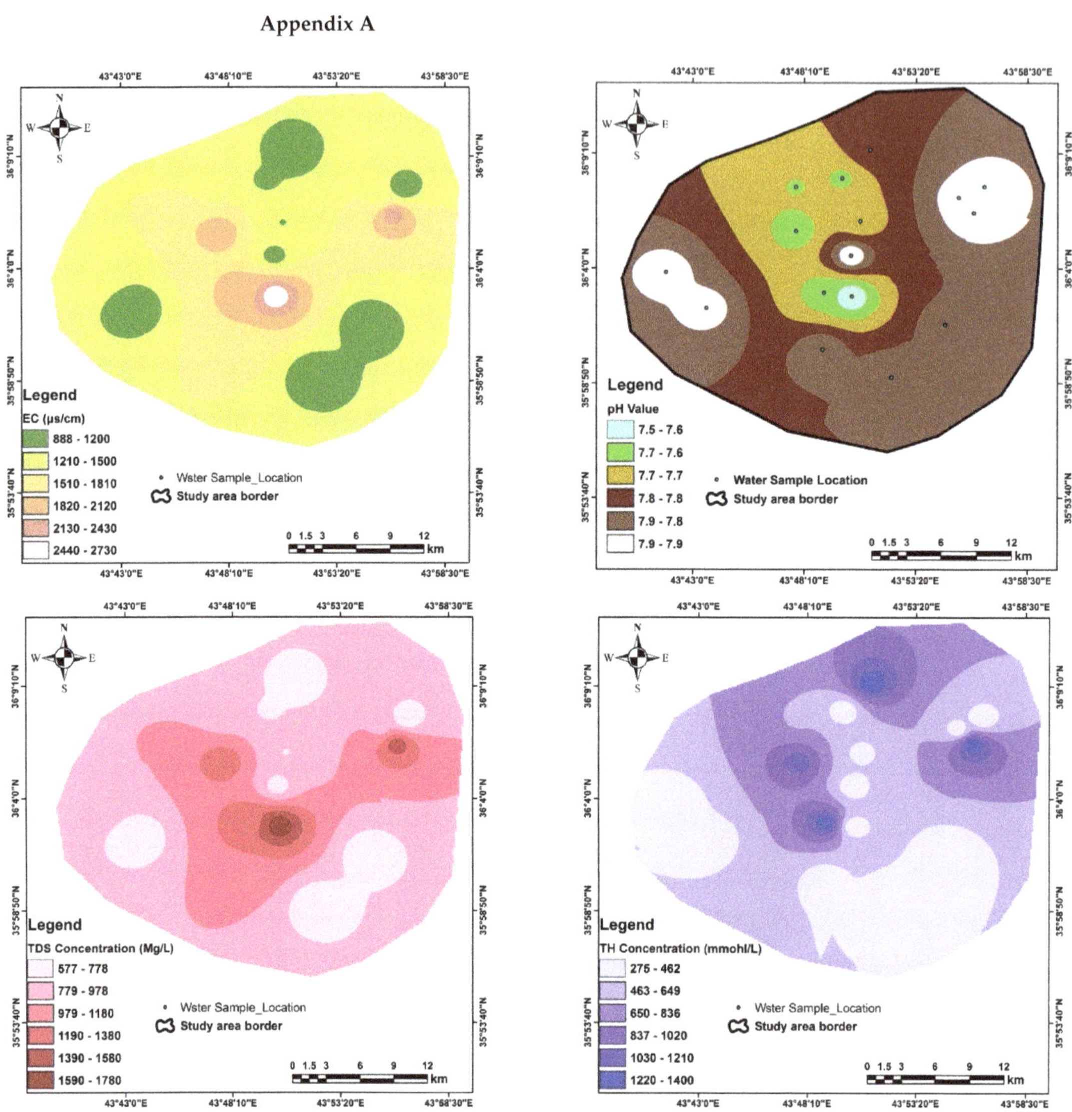

Figure A1. Special distribution of some physiochemical parameter in the study area.

Appendix B

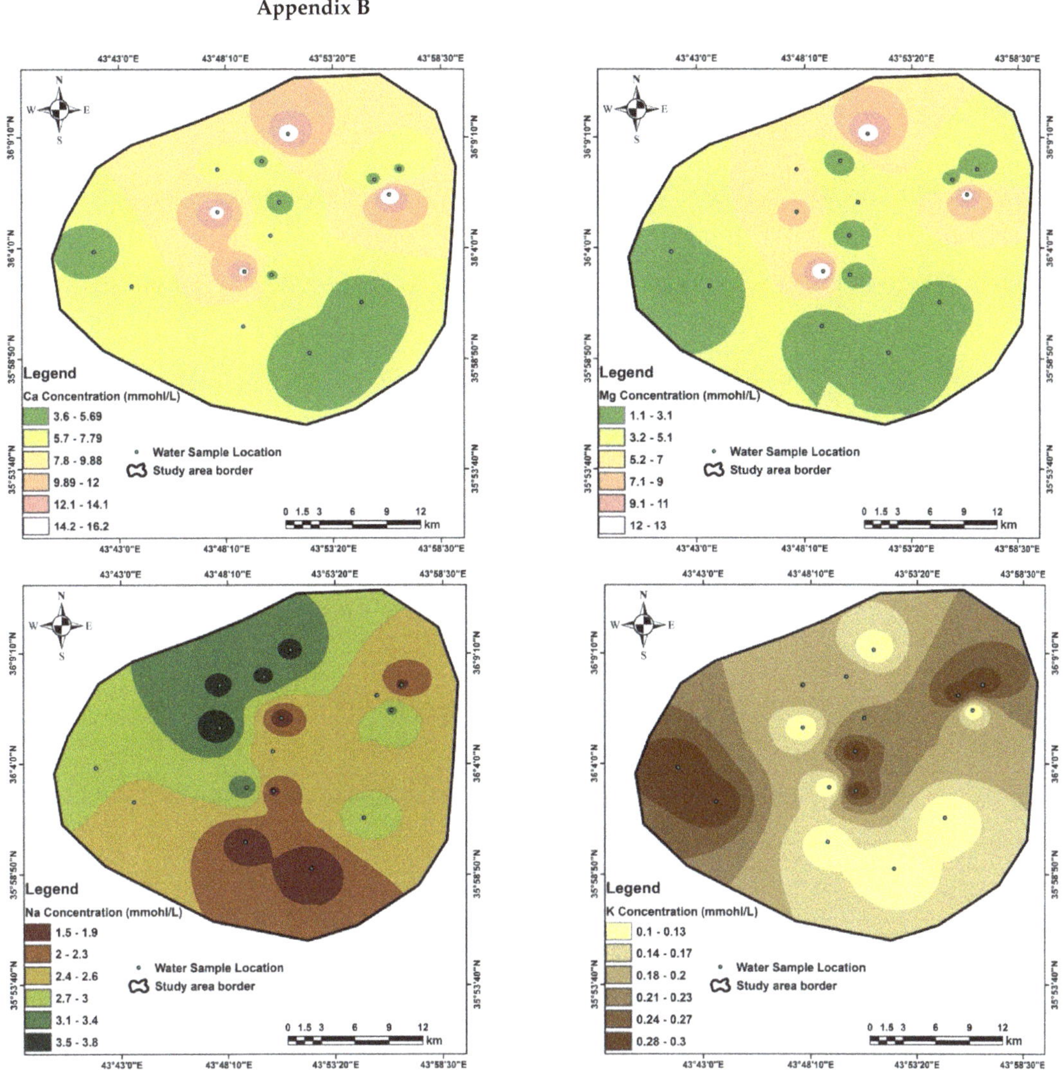

Figure A2. Special distribution of the cations in the study area.

Appendix C

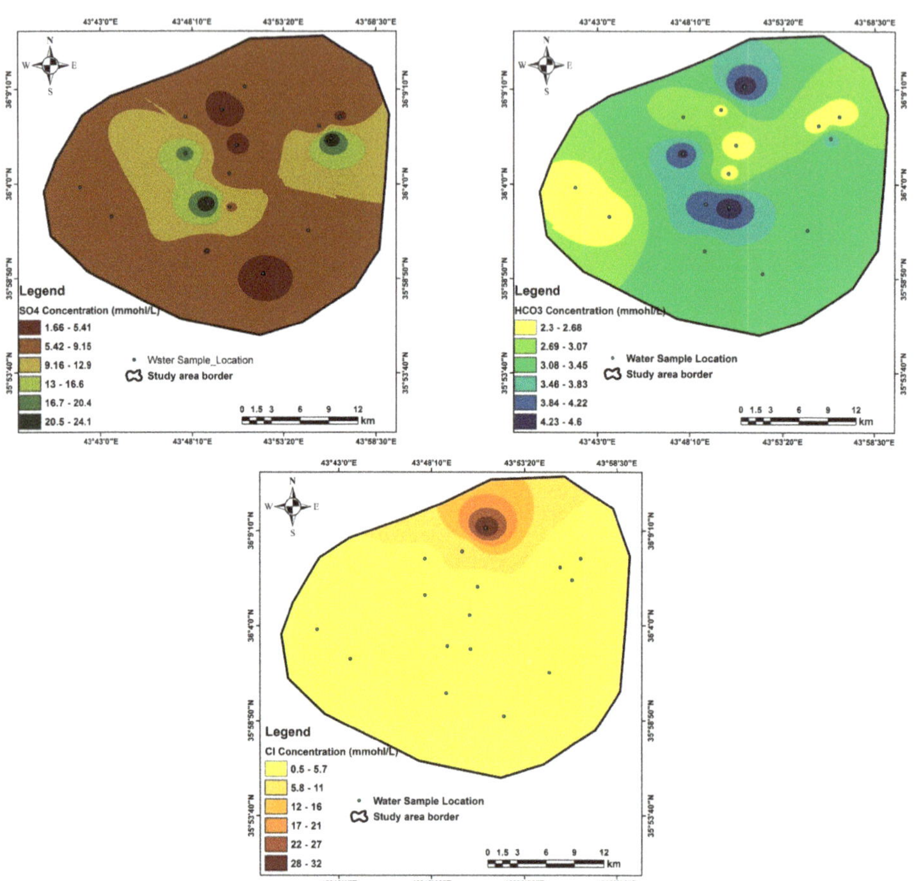

Figure A3. Special distribution of the anions in the study area.

Appendix D

Table A1. Calculated parameters of irrigation water quality in the study area.

SN	Well Name	Na%	SAR	Ps	RSC	MCAR	CROSS	MH	%PI
1	Tandura Village Well	16.1	1.83	4.4	−4.5	0.2	1.4	18.6	7.4
2	Mastawa Village Well	13.7	1.4	3.9	−1.9	0.2	1.1	18.5	8.1
3	Aliawa Shekh Village Well	9.7	1.15	3.3	−4.3	0.3	0.9	21.3	6.2
4	Dil uguleKhwaru Village Well	10.2	1.14	5.5	−5.7	0.7	0.9	42.5	5.4
5	Doosarafatah VillageWell	18.8	2.33	3.4	−3.2	0.6	1.8	38.5	9.1
6	Haza Village Well	16.3	1.83	4.6	−4.6	0.5	1.5	31.4	7.4
7	Shekh Sherwan Village Well-1	15.6	1.84	4.7	−5.2	0.2	1.4	16.5	7.0
8	YadiQizlar Village Well	12.2	1.26	2.5	−2.4	0.5	1.0	34.5	7.9
9	Dheivan Village Well-1	5.9	1.28	14.7	−23.3	0.9	1.0	47.3	2.8
10	Binberze Gichka Village Well	6.3	1.14	35.3	−22.5	0.8	1.1	44.4	3.0
11	Yarmja Village Well	22.0	2.74	6.7	−3.9	0.2	2.1	18.5	9.4
12	Lajan Harki Village Well-1	13.0	1.99	9.0	−9.8	0.8	1.6	44.0	5.5
13	Sardar Village Well	13.7	1.4	3.9	−4.1	0.2	1.1	16.9	6.8
14	Dhemat Village Well-1	7.8	2.67	13.4	−18.7	0.6	1.2	36.1	3.6
15	Awena Village Well-1	18.7	2.23	4.7	−4.6	0.6	1.8	35.7	8.1
16	Bryat Village Well	5.4	1.15	14.9	−24.3	0.7	0.9	42.1	2.6

Appendix E

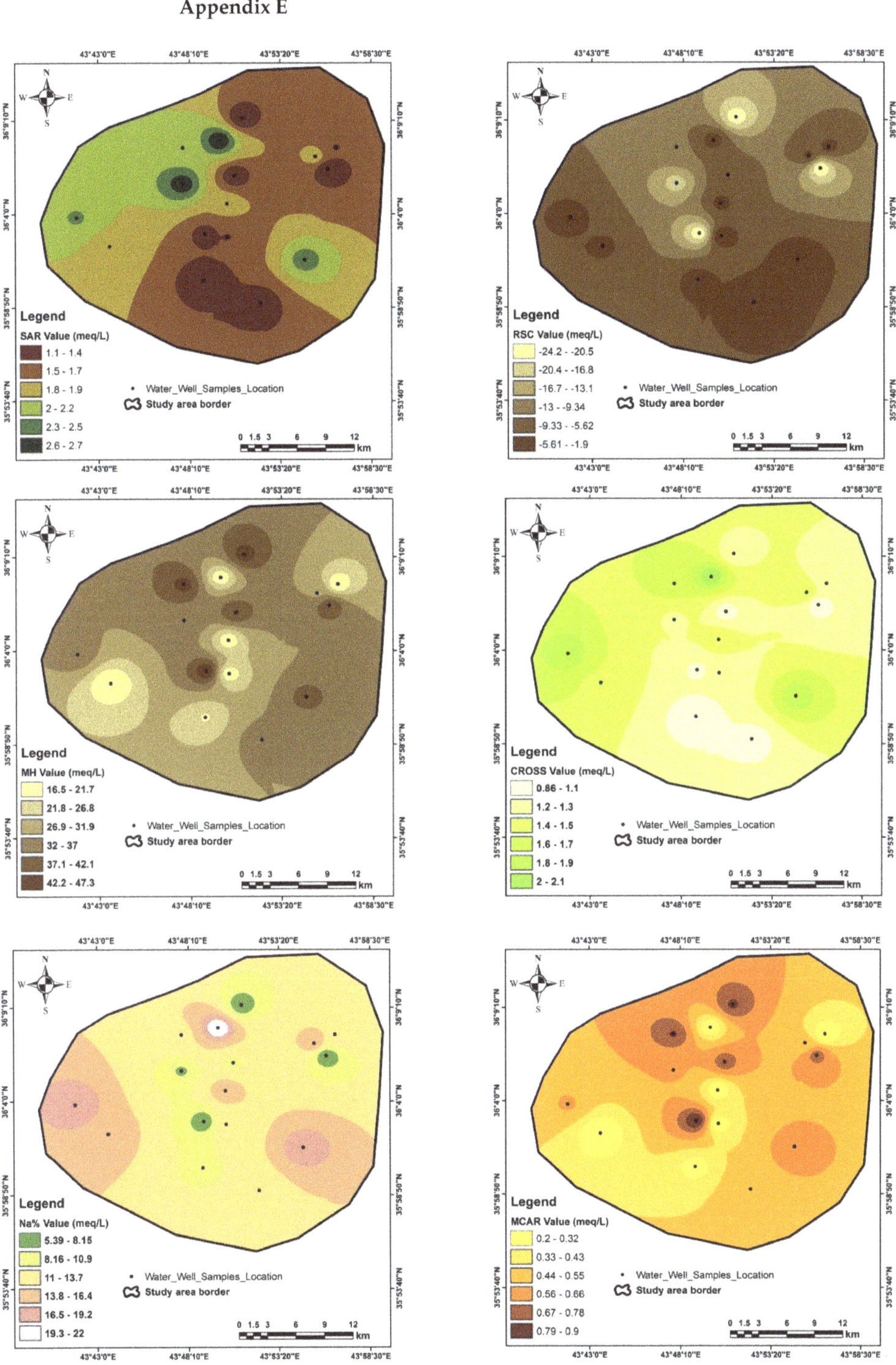

Figure A4. Special distribution of the quality assessment parameters in the study area.

Appendix F

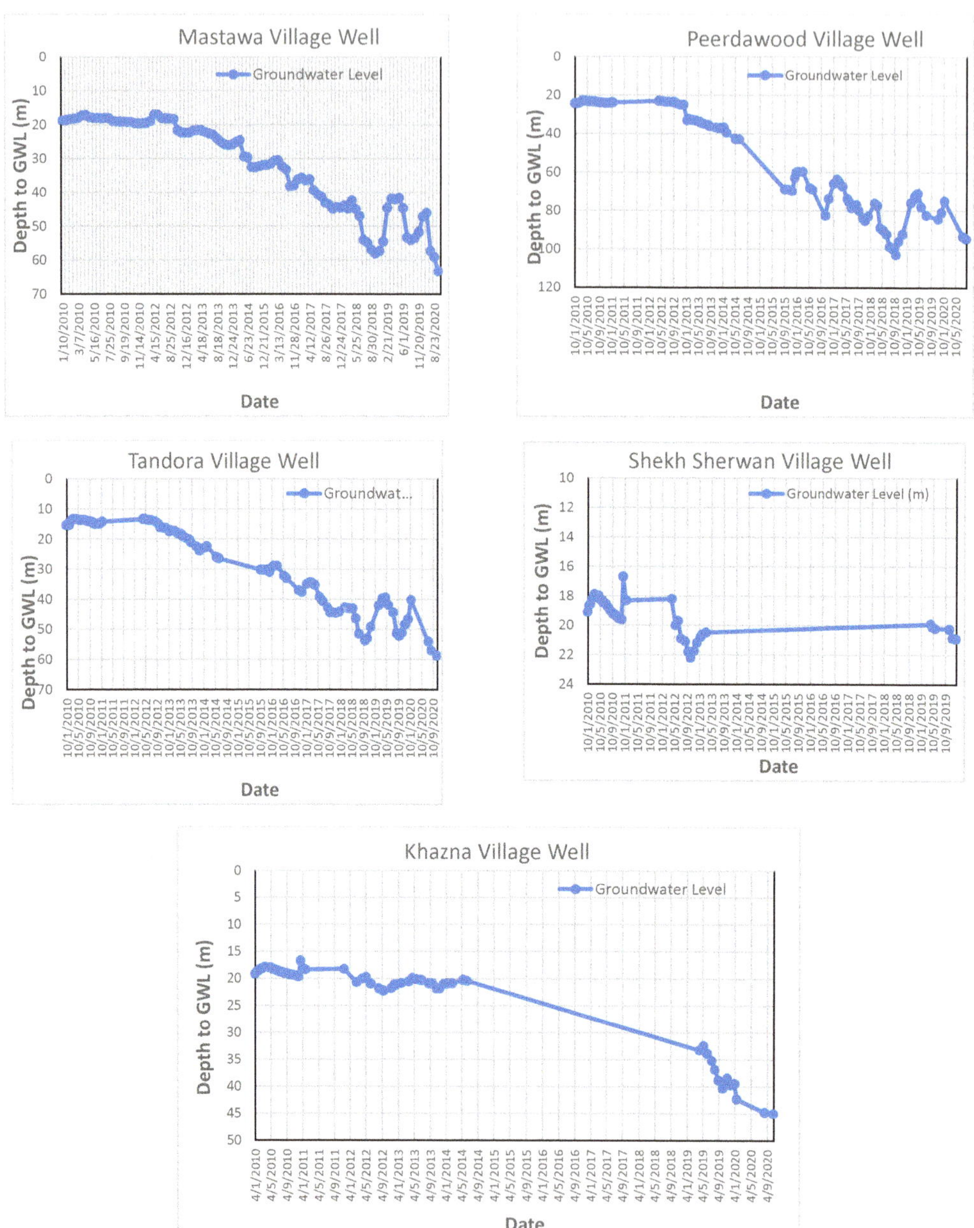

Figure A5. Groundwater level fluctuation curve in some of monitored wells.

References

1. Niemczynowicz, J. Urban hydrology and water management present and future challenges. *Urban Water* **1999**, *1*, 1–14. [CrossRef]
2. Seeyan, S.; Adham, A.; Mahdi, K.; Ritsema, C. Water Quality, Availability, and Uses in Rural Communities in the Kurdistan Region, Iraq. *Water* **2021**, *13*, 2927. [CrossRef]

3. Lamont, B. *Maintaining Drip Irrigation System*; Vegetable & Small Fruit Gazette; Penn State University Extension: Landisville, PA, USA, 2012; Volume 16.

4. Belkhiri, L.; Boudoukha, A.; Mouni, L.; Baouz, T. Application of multivariate statistical methods and inverse geochemical modeling for characterization of groundwater—A case study: Ain Azel plain (Algeria). *Geoderma* **2010**, *159*, 390–398. [CrossRef]

5. Al-Saffawi, A.Y.; Ibn Abubaka, B.S.; Abbass, L.Y.; Monguno, A.K. Assessment of groundwater quality for IWQ index in Al-Kasik sub-district Northwestern Iraq. *Niger. J. Technol.* **2020**, *39*, 632–638. [CrossRef]

6. Seeyan, S. Hydrochemical assessment and water quality of Koysinjaq area in Kurdistan Region-Iraq. *Arab. J. Geosci.* **2020**, *13*, 491. [CrossRef]

7. Lioyd, J.W. *Water Resources of Hard Rock Aquifers in Arid and Semi-Arid Zones*; Studies and Report in Hydrology 58; UNESCO: Paris, France, 1999; 184p.

8. Wilson, E.M. *Engineering Hydrology*, 3rd ed.; Bloomsbury Publishing Ltd.: London, UK, 1987; 309p.

9. Jassim, S.Z.; Goff, J.C. *Geology of Iraq*; Dolin, Prague and Moravian Museum: Brno, Czech Republic, 2006.

10. Hem, J.D. *Study and Interpretation of the Chemical Characteristics of Natural Water*, 3rd ed.; USGS Water Supply Paper: Reston, VA, USA, 1985; 263p.

11. Richard, L.A. *Diagnosis and Improvement of Saline and Alkali Soils*; Agricultural Hand Book 69; US Dept. Agric.: Washington, DC, USA, 1954; 160p.

12. Wilcox, L.V. The Quality of water for Irrigation USE. US Dept. *Tech. Bull.* **1955**, *962*, 40.

13. Szabolcs, I.; Darab, C. Influence of irrigation water of high sodium carbonate content of soils. In Proceedings of the 8th International Congress of ISSS, Tsukuba, Japan, 9 September 1964; pp. 803–812.

14. Doneen, L.D. Salination of Soil by Salts in the Irrigation Water. *Am. Geophys. Union Trans.* **1954**, *35*, 943–950. [CrossRef]

15. Doneen, L.D. *Notes on Water Quality in Agriculture. Published as a Water Science and Engineering*; Paper 4001; Department of Water Sciences and Engineering, University of California, Davis: Oakland, CA, USA, 1964.

16. Smiles, D.E.; Smith, C.J. A survey of the cation content of piggery effluents and some consequences of their use to irrigate soils. *Aust. J. Soil Res.* **2004**, *42*, 231–246. [CrossRef]

17. Qadir, M.; Sposito, G.; Smith, C.J.; Oster, J.D. Reassessing irrigation water quality guidelines for sodicity hazard. *Agric. Water Manag.* **2021**, *255*, 107054. [CrossRef]

18. Rengasamy, P.; Marchuk, A. Cation ratio of soil structural stability (CROSS). *Soil Res.* **2011**, *49*, 280–285. [CrossRef]

19. Ivanov, V.V.; Barvanon, L.N.; Plotnikova, G.N. The Main Genetic Type of Earth's Crust Mineral Water and their distribution in the USSR. In Proceedings of the International Geological Congress of 23rd Session, Prague, Czech Republic, 19–23 August 1968; Volume 12.

20. Grossman, Z. International Environmental Problems & Policy, a Class Website on Water Privatization and Commodification. Produced by Students of Geography at the University of Wisconsin-Eau Claire, USA. 2004. Available online: www.uwec.sdu/grossmzc/GEOG378.html (accessed on 12 August 2022).

21. Leavy, D.B.; Kearney, W.F. Irrigation of Native Rangeland Using Treated Waste Water from Institute Uralian processing. *J. Environ. Qual.* **1999**, *28*, 208–217. [CrossRef]

22. Train, R.E. *Quality Criteria for Water*; Castle House Publication, Ltd.: London, UK, 1979; 256p.

23. Bauder, T.A.; Waskom, R.M.; Davis, J.G. Irrigation Water Quality Criteria. Colorado State University Extension: Fort Collins, CO, USA, 2004; Available online: www.ext.colostate.eud (accessed on 12 August 2022).

24. Singh, K.K.; Tewari, G.; Kumar, S. Evaluation of Groundwater Quality for Suitability of Irrigation Purposes: A Case Study in the Udham Singh Nagar, Uttarakhand. *J. Chem.* **2020**, *2020*, 15. [CrossRef]

25. Rijtima, P.E. Quality Standards for Irrigation Waters. *Acta Hort.* **1981**, *119*, 25–35. [CrossRef]

26. Piper, A.M. A Graphical Interpretation of Water Analysis. *Eos Trans. Am. Geophys. Union* **1944**, *25*, 914–928. [CrossRef]

27. Davis, S.N.; Dewiest, R.J. *Hydrology*; John Wiley & Sons., Inc.: New York, NY, USA, 1966; 463p.

28. Shoeller, M. Edute Geochimique De La Nappe Des "Stables in fericurs" Du Bassin Daquitainse. *J. Hydrol.* **1972**, *15*, 317–328. (In French) [CrossRef]

29. Fetter, C.W. *Applied Hydrogeology*; Prentice-Hall, Inc., A Simon & Schuster Company: Upper Saddle River, NJ, USA, 1994; 691p.

30. Al-Jaleel, H.M. Effect of Industrial Wastes Discharge of Chemical Complex of Phosphate in Al-Qaim on Surface and Groundwater Pollution. Ph.D. Thesis, Baghdad University, Baghdad, Iraq, 2000; 17p. (In Arabic)

31. Chadha, D.K. A Proposed New Diagram for Geochemical Classification of Natural Waters and Interpretation of Chemical Data. *Hydrogeol. J.* **1999**, *7*, 431–439. [CrossRef]

32. Gibbs, R.J. Mechanisms Controlling World Water Chemistry. *Science* **1970**, *170*, 1088–1090. [CrossRef] [PubMed]

33. DeHayer, Z.; Gordon, J. *Classification of Irrigation Water Quality*; Springer: Cham, Switzerland, 2006.

34. Eaton, F.M. Significance of Carbonates in Irrigation Waters. *Soil Sci.* **1950**, *69*, 123–133. [CrossRef]

35. Nagaraju, A.; Sunil Kumar, K.; Thejaswi, A. Assessment of groundwater quality for irrigation: A case study from Bandalamottu lead mining area, Guntur district, Andhra Pradesh, South India. *Appl. Water Sci.* **2014**, *4*, 385–396. [CrossRef]

36. Rasouli, F.; Pouya, A.K.; Cheraghi, S.A.M. Hydrogeochemistry and Water Quality Assessment of the Kor-Sivand Basin, Fars Province, Iran. *Environ. Monit. Assess.* **2012**, *184*, 4861–4877. [CrossRef] [PubMed]

to consider different scenarios. The main goal of this study was to evaluate how much artificial recharge using a pond near the Kerbala wastewater treatment plant (WWTP) would raise water levels and improve water quality, as well as how large an area would be affected, as well as obtaining the values for the agricultural areas that would be added and made sustainable, resulting in an increase in the number of useful farmers and productive yields in Karbala Province, Iraq. This province has recently faced an increase in water consumption due to the rapid increase in population and climate changes.

2. Materials and Methods

2.1. Study Area

Figure 1 locates the study area in the middle of Iraq, between the cities of Karbala and Najaf. It is between latitudes 31°55′ and 32°45′ and longitudes 43°30′ and 44°30′. The Dibdibba aquifer is a shallow, unconfined groundwater aquifer with a homogeneous soil profile and is entirely recharged by rainfall. It has an area of 1100 km^2 and is surrounded by two ledges: Tar Al-Sayyed, which is inside the city limits of Kerbala, and Tar Al-Najaf, which is inside the city limits of Najaf province. The northern limit of the aquifer is close to Al-Razaza Lake, which is an open-surface reservoir. Quaternary sediments define the aquifer's eastern boundary. The topography shows elevations between 10 and 90 m above sea level. The primary direction of the flow of groundwater is typically from southwest to northeast, with hydrologic gradients varying between 0.0011 and 0.0005. The average temperature varies from 11 °C in winter to 37.6 °C in summer. The average rainfall varies between 90 mm, as recorded at the weather station in Karbala, and 112 mm at the weather station in Najaf. In addition, 80% of annual precipitation falls between November and March, which are the wettest months of the year. In this aquifer, water levels have dropped significantly since 2003, negatively impacting the quality of the available water. The withdrawal of water through overexploitation has also had a significant impact. According to previous research [26,27], salinization of groundwater is an indication of increased irrigation water use in agriculture, which also results in land degradation and the movement of pollutants to unconfined aquifers. More details about the area were given in a previous paper [3].

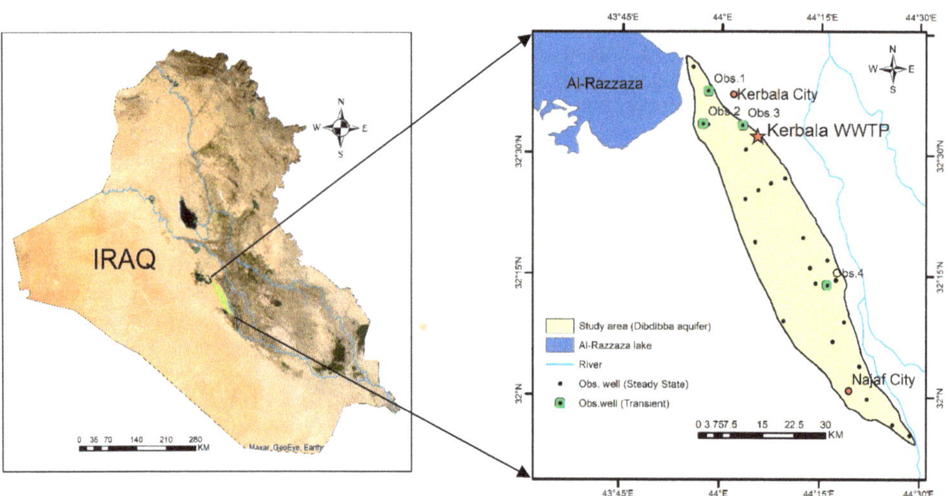

Figure 1. The study location with the Karbala WWTP.

2.2. Hydrogeological Characterization

The conceptual groundwater flow model and the subsequent progress of the calibrated model required an adequate explanation of the hydrologic conditions at the study site as a starting point. It is also difficult to choose an appropriate model or create an authoritative calibrated model without a clear description of the site. Only a few areas of the study site had hydrogeological properties available, such as the aquifer's characteristics. As a consequence, the Kriging approach was utilized to estimate information for the entire region, in order to create an approximation of the information required. Figure 1 depicts the sites of the wells chosen to estimate the properties of the aquifer. The accuracy of the calculation used to determine the artificial recharge of the aquifer was significantly impacted by these parameters. The Dibdibba formation is composed primarily of pebbly sandstone, sandstone, siltstone, and lime. Secondary gypsum is also present. The formation is between 45 and 60 m thick.

2.3. Modeling Approach

A conceptual numerical model was built using the Processing MODFLOW package. A proper representation of the hydrogeological characteristics in the research area was necessary for the development of both the conceptual groundwater movement simulation and for an accurately calibrated model. The model was calibrated in steady and transient modes, based on historical data observed from 2016 to 2017. The steady state calibration was necessary to understand the transient situation, thereby serving as the basis for the transient models. The steady state simulation model was validated utilizing fifteen wells with two adjustable elements. The well depths varied from 20 to 50 m, and the pump rate was approximately 25 to 30 m^3/h. The study area was divided into seven regions, using hydraulic conductivity parameters predicted as a result of the pumping tests for 20 wells [28], and three sub-catchment regions that were naturally recharged based on the parameters of the aquifer (Figure 2). Hydraulic conductivity and natural recharge were considered separately for each homogeneous area in the study area. This was done to reduce the uncertainty caused by the projected variation over the wide area studied. The GMS program's PEST (parameter estimation) calibration software was used to automatically complete the parameter estimation process. In this study, the spatial statistical method (Kriging) was utilized to interpolate groundwater table values, as well as all other aquifer properties, including the hydraulic conductivity and the top and bottom of the aquifer bounds. Kriging is a method for obtaining optimal and impartial estimates of geographic characteristics at non-sampled locations, by utilizing structural features of a semivariogram and initial group data values. It takes the spatial structure of the variable into account, so it was chosen over the nearest neighbor approach, the arithmetic average technique, the polynomial approximation, and the inverse distance-weighted approach. Kriging can also provide variance distribution estimates for each estimated location, which can serve as a primary indicator of estimated value precision or uncertainty. The conventional Kriging method was employed in this work, to generate expected values for various beginning input data at every unsampled position in the region. It is recommended that the input parameters are regularly distributed for optimal performance of the Kriging interpolation method (bell curve). To determine the degree to which the input variables (aquifer characteristics) matched the bell curve, two testing procedures were used: the first was drawing data histograms, and the other was a normal likelihood chart. A standard normal likelihood chart was produced by graphing the input values of the information versus the standard normal distribution at the point where their summed distributions were similar. The data distribution was normal if the points clustered around a straight line.

Using trial and error, the best grid size for the models was determined, and the grid subsequently used was composed of 3600 active cells. The width of each cell in both the x and y directions (rows and columns) was 500 m. The model area was developed in a 3D grid on the horizontal axis, with single unbounded layers on the vertical axis. Groundwater flow patterns in the Dibdibba aquifer were used to establish boundary conditions. A constant-head boundary was applied along the study area's eastern and western edges, with values of 5 and 35 m provided by measurements taken from various observation wells. Tar Al Sayyed and Tar Al Najaf are specified no-flow borders, located on the northwestern and southwestern boundaries of the study region. For the predicted values of the precipitation for the future period (2022–2030) in the study area, the results of expected precipitation [29] were used in this study in the conceptual model.

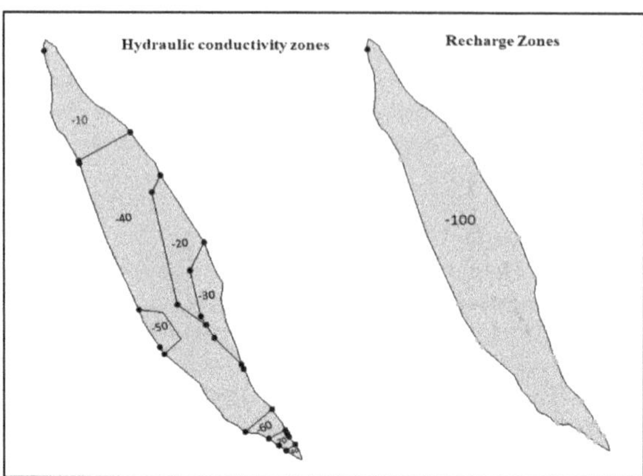

Figure 2. Recharge and hydraulic conductivity zones of Dibdibba aquifer.

Using the calibrated models, the effects of artificial recharge on the groundwater table and the quality of the water in the aquifer, especially in the surrounding area, were estimated. This was done by adding a hypothetical pond with several scenarios of pumping rates close to the WWTP. Different scenarios were considered based on the idea that groundwater pump rates and both artificial and natural recharge will increase between 2022 and 2030. Observations of actual groundwater levels were used to check the results of the model for the time period 2016–2017. In this study, treated wastewater was used as an artificial source of recharge, because there is high output of treated water (100,000 m^3/day) and it is close to a wastewater treatment plant.

3. Results and Discussion

3.1. Calibration and Validation Model

The steady state calibration of the flow model allowed for the estimation of natural groundwater recharge varying from 9 to 12 percent of the mean annual precipitation for the Dibdibba aquifer. These ratios were similar to those employed in earlier research [3,28]. For the steady state, the results of the calibrated model were acceptable and matched the observed data with projected groundwater table, and the determination coefficient (R^2) was around 0.96 (Figure 3). Figure 3 illustrates the calibration results of the model with observation ground water levels in the selected wells. A scatter diagram of these results was plotted and is shown in Figure 3. The observed value is shown by the center line. The whiskers on the top and bottom edges represent the measured data plus the period and the measured values minus the period, respectively. The full line displays the error, where a green line signifies that the error is contained inside the designated range (lower than 0.5 m). Red lines indicate errors greater than 1 m, while yellow bars indicate errors between 0.5 and 1 m. The Parameter ESTimation (PEST) tool within the GMS software was used to obtain the optimal solution mode compatible with the minimum errors during the calibration process. Groundwater levels from 15 wells were used for the numerical model calibration and validation under steady-state conditions. These wells were selected in order to cover the entire research area and reduce the length of the simulation run during the calibration period. In order to achieve an acceptable degree of accuracy, 15 calibrating targets were used to simulate levels of water in the models; 14 of the lines were green (error below 0.5 m), but one line was yellow (error over 0.5 but lower than 1.0 m). During the conceptual model's calibration procedure, the research area was separated into a number of zones by hydraulic conductivity and expected natural recharge rates.

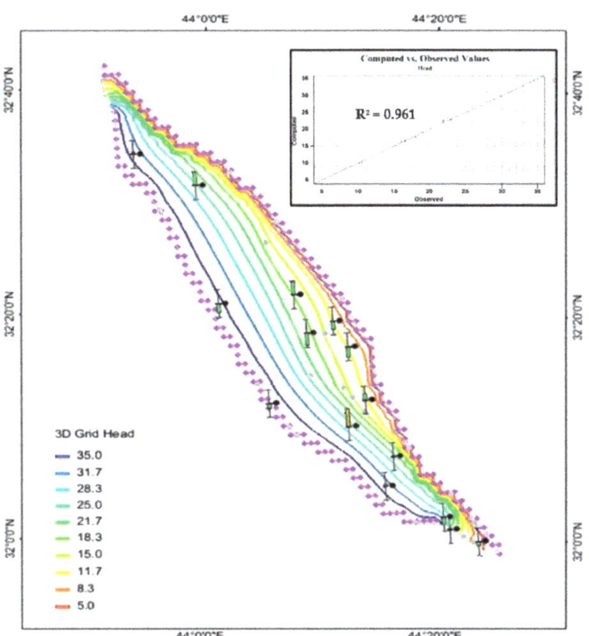

Figure 3. Simulated heads after calibration of the study aquifer under steady state.

At the end of the simulation, the numerical model produced calibrated values of these two crucial characteristics for all the study areas. These results were an indication of high confidence in the simulation model's estimates, meaning that this data could be used as the initial condition when calibrating the transient model. It was not possible to calibrate the salinity, partly because there were not enough measurements and mostly because the processes were too complicated [30].

3.2. Validation of the Transient Model

In order to investigate the validity of the numerical model results for the future time period, the transient numerical model results were examined with historically observed groundwater levels from four monitoring wells during the period 2016–2017. Building a transient simulation mode requires the handling of a huge quantity of transitory information from various sources, such as data on the depth of water in observation wells and on recharging and pumping wells. The transient state was simulated by initiating a simulated steady-state groundwater table. The aim of this operation was to measure the groundwater table. The Dibdibba unconfined aquifer's specific yield (Sy-value) was determined using transient models as preliminary projections. Previous studies' pumping experiments gave Sy values ranging from 0.001 to 0.05 [28]. According to the agricultural needs in the research region, different monthly pumping rates were used. Designing a transient model requires careful consideration of the simulation time step used, because this has a significant impact on the results of the numerical model [31]. The time period was split into 20 time periods and four control wells, with a start time of 1 January 2016, and an end time of 1 December 2017. The selected wells were close to the Karbala treatment plant, especially Obs1 and Obs3 (Figure 1). Specific yield factors were chosen to serve as the simulated results in the PEST operations during the calibration of the transient state. This parameter varied continuously, until a satisfactorily acceptable difference between the measured and simulated water depth was obtained from January 2016 to December 2017. The four calibration objectives in this model are indicated by the color green (Figure 4). The transient numerical model was also examined using a transient scatter plot, which compared the groundwater levels against the simulations. These comparisons are displayed in Figure 4, and the models accurately predicted the measured levels with a determination coefficient (R^2) of 0.923.

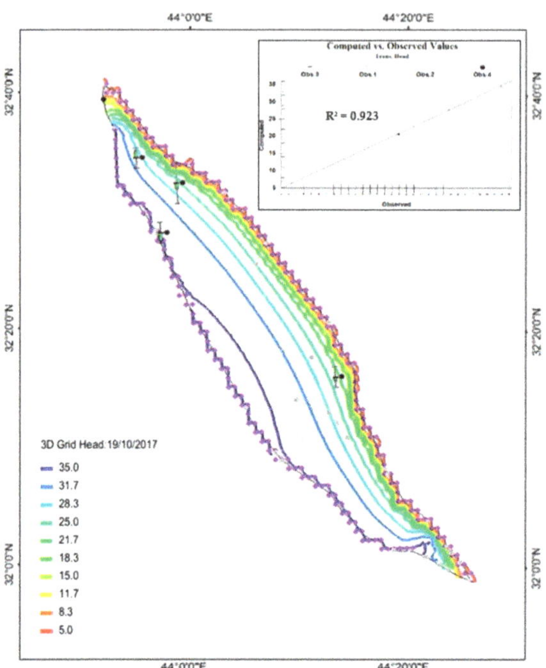

Figure 4. Calibration simulated heads of the steady area during the transient simulation (19-Oct.-2017).

The groundwater flow simulation model was validated from January 2016 to December 2017. Figure 5 shows how it nearly recreated the groundwater table fluctuations of various earlier observation readings. Figure 5 compares the observed and estimated fluctuations in the groundwater table in the observation wells. The transient simulation of the conformity between the predicted and observed groundwater tables in the monitoring wells demonstrated the simulation's excellent performance. During the monitoring periods, all of the figures show a decreasing trend in groundwater levels. According to these assessments, the model was adequately calibrated. Figure 5 shows that the simulation model outputs for observation well No. 1 (Obs.1) were always lower than the measured groundwater levels, while well No. 2 (Obs.2) always exceeded the measured groundwater table throughout the validation time period but was still within agreeable limits. The projected values for the other wells measured, both overestimated and underestimated the value for the various observation periods. These patterns could be explained by the varying prediction values for the hydrogeology characteristics of the steady area, as well as the impact of the well's distance from the study area's hydraulic boundaries.

The validated numerical models were applied to project the effect of the proposed groundwater artificial recharge using a pond on the level and quality of groundwater, with a focus on the Kerbala WWTP zone. The amount of groundwater extraction by pumps between 2016 and 2030, as well as both natural and artificial recharge, were taken into account in the various scenarios.

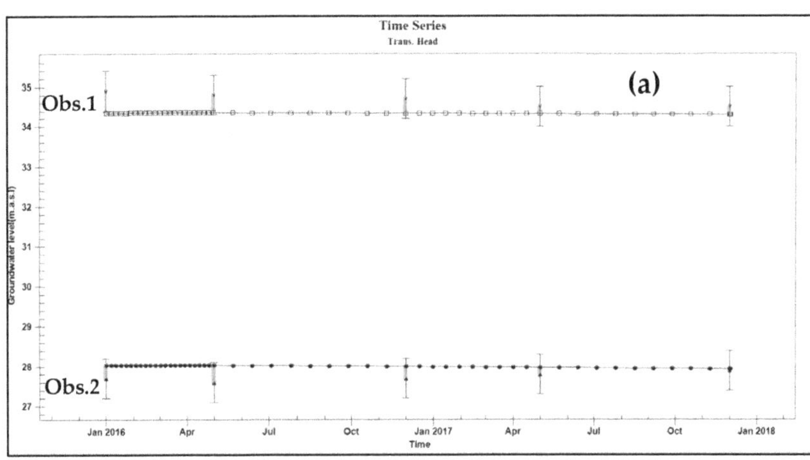

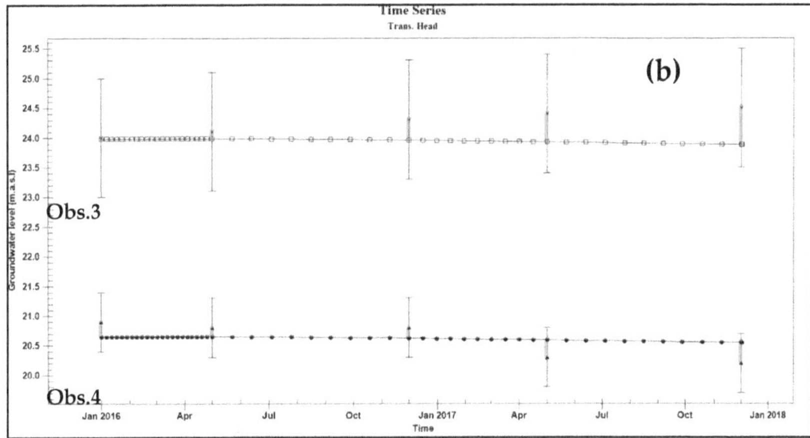

Figure 5. Validation in four observation wells throughout the transient duration 2016 and 2017, (**a**) validation in observation wells 1 and 2, (**b**) validation in observation wells 3 and 4.

3.2.1. Quantitative Evaluation

In this study, three different simulation scenarios (SIM1, SIM2, and SIM3) were applied to the prediction period 2016–2030. SIM1 used the same initial conditions, making the assumption that the existing recharge and extraction rates would stay the same as during 2016, excluding the additional rate of artificial recharge from treated water. Based on this first scenario (SIM1), the simulation model projected that the groundwater level would drop by more than 1 m near the Kerbala WWTP location by 2030, and the water quality would become worse. According to this scenario, there would be a loss of more than 2 m of groundwater in the vicinity of the Kerbala WWTP by 2030 (Figure 6). This decline was justified as a result of the excessive use of ground water, the decrease in natural recharge due to the lack of precipitation, and the increased temperatures in the past few years due to climatic changes in the region. Many agricultural areas have suffered in the last few years due to this decline in groundwater levels, which increased withdrawal costs and degraded water quality. Most climate studies have shown that the effects of climate change in the region could become worse at a fairly high rate [32]. Due to this, it is expected that this drop in water level and quality will have a greater effect on farmers and the region's ecosystem in the near future.

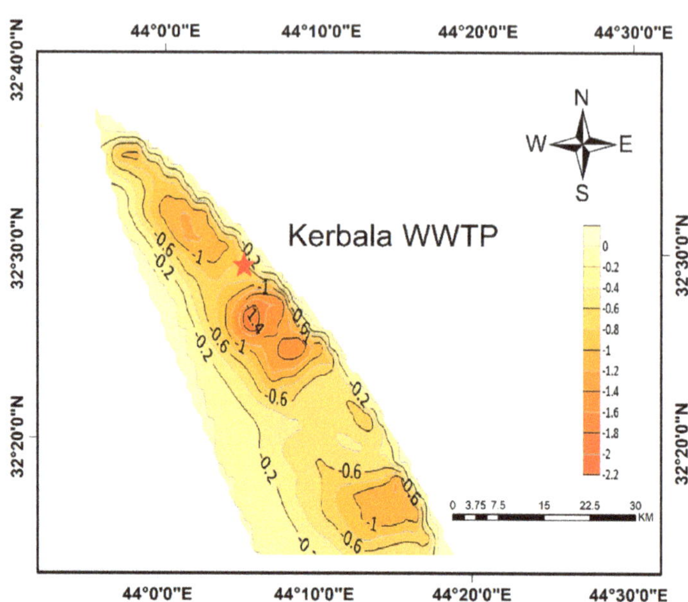

Figure 6. The predicted groundwater elevation drops in 2030 over the study area without the use of artificial recharge.

As a result of its closeness to the treatment plant, the significant effects of extraction processes in this region, and the lack of natural recharge, this site is regarded as one of the best for establishing an artificial recharge pond. These estimates of the predicted drop in groundwater levels are sensitive to the assumptions made regarding extraction values (11,000 m^3/day), which depend on a number of other assumptions, because there have not been enough observations in the field.

In SIM2 and SIM3, the same initial conditions as in SIM1 were implemented, with only the addition of an artificial recharge flow that was pumped into the recharge pond at a rate of 5000 and 10,000 m^3/day, respectively. In the simulation model, the recharge pond was represented by one cell. For operational and economic reasons, the site selected for the recharge pond was near the WWTP. When the projected groundwater levels in 2030 for SIM1 and SIM2 were compared, it was found that an increase of up to 0.8 and 1.4 m would be possible with an artificial recharge rate of 5000 and 10,000 m^3/day, respectively (Figure 7a,b). Taking into account a minimum rise of 0.2 m, the impacted region around the Kerbala WWTP would be about 78.2 km^2 for SIM2 and 110 km^2 for SIM3. During the study period, this expected increase would potentially be very important for lowering withdrawal costs and adding more farmland to the region.

The results related to the monitoring well (Obs.3) situated close to the artificial recharge zone were used to assess the temporal variation of the groundwater elevation, as illustrated in Figure 8. As shown in Figure 8a,b, groundwater levels rose over the eight years of the simulation by 7.5 cm and 12.3 cm each year for the SIM2 and SIM3 artificial recharge modeling scenarios, respectively. It is important to note that the rate of rise differed between the two scenarios because of the different pumping rates of the recharge pond. At the end of the time simulation period (2030), the groundwater table in the observation well (Obs.3) had increased by more than 0.67 m and 1.2 m under SIM2 and SIM3, respectively.

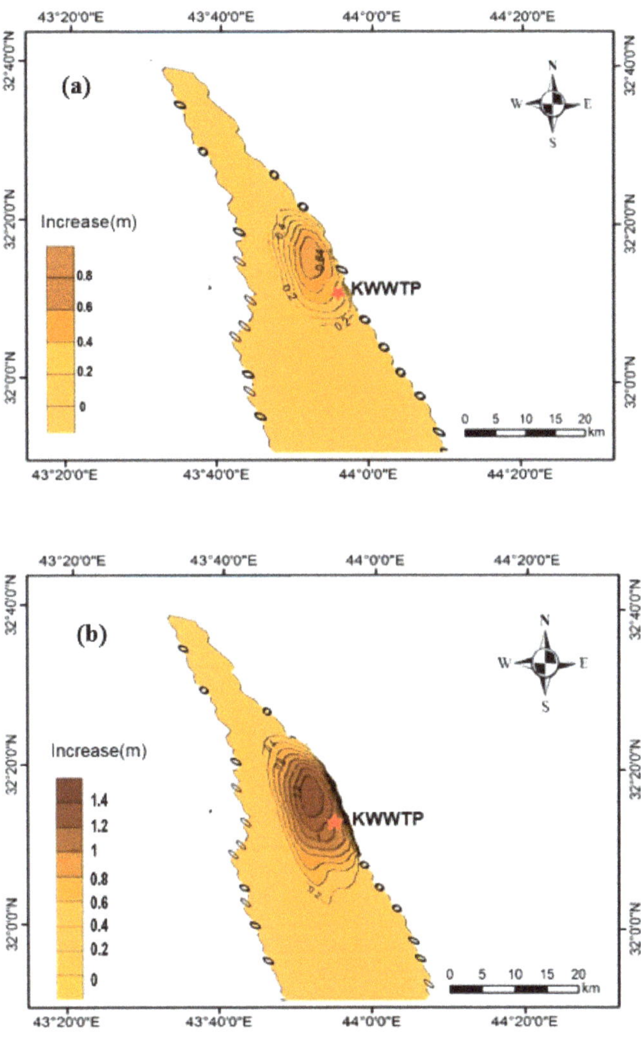

Figure 7. The impact of artificial recharge using a pond in 2030: increases in groundwater levels (**a**) between SIM2 and SIM1 and (**b**) between SIM3 and SIM1.

The prediction of the water requirements in the study region is dependent on two important aspects. The first is the weather, which includes precipitation, temperature, sunlight, wind velocity, and moisture. The second component is cultivation type, which impacts the irrigation water demand and represents the plant's modulus [33]. According to the Iraqi Ministry of Water Resources and the Ministry of Agriculture, the maximum irrigation water requirement for an agricultural plan is 3 m^3/donum/day (one donum is 2500 m^2). According to the modeling findings for the coming eight years, a new agricultural region of more than 8950 donum (22.4 km^2) could be added to the area if 5% of the reclaimed wastewater output from the WWTP was utilized for the artificial recharge operation with a pond. Moreover, when 10% of the treated wastewater output is used, the additional area might be increased to 16,200 donum. This anticipated increase in farmed area would be extremely beneficial in combating desertification, global climate change, and enhancing the ecosystem in the study region.

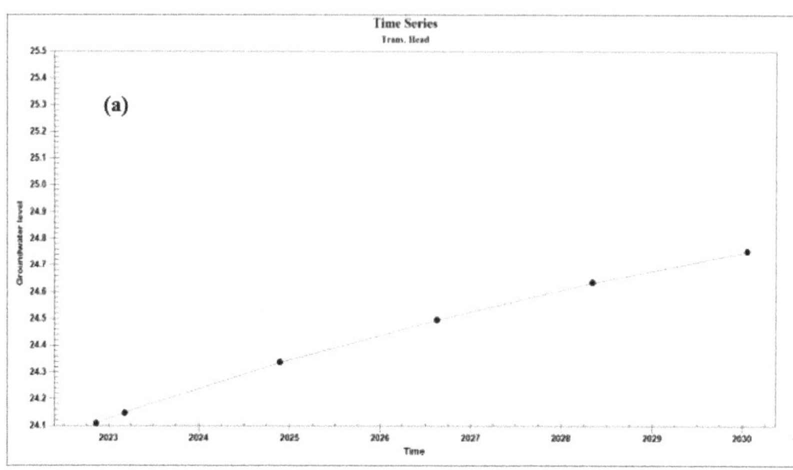

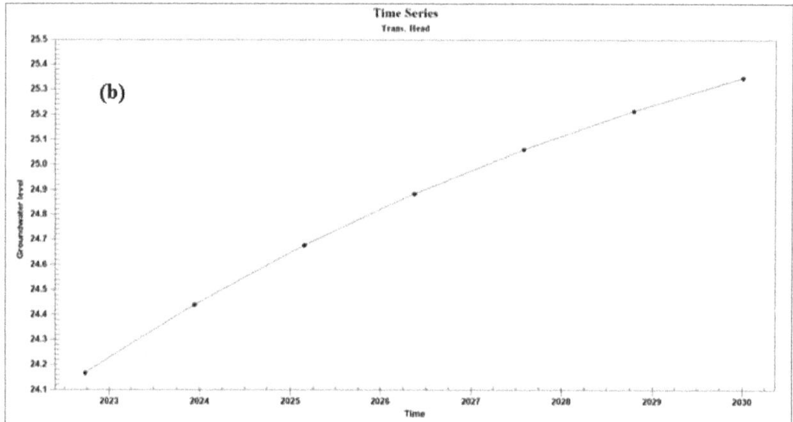

Figure 8. Changes in groundwater table (m.a.s.l) at the observation well No.3 (**a**) for SIM2 and (**b**) for SIM3 during the simulation period.

3.2.2. Improving Groundwater Quality

The MT3DMS model within GMS Software was used to predict groundwater quality based on the TDS and EC values of groundwater and treated water. The results of the first scenario in this study indicated that the water quality in the aquifer will continue to deteriorate during the coming years, as a result of the excessive withdrawal of groundwater. Artificial recharge may be the best and most practical way to improve the quality of water and reduce salinity. According to the laboratory results of the field samples of groundwater taken from the operation wells near the site of the Kerbala WWTP, the groundwater was salty, with a TDS of more than 4320 ppm and an EC of more than 4780 ppm, while the salinity of the treated wastewater that would be injected in the artificial pond reached a TDS of less than 1100 ppm and an EC of less than 1185 μ.S/cm.

Under the artificial recharge situation of SIM2, with a daily pumping rate of 5000 m³/d, groundwater salinity could be lowered by up to 2400 ppm TDS close to the pond site, with a reclaimed area of approximately 32 km² (Figure 9a). This would extend to a 51 km² recovery region with a TDS change of less than 3000 ppm. For the EC, the artificial recharge pond could decrease the concentration from 4779 μ.S/cm to less than 1600 μ.S/cm near the recharge pond. With a decreasing ratio close to 50%, the recovery area reached 68.4 km^2 (Figure 9b).

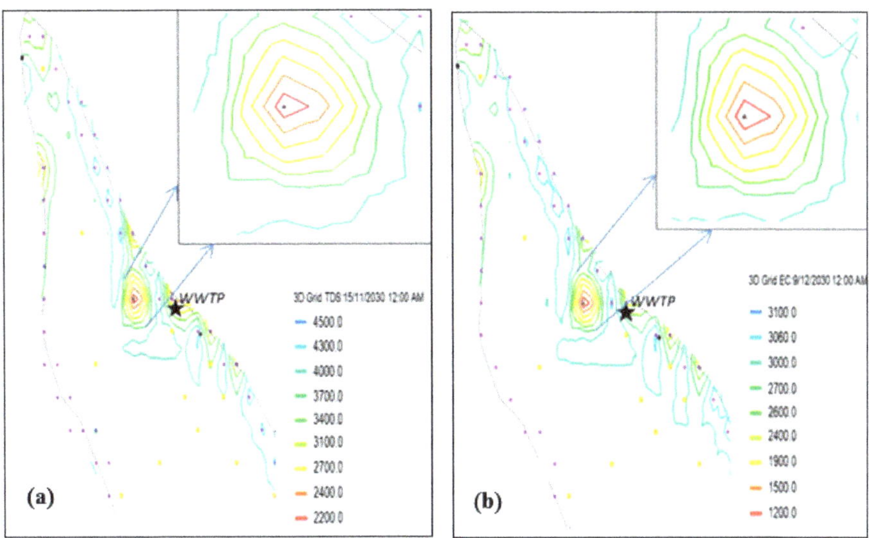

Figure 9. Effect of the artificial recharge on water quality under SIM2 simulation model (**a**) TDS and (**b**) EC for the period 2022–2030.

For the third scenario, SIM3, with an increased pumping rate of 10,000 m^3/d, as a consequence of this increase, a greater effect was obtained on the quality of groundwater (Figure 10). In fact, the reclaimed area was increased to 62.7 km^2 with a TDS of 2600 ppm, and the maximum reduction in TDS reached 1900 ppm near the pond (Figure 10a). For the EC, the region influenced was t increased to almost 77.4 km^2 with an EC of 1800 μ.S/cm and a maximum reduction of 1200 μ.S/cm near the recharge pond, as illustrate in Figure 10b.

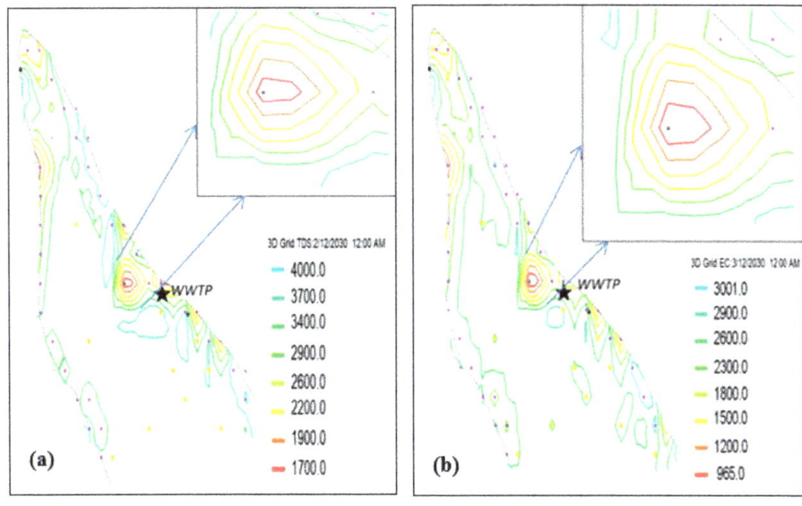

Figure 10. Effect of the artificial recharge on water quality under SIM3 simulation model (**a**) TDS and (**b**) EC for the period 2022–2030.

3.3. Sensitivity Analysis

Sensitivity analysis is the method used to quantify the degree of uncertainty in a calibrated model as a result of unknown aquifer characteristics. The aim of sensitivity analysis was to comprehend how different model variables and hydrogeological stressors affect the aquifer and to identify the parameters which were most sensitive, thereby requiring special attention in future investigations. At the conclusion of the PEST iteration, sensitivity analyses for each of the parameters were carried out. As shown in Figure 11, hydraulic conductivity and natural recharging values at each site were examined for the steady state. In comparison to changes in hydraulic conductivity, the model was much more sensitive to variations in natural recharge. Compared to the other input characteristics, the largest zone (RECH-1) natural recharge had the greatest influence on groundwater levels. The simulation model results were significantly influenced by hydraulic conductivity parameters for the regions HK-30, HK-40, and HK-70. In comparison with other regions, the relatively small regions (HK-10, HK-20, HK-50, and HK-80) were less sensitive.

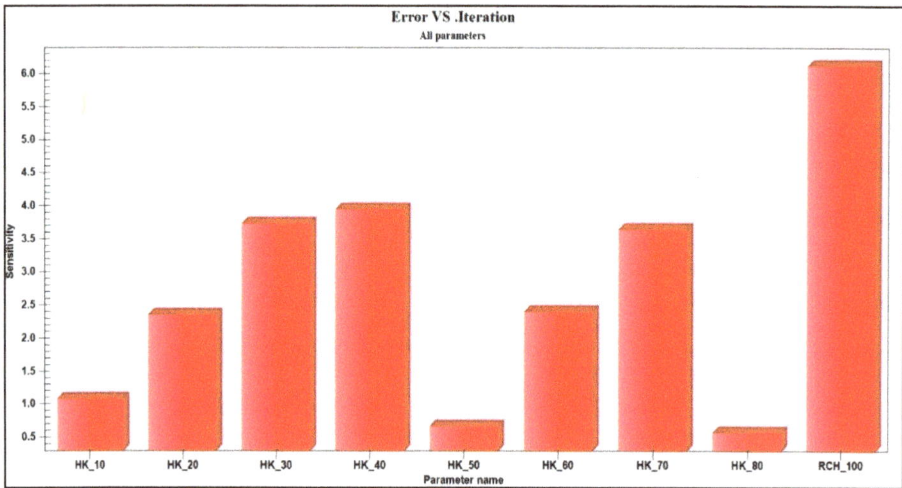

Figure 11. Model parameter sensitivity analysis.

4. Conclusions

The artificial recharge of groundwater is an effective way to alleviate the current water resource crises and to improve the quality of water supplies. The effect of artificial recharge on groundwater levels in the Dibdibba unconfined aquifer, Iraq, was investigated using treated wastewater (tertiary treatment) from the WWTP in Kerbala. Groundwater flow and solute transport were simulated using a 3D numerical model (MODFLOW and MT3DMS) with GMS 10.6 software. The PEST tool was used for the automatic calibration of the created models. The steady-state and transient groundwater levels simulated for 2016–2017 were in agreement with the observed groundwater levels. Three scenarios were simulated using calibrated models. The first scenario applied the current situation circumstances without an artificial recharge, while the other scenarios applied artificial recharge (5000 and 10,000 m^3/day) injected into a pond with an area of 0.25 km^2, to demonstrate how the aquifer would perform between 2022 and 2030. The results revealed that during this time, a pond that was artificially recharged at rates of 5000 and 10,000 m^3/day would result in annual increases in groundwater levels of more than 12.3 cm. This increase would result in the recovery of groundwater levels of up to 40 km^2 for the new agricultural area. Consequently, a decrease in TDS and EC concentrations in groundwater could also be observed at approximately 1900 ppm for TDS and 1500 μ.S/cm for EC near the pond. The reclaimed area increased by 62.7 km^2 for TDS and 77.4 km^2 for EC. These results indicated that eight years of artificial recharge would result in a minimum recovery of groundwater levels of 20 cm, over a recovery area measuring 110 km^2. Therefore, the application of the artificial recharge site should increase groundwater reserves and enhance water quality; the numerical modeling used here is indicative of artificial recharge being an effective approach for the conservation of groundwater. Furthermore, the artificial recharge had a considerable impact on salinity. The main limitations of this study were the hydrogeological features of the aquifer and the future estimates of the natural recharge and discharge that were used. In addition, in this study, only salinity was considered in the water quality assessment.

The findings of this study can assist decision-makers in developing strategies to reduce water scarcity and adapt to climate change. Moreover, other researchers can use it at full scale and implement further field tests to decrease the uncertainty of the study, as well as increase the number of water quality parameters, such as for heavy metals.

Author Contributions: Conceptualization, W.H.H. and B.K.N.; methodology, K.M.; software, W.H.H. and F.A.M.; validation, M.R., C.R. and A.A.J.G.; formal analysis, A.A.; investigation, W.H.H.; resources, B.K.N.; data curation, F.A.M.; writing—original draft preparation, F.A.M.; writing—review and editing, W.H.H., K.M. and C.R.; visualization, M.R.; supervision, W.H.H.; project administration, K.M. and B.K.N.; funding acquisition, K.M., M.R. and C.R. All authors have read and agreed to the published version of the manuscript.

Funding: This research received no external funding.

Data Availability Statement: All data are contained within the article.

Acknowledgments: This study has been funded by NUFFIC's Orange Knowledge Programme, through the OKP-IRA-104278 project entitled "Efficient water management in Iraq switching to climate smart agriculture: capacity building and knowledge development" coordinated by Wageningen University & Research, the Netherlands. The authors extend their thanks to University of Warith Al-Anbiyaa, Iraq for financial support for this study.

Conflicts of Interest: The authors declare no conflict of interest.

References

1. Shahid, S.; Alamgir, M.; Wang, X.J.; Eslamian, S. Climate change impacts on and adaptation to groundwater. In *Handbook of Drought and Water Scarcity*; CRC Press: Boca Raton, FL, USA, 2017; pp. 107–124.
2. Hassan, W.H.; Hussein, H.H.; Nile, B.K. The effect of climate change on groundwater recharge in unconfined aquifers in the western desert of Iraq. *Groundw. Sustain. Dev.* **2022**, *16*, 100700. [CrossRef]
3. Hassan, W.H.; Nile, B.K.; Mahdi, K.; Wesseling, J.; Ritsema, C. A feasibility assessment of potential artificial recharge for increasing agricultural areas in the Kerbala desert in Iraq using numerical groundwater modeling. *Water* **2021**, *13*, 3167. [CrossRef]
4. Seeyan, S.; Akrawi, H.; Alobaidi, M.; Mahdi, K.; Riksen, M.; Ritsema, C. Groundwater Quality Evaluation and the Validity for Agriculture Exploitation in the Erbil Plain in the Kurdistan Region of Iraq. *Water* **2022**, *14*, 2783. [CrossRef]
5. Baaloudj, O.; Badawi, A.K.; Kenfoud, H.; Benrighi, Y.; Hassan, R.; Nasrallah, N.; Assadi, A.A. Techno-economic studies for a pilot-scale Bi12TiO20 based photocatalytic system for pharmaceutical wastewater treatment: From laboratory studies to commercial-scale applications. *J. Water Process Eng.* **2022**, *48*, 102847. [CrossRef]
6. Hassan, W.H.; Hashim, F.S. The effect of climate change on the maximum temperature in Southwest Iraq using HadCM3 and CanESM2 modelling. *SN Appl. Sci.* **2020**, *2*, 1494. [CrossRef]
7. Gaznayee, H.A.A.; Al-Quraishi, A.M.F.; Mahdi, K.; Messina, J.P.; Zaki, S.H.; Razvanchy, H.A.S.; Hakzi, K.; Huebner, L.; Ababakr, S.H.; Riksen, M.; et al. Drought Severity and Frequency Analysis Aided by Spectral and Meteorological Indices in the Kurdistan Region of Iraq. *Water* **2022**, *14*, 3024. [CrossRef]
8. Hassan, W.H. Climate change impact on groundwater recharge of Umm er Radhuma unconfined aquifer Western Desert, Iraq. *Int. J. Hydrol. Sci. Technol.* **2020**, *10*, 392–412. [CrossRef]
9. Jarraya Horriche, F.; Benabdallah, S. Assessing aquifer water level and salinity for a managed artificial recharge site using reclaimed water. *Water* **2020**, *12*, 341. [CrossRef]
10. Abraham, M.; Mathew, R.A.; Jayapriya, J. Numerical Modeling as an Effective tool for Artificial Groundwater Recharge Assessment. *J. Phys. Conf. Ser.* **2021**, *1770*, 012097.
11. Ranganathan, P.C.; Chuluke, D.; Chena, D.; Senapathi, V. Artificial recharge techniques in coastal aquifers. In *Groundwater Contamination in Coastal Aquifers*; Elsevier: Amsterdam, The Netherlands, 2022; pp. 279–283.
12. Islam, H.; Abbasi, H.; Karam, A.; Chughtai, A.H.; Ahmed Jiskani, M. Geospatial analysis of wetlands based on land use/land cover dynamics using remote sensing and GIS in Sindh, Pakistan. *Sci. Prog.* **2021**, *104*, 00368504211026143. [CrossRef]
13. Bouwer, H. Artificial recharge of groundwater: Hydrogeology and engineering. *Hydrogeol. J.* **2002**, *10*, 121–142. [CrossRef]

14. Al-Assa'd, T.A.; Abdulla, F.A. Artificial groundwater recharge to a semi-arid basin: Case study of Mujib aquifer, Jordan. *Environ. Earth Sci.* **2010**, *60*, 845–859. [CrossRef]

15. Mohammed, M.H.; Zwain, H.M.; Hassan, W.H. Modeling the impacts of climate change and flooding on sanitary sewage system using SWMM simulation: A case study. *Results Eng.* **2021**, *12*, 100307. [CrossRef]

16. Voudouris, K.; Diamantopoulou, P.; Giannatos, G.; Zannis, P. Groundwater recharge via deep boreholes in the Patras Industrial Area aquifer system (NW Peloponnesus, Greece). *Bull. Eng. Geol. Environ.* **2006**, *65*, 297–308. [CrossRef]

17. Rambags, F.; Raat, K.J.; Zuurbier, K.G.; van den Berg, G.A.; Hartog, N. *Aquifer Storage and Recovery (ASR): Design and Operational Experiences for Water Storage through Wells*; PREPARED 2012.016, the Seventh Framework Programme; European Union/European Commission: Amsterdam, The Netherlands, 2013; 40p.

18. Arya, S.; Subramani, T.; Karunanidhi, D. Delineation of groundwater potential zones and recommendation of artificial recharge structures for augmentation of groundwater resources in Vattamalaikarai Basin, South India. *Environ. Earth Sci.* **2020**, *79*, 102. [CrossRef]

19. Bouri, S.; Dhia, H.B. A thirty-year artificial recharge experiment in a coastal aquifer in an arid zone: The Teboulba aquifer system (Tunisian Sahel). *Comptes Rendus Geosci.* **2010**, *342*, 60–74. [CrossRef]

20. Kareem, I.R. Artificial groundwater recharge in Iraq through rainwater harvesting (Case Study). *Eng. Tech. J* **2012**, *31*, 1069–1080.

21. Ali, M.T.; Saleh, S.A.; Gweer, A.N. Experimental Study of Artificial Recharge of Unconfined Groundwater Aquifer for Elected Sites in Salahaddin, Iraq. *Iraqi Geol. J.* **2022**, *55*, 162–175. [CrossRef]

22. Vandenbohede, A.; Van Houtte, E.; Lebbe, L. Groundwater flow in the vicinity of two artificial recharge ponds in the Belgian coastal dunes. *Hydrogeol. J.* **2008**, *16*, 1669–1681. [CrossRef]

23. Zhang, H.; Xu, Y.; Kanyerere, T. Site assessment for MAR through GIS and modeling in West Coast, South Africa. *Water* **2019**, *11*, 1646. [CrossRef]

24. El Ayni, F.; Manoli, E.; Cherif, S.; Jrad, A.; Assimacopoulos, D.; Trabelsi-Ayadi, M. Deterioration of a Tunisian coastal aquifer due to agricultural activities and possible approaches for better water management. *Water Environ. J.* **2013**, *27*, 348–361. [CrossRef]

25. Hassan, W.H.; Khalaf, R.M. November. Optimum Groundwater use Management Models by Genetic Algorithms in Karbala Desert, Iraq. In *IOP Conference Series: Materials Science and Engineering*; IOP Publishing: Bristol, UK, 2020; Volume 928, p. 022141.

26. Al-Sudani, H.I.Z. Groundwater system of Dibdibba sandstone aquifer in south of Iraq. *Appl. Water Sci.* **2019**, *9*, 72. [CrossRef]

27. Abdulameer, A.; Thabit, J.M.; Kanoua, W.; Wiche, O.; Merkel, B. Possible Sources of Salinity in the Upper Dibdibba Aquifer, Basrah, Iraq. *Water* **2021**, *13*, 578. [CrossRef]

28. Al-Ghanimy, M.A. Assessment of Hydrogeological Condition in Karbala—Najaf Plateau, Iraq. Ph.D. Thesis, University of Baghdad, Baghdad, Iraq, 2018, *unpublished*.

29. Nile, B.K.; Hassan, W.H.; Alshama, G.A. Analysis of the effect of climate change on rainfall intensity and expected flooding by using ANN and SWMM programs. *ARPN J. Eng. Appl. Sci.* **2019**, *14*, 974–984.

30. Konikow, L.F. The secret to successful solute-transport modeling. *Groundwater* **2011**, *49*, 144–159. [CrossRef] [PubMed]

31. Anderson, M.P.; Woessner, W.W.; Hunt, R.J. *Applied Groundwater Modeling: Simulation of Flow and Advective Transport*; Academic Press: Cambridge, MA, USA, 2015.

32. Hassan, W.H. Climate change projections of maximum temperatures for southwest Iraq using statistical downscaling. *Clim. Res.* **2021**, *83*, 187–200. [CrossRef]

33. FAO. *The State of World Fisheries and Aquaculture, 1998*; Food and Agriculture Organization, Fisheries Department: Rome, Italy, 1999.

Article

A Feasibility Assessment of Potential Artificial Recharge for Increasing Agricultural Areas in the Kerbala Desert in Iraq Using Numerical Groundwater Modeling

Waqed H. Hassan [1,*], Basim K. Nile [1], Karrar Mahdi [2], Jan Wesseling [2] and Coen Ritsema [2]

[1] Civil Engineering, College of Engineering, University of Kerbala, Kerbala 56001, Iraq; dr.basimnile@uokerbala.edu.iq
[2] Soil Physics and Land Management Group, Wageningen University & Research, 6708 PB Wageningen, The Netherlands; Karrar.mahdi@wur.nl (K.M.); jan.wesseling@wur.nl (J.W.); coen.ritsema@wur.nl (C.R.)
* Correspondence: waaqidh@uokerbala.edu.iq or waqed2005@yahoo.com

Abstract: Groundwater in Iraq is considered to be an alternative water resource, especially for areas far away from surface water. Groundwater is affected by many factors including climate change, industrial activities, urbanization, and industrialization. In this study, the effect of artificial recharge on the quantity of groundwater in the Dibdibba unconfined aquifer in Iraq was simulated using a groundwater modeling system (GMS). The main raw water source used in the artificial recharge process was the reclaimed water output (tertiary treatment) from the main wastewater treatment plant (WWTP) in Kerbala, with 20 injection wells. After calibration and validation of the three-dimensional numerical model used in this study and taking wastewater recharge rates into account, two different scenarios were applied to obtain the expected behavior of the aquifer when the groundwater levels were augmented with 5% and 10% of the daily outflow production of the WWTP in Kerbala. The model matched the observed head elevations with $R^2 = 0.951$ for steady state and $R^2 = 0.894$ for transient simulations. The results indicate that the injection of treated water through 20 wells raised the water table in more than 91 and 136 km^2 for 5000 and 10,000 m^3/day pumping rates, respectively. Moreover, increasing the volume of water added to the aquifer could lead to establishing new agricultural areas, spanning more than 62 km^2, extending about 20 km along the river.

Keywords: artificial recharge; Dibdibba aquifer; groundwater modeling system; GMS

Citation: Hassan, W.H.; Nile, B.K.; Mahdi, K.; Wesseling, J.; Ritsema, C. A Feasibility Assessment of Potential Artificial Recharge for Increasing Agricultural Areas in the Kerbala Desert in Iraq Using Numerical Groundwater Modeling. *Water* **2021**, *13*, 3167. https://doi.org/10.3390/w13223167

Academic Editor: Adriana Bruggeman

Received: 1 October 2021
Accepted: 8 November 2021
Published: 10 November 2021

Publisher's Note: MDPI stays neutral with regard to jurisdictional claims in published maps and institutional affiliations.

1. Introduction

Water and groundwater play a pivotal role in food security and economic evolution all over the world. Unfortunately, in recent decades, the ever-increasing demand for water due to urbanization, economic development, population growth, and climate change has caused water scarcity and restricted economic evolution in many countries. Groundwater is used as an alternative source of water when there is a shortage of surface water. Groundwater is the major source of irrigation water in countries with arid and semiarid climates. In some Middle Eastern countries, the domestic water supply depends completely on groundwater [1,2]. Efficient management of groundwater resources is necessary to meet the growing water demand. Climate change and global warming can influence groundwater resources in many ways, either directly or indirectly. Increasing temperatures and changing patterns of precipitation will directly impact groundwater recharge, discharge, water levels, and annual storage. Moreover, the rising level of the sea, increased demand for irrigation water, and changes in vegetation cover can indirectly affect the quality of groundwater resources. Global warming will lead to changes in plant transpiration and evaporation rates, which denote soil dryness, causing higher soil moisture losses and reducing natural groundwater recharge [3].

Overexploitation of groundwater has led to a rapid decline in groundwater levels in many parts of Iraq [4]. One way to control the drop in the groundwater table is to artificially recharge water using wells [5,6]. Many studies have shown an increasing trend in artificial recharge methods in numerous regions all over the world such as Finland, The Netherlands, Belgium, Greece, Denmark, and Spain. Other regions in which artificial recharge techniques are used include Iran, Tunisia, Morocco, and South Africa [7]. The concept of artificial groundwater recharge technology is gaining momentum day by day because groundwater is such a variable and precious natural resource. Artificial recharge not only provides an efficient method of water storage allowing better management of the available resources, but it can also affect water quality indexes [8].

The utilization of treated wastewater is an important part of a water management planning strategy, particularly for the artificial recharge of groundwater resources and agriculture. Consequently, reclaimed wastewater is used to improve the quantity and quality of groundwater in an aquifer. Several studies have proven that such alternatives are plausible when traditional fresh raw water sources become severely limited [9–13]. The use of artificial recharge techniques has already been recognized as a way of increasing groundwater levels and improving water quality in different groundwater aquifers [14,15]. For example, Bouri and Ben Dhia [16] reported that the groundwater level of the Teboulba aquifer located in Tunisia rose nearly 30 m after six years of artificial well recharge.

Iraq is located in a region that suffers from water scarcity with only a few special water resources of its own. It is facing considerable interdependent political, economic, environmental, and security challenges [17]. Among the most important of these challenges is the deterioration of the Euphrates and Tigris rivers, which are essential to agriculture and water security in Iraq. Both rivers originate outside Iraqi borders in Turkey and Iran. Iraq has no control or authority over them. The adverse impacts from the installation of hydraulic structures such as big dams upstream on the Euphrates and Tigris rivers coupled with climate change will further undermine the agricultural sector in Iraq, causing further environmental deterioration and increasing desertification.

The current study focuses primarily on the unconfined Dibdibba aquifer in southeastern Iraq. Over the past 20 years in this aquifer, there has been a severe decline in both groundwater levels and water quality caused by the excessive use of groundwater for cultivation [18]. This problem has motivated numerous researchers to study hydrodynamic systems using numerical modeling [19–21] to obtain the spatial distribution of groundwater levels and the directions of groundwater flow or to identify the amount of re-usable recharge by rainfall. Unfortunately, there have been no previous studies regarding artificial recharge in the region, perhaps due to the lack of treated water or any water sources other than rain in the study area. The first wastewater treatment plant (WWTP) in Kerbala, which is within the boundaries of the aquifer formation, was opened in 2020.

Numerical models such as MODFLOW for simulating the flow of groundwater have been used in several previous studies over the last decade. These types of programs are connected to GIS technology and play a vital role in the management and evaluation of groundwater in many regions [22,23]. Numerical modeling investigations have been frequently used by researchers to predict the potential of an aquifer or to implement recovery of groundwater levels depending on numerical models with multi-scenario development [7,12,13,24–27]. In this paper, the impact of artificial recharge by wells on the groundwater level behavior in an unconfined aquifer (Dibdibba) was evaluated based on several scenarios using 3D calibrated numerical models constructed by the MODFLOW application using GMS 10.4 software. The main objective of the current study is to obtain the expected increase in groundwater levels and the range of the affected area as a result of applying artificial recharge by selected wells. The present study focuses on the range area near the Kerbala wastewater treatment plant (WWTP). It is the first scientific study dealing with the process of artificial recharge in the study area (Dibdibba unconfined aquifer) and one of the first studies in Iraq on the application and sustainability of treated water.

2. Materials and Methods

2.1. Study Area

Iraq is a country in a semi-arid/arid region of the Middle East. Geographically, the study area (Dibdibba aquifer) is located in central Iraq, between the cities of Najaf and Kerbala, with coordinates between 31°55′ N–32°45′ N latitude and 43°30′ E–44°30′ E longitude, as shown in Figure 1. The Dibdibba aquifer is an unconfined shallow aquifer with a single soil layer and fully depends on rainfall for recharge. It covers an area of 1100 km^2 and is limited by two cliffs: the Tar AlSayyed within Kerbala city boundaries in the northwest and the Tar Al-Najaf within the boundaries of Najaf city in the south and southwest. The Al-Razzaza lake, located near the northern boundary of the aquifer, is an open surface reservoir. The eastern side of the aquifer is bounded by quaternary sediments. The topography elevation ranges from 10 m to 90 m above sea level. The study area is considered to be the most important aquifer in Iraq. The region is known for its significant cultivation activity, which has led to a growing demand for irrigation water and thus an increase in groundwater withdrawal. For the study area, the main direction of groundwater flow is generally towards the Euphrates River, from the southwest to the northeast, and the hydraulic gradient value ranges from 0.0011 to 0.0005. The mean temperature ranges from 11 °C during the winter season to 37.5 °C during the summer. The mean annual rainfall in the study area ranges from 90 mm at the Kerbala meteorological station to 112 mm at the meteorological station in Najaf. The wettest months of the year are between November and March; 80% of the yearly precipitation falls in this period. Summer is the driest season, from June to September. Scant and irregular rain falls in May and October. Since 2003, the groundwater level has decreased and the quality of the water has been degraded. This area has been greatly influenced by water withdrawal, especially in places that were subject to overexploitation [28]. Moreover, other studies [29,30] have referred to the salinization of groundwater as an indicator of the increasing use of irrigation water for agriculture, which has led to soil leaching and the transfer of fertilizers to the unconfined aquifer. To address the issues of the low groundwater table and the deteriorating water quality in the Dibdibba unconfined aquifer, this study used treated municipal wastewater as an artificial recharge source. The main reasons for choosing this type of water were its availability in relatively large quantities (100,000 m^3/day) and the location of the nearby wastewater treatment plant.

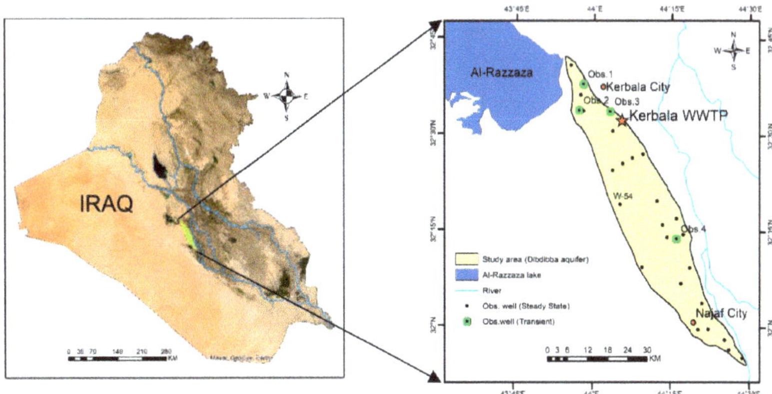

Figure 1. Location of the study area with the Kerbala WWTP.

2.2. Hydrogeological Characterization

The correct description of the hydrogeological situation in the aquifer under study is essential for understanding the value of the pertinent flow operations. Without a correct description of the area, it is impossible to choose a suitable model or develop an authoritative calibrated model. Hydrogeological characteristics of the modeled area, such as the parameters of the aquifer, are only available for a few sites in the study area. In order to obtain an estimate of the information needed, the Kriging method was used to predict data across the entire region. Figure 1 shows the locations of the selected production wells used to predict the aquifer characteristics in the study area.

These characteristics have an important impact on the accuracy of calculating the artificial recharge of the aquifer. The most important soil attributes are the main parameters that control the flow rate of the downward percolation and infiltration, especially with this type of technique. Figure 2 shows the lithology formation that was identified utilizing deep well logs and results from the infiltration test. The stratigraphic column in the area of study includes many formations (from old to young): middle late Eocene (Al-Dammam), late lower Miocene (Euphrates), middle Miocene (Fatha and Nfayil), upper Miocene (Injana), and upper Miocene-Pliocene (Dibdibba). The last formation (Al-Dibdibba) represents the top of the main unconfined aquifer and covers 1100 square kilometers of the Najaf-Kerbala plateau. The aquifer is recharged by the seasonal stormwater, with water coming from the eastern and northeastern areas as well as directly from rainfall falling on the plateau [31,32]. The formation of Dibdibba consists of pebbly sandstone and sandstone with some claystone, siltstone, and marl associated with secondary gypsum. The thickness of the formation ranges from 45 to 60 m. The Al-Dibdibba formation is exposed at both ridges of the Tar Al-Najaf and the Tar Al-Sayyed, taking the topmost of the uncovered sequence, hence making up the bedrock of the desert plain between Najaf and Kerbala.

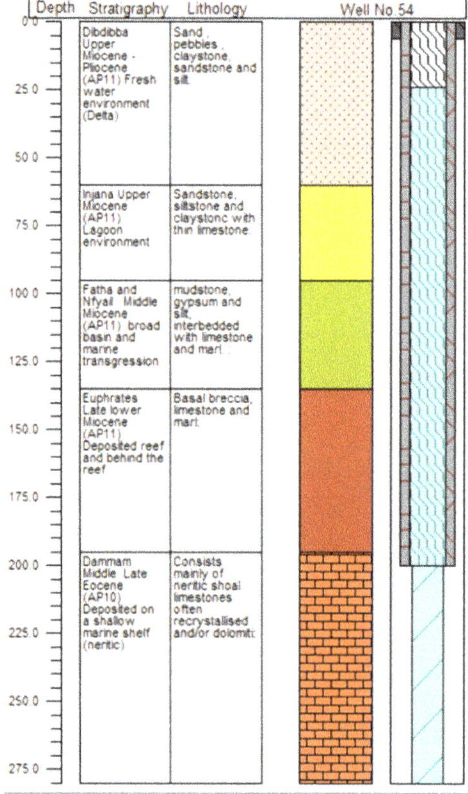

Figure 2. The stratigraphy and lithology of the formations in the study area.

2.3. Groundwater Flow Simulation

In order to meet the objectives of the numerical modeling, a conceptual model was constructed. The conceptual model represents an idealistic and simplified representation of the problem, consisting of the spatial distribution of hydrogeological and geological units, types and location of boundaries, and the values of the aquifer parameters. A suitable description of the hydrogeological conditions in the study area is essential for building the conceptual groundwater flow model as well as for developing the reliably calibrated model. To obtain the input data for the conceptual model, a spatial statistical technique was used for all data. The geostatistical technique (Kriging) was used in this study to interpolate the groundwater level data and all other aquifer characteristics such as hydraulic conductivity and the upper and lower limits of the unconfined aquifer. These terms represent the major input data for the MODFLOW under the GMS simulation program.

In this study, the ordinary Kriging method was used to provide the estimated values for different initial input parameters at any unsampled points in the aquifer. For the best performance of the Kriging interpolation method, it is recommended that the input data are normally distributed (bell curve). Two testing methods were utilized to investigate whether or not the input data (aquifer parameters) followed a normal distribution; the first was the drawing of histograms of the data, and the second was a normal probability (Q-Q) plot. A normal Q-Q plot is generated by plotting the values of input data against the value of the standard normal distribution where their cumulative distributions are equal. If the points cluster around a straight line, the data distribution matches the normal distribution. The experimental semivariograms and the best-fitted theoretical models for the data of groundwater levels (head) are shown in Figure 3. The three nonlinear main different semi-variogram models (exponential, spherical, and Gaussian) for observed groundwater head levels were plotted as shown in Figure 3a–c, respectively. The semivariogram parameters (i.e., nugget, partial sill, and range) were obtained for each model. To check the validity of all the assumptions made in the development of the theoretical model and estimation of model parameters, cross-validation was carried out on the data. It is clear from Figure 3 that the best fit of the data was the Gaussian model compared to other models. Therefore, the Gaussian model was chosen as the final model to be used in Kriging for the initial groundwater level data.

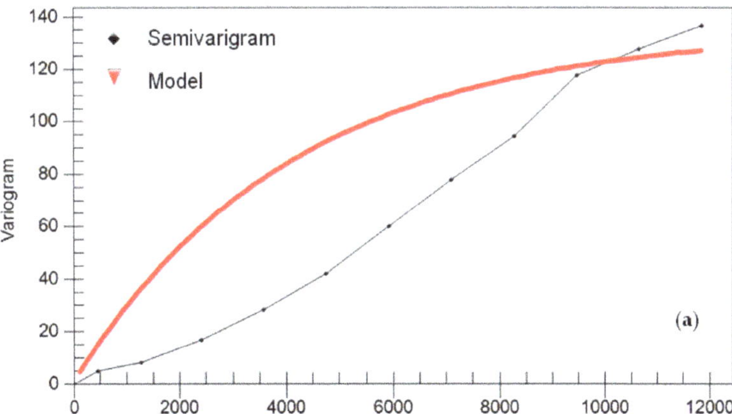

Figure 3. *Cont.*

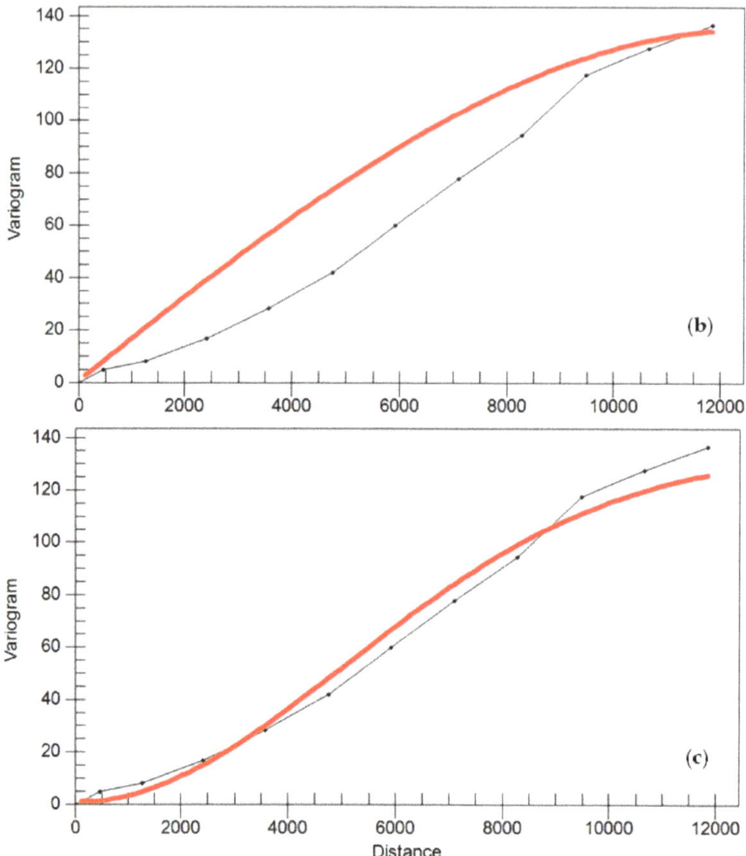

Figure 3. Experimental and fitted semivarigrams for the groundwater head level: spherical (**a**), exponential (**b**), and Gaussian (**c**).

The second stage was the calibration of the model. Through the calibration process, the fine-tuning of the parameters was carried out in such a way that the numerical model could simulate groundwater levels that match the field measured values in the best possible way. The parameter estimation process was performed automatically using the PEST (=parameter estimation) calibration software under the GMS program. During the calibration process, this tool operates automatically to decrease the difference between measured and computed values by changing different aquifer parameter values such as hydraulic conductivity and transmissibility until minimum differences are reached. The tool uses mathematical optimization algorithms.

During the calibration process, sensitivities of the computed groundwater levels with respect to the aquifer parameters were calculated. These statistical indexes and others were used when deciding which aquifer parameters to consider and to detect amended calibration. Calibration was performed for two main states of the numerical model; the first was the steady state and the second was the transient state to investigate the performance of models to simulate the aquifer for selected periods in the future. The calibrated steady state is very important for the transient state as it represents the initial condition of the transient models.

The model calibration process is an important part of any groundwater modeling process. For the calibration under a steady-state condition, the model variables of the aquifer's natural recharge and hydraulic conductivity were predicted. The steady state was calibrated using 15 observation wells and 2 adjustable variables. The calibration of the steady-state condition was obtained by reducing the variation between the observed groundwater table and the simulated groundwater table, where the observed heads automatically compared with heads computed by the model. The final calibrated model was created using the parameter estimation tools (PEST) application in GMS software. The measured values were registered, and the intervals of confidence (95%) and observation head interval (0.5 m) were selected.

For each homogeneous region in the study area, the natural recharge and hydraulic conductivity needed to be estimated separately in order to decrease the uncertainty resulting from the expected variance of the relatively large area studied. Therefore, the study region was divided into seven zones with estimated hydraulic conductivity parameters based on the results of the pumping test of twenty wells [33] and three sub-catchment areas for natural recharge depending on the aquifer characteristics (Figure 4).

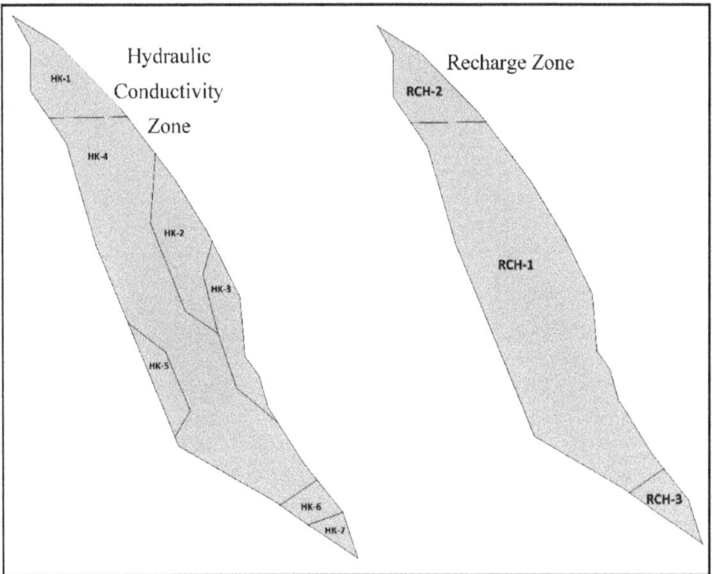

Figure 4. The divided zones of the hydraulic conductivity and natural recharge of the study area.

The model was validated using well data for the transient condition. The calibrated models were used to estimate the effect of the suggested artificial recharge by the wells on the aquifer groundwater table with a focus on the surrounding area. Different scenarios were applied based on the hypothesis of groundwater pumping rates and artificial and natural recharge over the future period between 2022 and 2030. The results of the model were validated through the historical period 2016–2017 using realistic recharge rates drawn from the observations.

2.3.1. Grid Design

The domain of the model was chosen to cover 1100 km^2. The optimum grid size for the model was found by trial and error. Two main criteria were considered. The first was the stability of the results for the last three trials at least. The second was the time required to complete the run, especially with 3D models and the use of the automatic calibration tool (PEST), as increasing the grid size to more cells than the optimal amount caused time-delays and irregularities in the solution, especially in the transient case. The grid of the model was composed of 7800 active cells. The cell width along rows and columns (x and y directions) was set at 500 m. The model region was created horizontally on a two-dimensional grid and vertically as a single unconfined layer. It is also important to note that independent results of the grid were obtained. Upper elevation values of the aquifer were determined based on a map of the topographic contour lines of the region, Figure 5, and the aquifer's low elevation, which comprised the top elevations minus the depth of formation. The average depth of the geological formation was 40 m. Due to the MODFLOW within the GMS program, we adopted the topography as a top layer and it was determined by meters above sea level (m.a.s.l.). Consequently, all the other input and output contour map layers were produced with the same units.

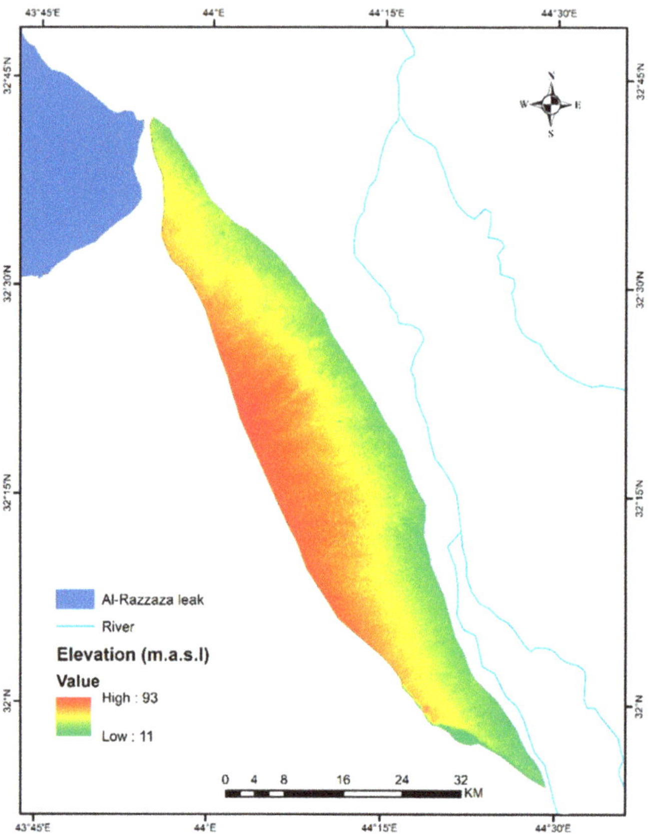

Figure 5. DEM of the study area.

2.3.2. Boundary Conditions

One of the most important requirements needed to solve the governing equations describing the flow in porous media like groundwater flow is determining the boundary conditions of the domain (study area). Boundary conditions refer to hydraulic conditions along the perimeter of the problem domain and mathematically can be classified into three types: constant head boundary, specified flow boundary, and head-dependent boundary. For the study area, boundary conditions were determined based on the flow pattern of groundwater in the Dibdibba unconfined aquifer and the observations of the groundwater level in the wells. A constant head boundary was applied at the eastern and western sides of the area. These head values were 35 m and 5 m, respectively, as obtained from measurements taken at observation wells. Moreover, the two features in the study area (i.e., Tar Al Sayyed and Tar Al Najaf) were defined as a no-flow boundary on the study area's northwest and southwest edges since it matched the flow directions (streamlines).

2.3.3. Recharge Estimation

Because field recharge values are difficult to determine, calibrated recharge modeling was used. The spatially calibrated recharge was first distributed according to the water budget analysis and then adjusted until a good match was obtained between the calculated and observed groundwater levels. The amount of groundwater recharge in the region was fundamentally computed from rain infiltration, assumed to be approximately 10% of the mean annual rainfall, as was found by other researchers [33]. The groundwater pumping rate through produced wells to meet the water demand for irrigation in the study region constituted a main component of the outflow of the system. There were about 3000 wells operating in the study area aquifer until recently when the number of operating wells was significantly decreased. According to the General Commission of Groundwater reports, there are only 500 to 600 wells still operating in the study area. In general, well depths ranged from 20 to 90 m and the pumping rates were between 25 and 30 m^3/h. The specific capacity of wells ranged from 5 to 220 m^3/h [33].

Based on the pumps' withdrawal rates and average well operating times, the value of the total abstraction rate could be calculated as follows:

$$\text{Total withdrawal} = (\text{pumping rate} \times \text{ operation time} \times \text{ days operation/year}) \times \text{number of production wells}/365 \qquad (1)$$

When assuming that the operation time ranges from 6 to 8 h/day with an average discharge of 8 L/s and 145 days of operation per year, the annual pumping rate is 11,000 m^3/day. These calculated values were applied as input for the simulation model due to the lack of direct field measurements.

3. Results and Discussion

By using the calibration and validation process for the numerical model and assurance of its consistency with real aquifer conditions, the simulation model can be used for the management of the aquifer. For this paper, the calibrated numerical model was used to investigate the impact of artificial recharge by wells in the Al-Dibdibba region on the groundwater levels of the unconfined aquifer. Therefore, constructing and validating an accurate simulation model was fundamental to reaching the objectives of the research. With the simulation model, the impact of the artificial recharge of the wells was evaluated using several scenarios.

3.1. Steady-State Model Calibration and Validation

For the calibration and validation processes, the observation coverage in GMS was used, where the observed values from the field were automatically compared with the values computed by the model. The measured values were registered, and a confidence interval (95%) and observation head interval (0.5 m) were specified. The results of the calibration along with the calibration target bars are shown in Figure 6. The middle line represents the observed value. The top and bottom ends of the whiskers indicate the observed values plus the interval and the observed value minus the interval, respectively. The filled bar shows the error; the green color indicates that the error is within the specified interval (less than 0.5 m). Yellow bars denote that the error is between 0.5 and 1 m, and for the red bar, the error is more than 1 m. In the calibration process, effort was put into reducing the error or colored bar. For steady-state conditions, the calibration and validation of the numerical model were performed with available measured groundwater levels for 15 observed wells only. These 15 wells were chosen in such a way that they covered the entire study regions as well as reducing the duration time of simulated run in the calibration process (which was done automatically by using PEST tools within GMS software).

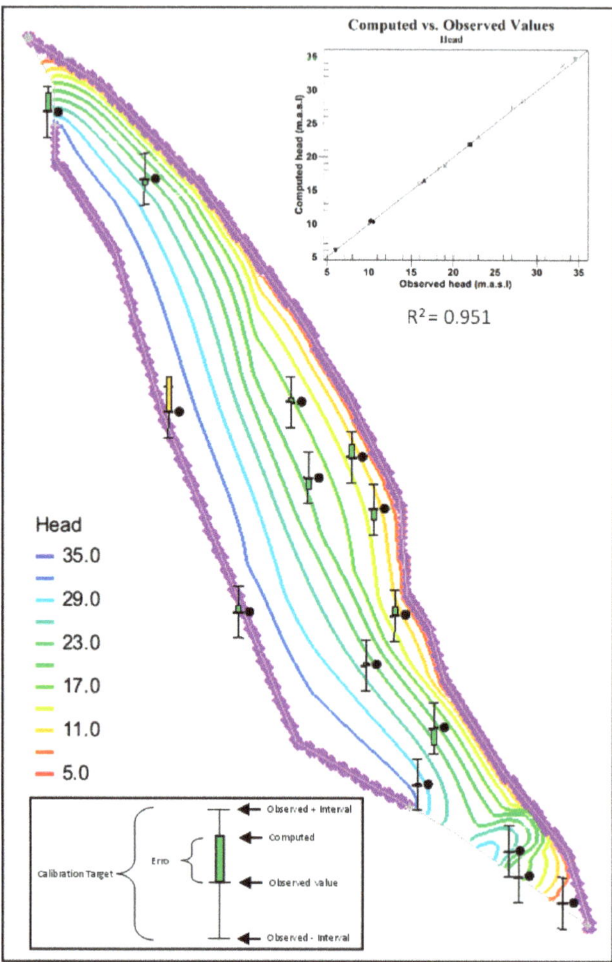

Figure 6. Contour map of the simulated groundwater level in meters above sea level (m.a.s.l.) after calibration of the Dibdibba aquifer under steady-state conditions.

Figure 6 shows the calibration results and contour map of simulated groundwater levels of the Dibdibba aquifer for steady state conditions. In Figure 6, 15 of the calibration targets were used to represent the water level in the model, 14 bars are green (error less than 0.5 m), and one bar is yellow (error more than 0.5 and less than 1 m), so that the obtained values matched with the measured values at an acceptable accuracy level. Moreover, the results of the calibration model were deemed acceptable after comparing the computed groundwater elevations with the measured values as shown in Figure 6. The model matched the observed head with a determination coefficient (R^2) of 0.951. This relatively high correlation value was obtained by using the PEST auto-calibration tool in the GMS software. In addition to the process of estimating hydraulic conductivities and recharge separately for each homogeneous zone, the study area was divided into several zones for these two important parameters during the calibration process of the conceptual model. These validated results are an indicator for good confidence for the estimated results for the simulated model. Consequently, the results of this steady-state calibrated model can be used as an initial condition in the process of calibrating the transient model.

3.2. Transient Model Validation

In the case of transient conditions, groundwater levels are a function of time. Simulated steady-state aquifer levels were used as initial aquifer levels to simulate the transient state. Measuring groundwater levels was an important calibration objective of this operation. Moreover, the transient models determined preliminary predictions of the specific yield (Sy-values) of the Dibdibba unconfined aquifer. Values of Sy, computed from pumping tests in previous studies, ranged from 0.001 to 0.05 [33]. Constructing a transient mode ideally means managing large amounts of transient data from a diversity of sources, including water levels in observation wells as well as recharge and pumping well data. The values of pumping rates per month varied according to the requirements of the cultivations in the study area. The choice of a simulation time stage is a crucial part of designing a transient model because the determination of space and time strongly affects the numerical model results [34].

The transient calibration model had a starting date of 1 January 2016 and the end date was set to 1 December 2017, while the time period was divided into 23 time steps. Data used for the calibration and validation processes included the corresponding measured groundwater levels from 4 control wells (Figure 1). The locations of these wells approximately represented the region, especially the area near the Kerbala WWTP (Obs.1 to Obs.3) because the results of the study will focus on it. The corresponding variable was computed from the steady-state calibration model. The relevant discharge rates were estimated from the realistic consumption in the study region.

In the process of calibrating the transient state, specific yield values were selected to be the variable parameter within the PEST operation; this changed automatically until a good match between the observed and calculated groundwater levels from January 2016 to December 2017 was achieved (Figure 7). Of the 4 calibration targets represented in the model, only one bar was yellow and 3 were green. A transient scatter plot of observed versus simulated groundwater levels was used to investigate the transient numerical model. The results are shown in Figure 6. The scatter plot shows a coefficient of determination $R^2 = 0.894$ at the end of simulation in 1 December 2017. As expected, the congruence in the transition state is less than the congruence in the steady state, where the determinant coefficient (R^2) was less than the value obtained in the steady state. The simulation model of groundwater flow was validated for the period from January 2016 to December 2017. It reproduced close groundwater level variation for different previous monitoring readings as displayed in Figure 8. A comparison between the measured and the computed variations in groundwater levels in the monitoring wells is presented in Figure 8. The excellent performance of the simulation was shown by the transient simulation of the concordance between the modeled and measured groundwater levels in the observation wells. All the figures show a declining trend in elevation during the monitoring periods. According to these evaluations, the model was well calibrated. It is clear from Figure 8 that, for observation well No.2 (Obs.2), the simulation model results always overestimated observed groundwater levels during the validation period but within an acceptable range (green color bar). However, for the other observed wells, the estimated values both over- and underestimate the values for different monitoring times. These behaviors may be due to the different estimation ranges for the hydraulic parameters of the aquifer in addition to the effect of the well's distance from the boundaries of the study area.

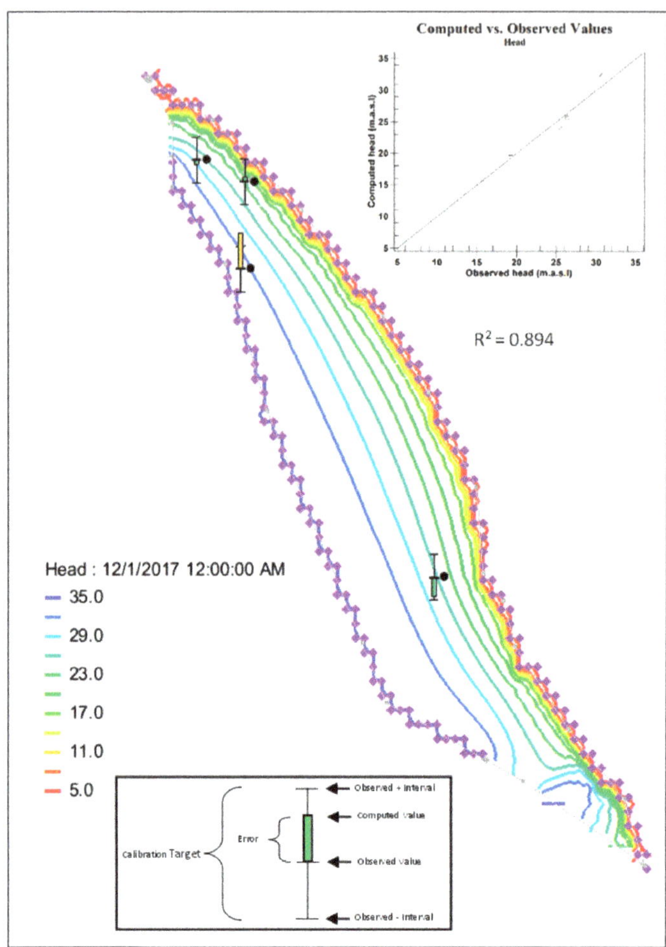

Figure 7. Contour map of the simulated aquifer groundwater level on 1 December 2017.

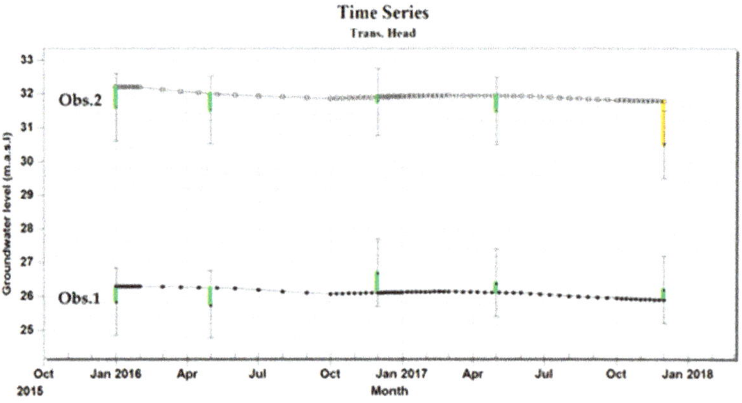

Figure 8. *Cont.*

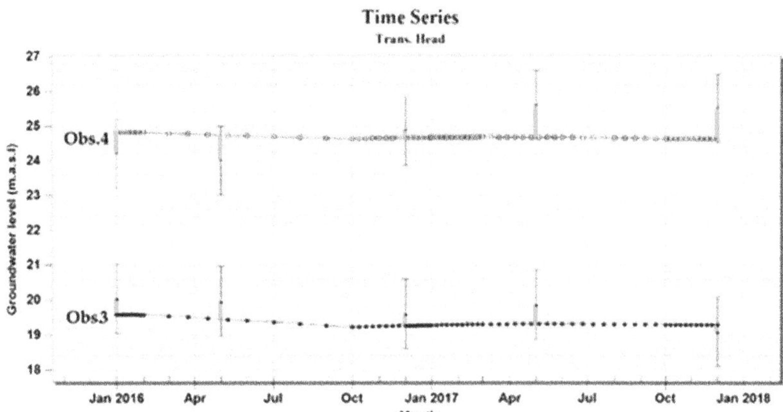

Figure 8. Validation in four observation wells during the transient period 2016–2017.

3.3. Evaluation of Artificial Recharge

The calibrated models were used to estimate the influence of the suggested artificial recharge by wells on the elevation of groundwater, focusing on the region surrounding the Kerbala WWTP. Different scenarios were applied, which considered the natural and artificial recharge and the rate of groundwater withdrawal by pumps between 2016 and 2030. This was done with two simulation models (MOD1 and MOD2) that performed calculations over the prediction time interval (2016–2030). In MOD1, the same boundary conditions were maintained during the entire simulation period, assuming the continuity of the present extraction and natural recharge rates of 2016 and the exclusion of the artificial recharge rate from the WWTP. Based on this scenario, the simulated groundwater levels would be more than 2 m lower at the Kerbala WWTP by the year 2030 (Figure 9). Therefore, this site is considered to be one of the best sites to establish artificial recharge wells due to its proximity to the treatment plant and the severe impact from withdrawal operations in this area as well as the lack of natural recharge. These results of the expected decline in groundwater levels have a sensitivity of assumptions made about the extraction values calculated by Equation (1) (11,000 m^3/day), which depend on some other assumptions due to a lack of observation data.

In MOD2, the same boundary conditions as in MOD1 were applied with the addition of a recharge flow of 5000 m^3/day injected into 20 selected recharge wells. These recharge wells were regularly distributed over the area near the treatment plant for economic and operational reasons. The comparison of the simulated groundwater elevation for 2030 based on MOD1 and MOD2 illustrated that an increase of close to 1.6 m (maximum) could be achieved with an artificial recharge pumping rate of 5000 m^3/day (Figure 10a). Considering the minimum change of 0.2 m, the affected area around the WWTP site under recharge conditions would be about 91.6 km^2, divided into two pools, one located on the left side of the Kerbala WWTP with an area of 45.4 km^2 and the other located on the right side with an area of 45.2 km^2, as shown in Figure 10a. It would spread over 18.5 km along the river and include a 36 km^2 recovery area with a groundwater level increase of more than 0.5 m from MOD1.

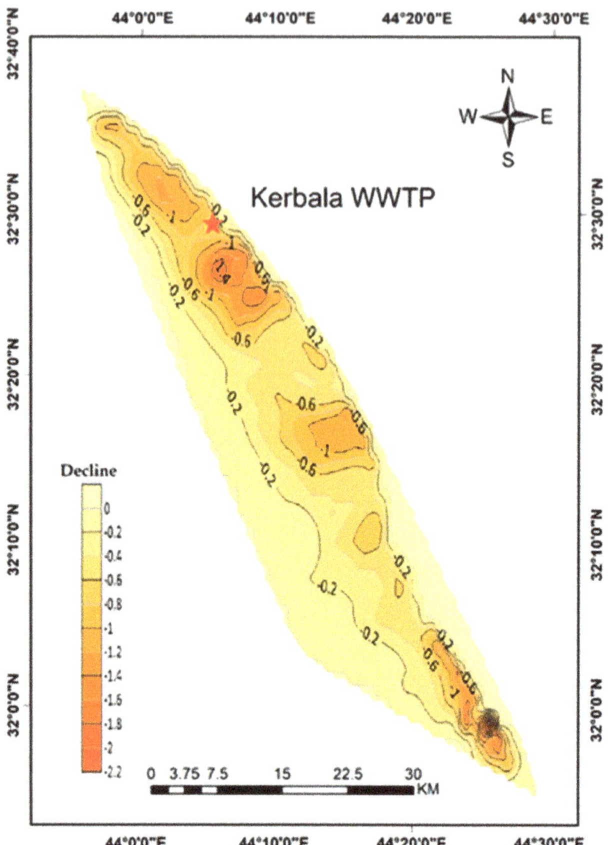

Figure 9. The expected groundwater level decline in 2030 in the study area without recharge.

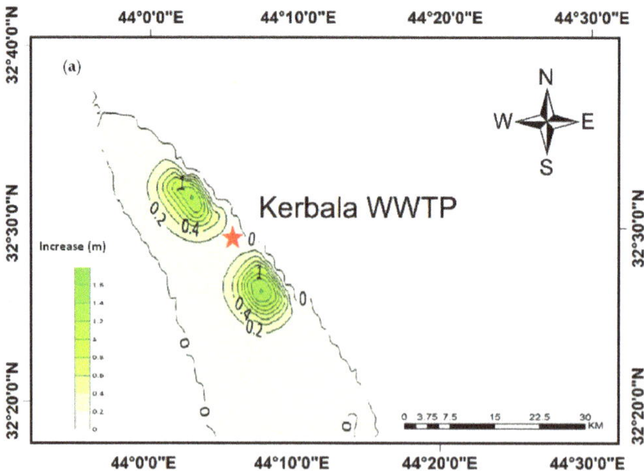

Figure 10. *Cont.*

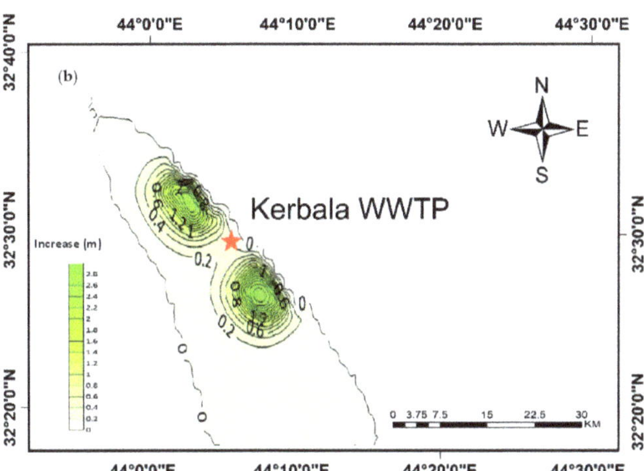

Figure 10. Effect of artificial recharge in 2030: differences in groundwater elevation increases (**a**) between MOD2 and MOD1 and (**b**) between MOD3 and MOD1.

An additional scenario, MOD3, was considered, in which 10% of the WWTP outflow (10,000 m^3/day) was used as an artificial recharge rate for the aquifer groundwater table. As a consequence of this increase, a higher impact was obtained on the groundwater level (Figure 10b). The positively influenced area (0.2 m) increased to 136.8 km^2 and the maximum increase in groundwater levels reached 2.8 m. The affected area extended 23 km along the river, including an 81 km^2 recovery area and a rise of at least 0.5 m in the groundwater level.

Results relative to the observation well (Obs.3) located near the artificial recharge site as shown in Figure 1 were used to compare the temporal development of the groundwater level. Figure 11a illustrates that the artificial recharge would increase the groundwater level, especially near observation well No.3 (Obs.3). However, this effect diminished farther away from the recharge wells. A groundwater level increase occurred during the next decade of the modeling with an annual increase of 7 cm and 20 cm for 5000 and 10,000 m^3/day recharge pumping rates, respectively, as shown in Figure 10a,b. It is clear from Figure 11 that there is a difference in the rate of increase due to the difference in the pumping rates of the recharge wells between the two scenarios. There is also a clear increase in the tendency of the expected rise in the groundwater level after 2026 for both scenarios. This may be due to the expected increase in the permeability of the soil surrounding the pumping well due to continuous pumping during the period before 2026.

At the end period of the simulation, the maximum expected groundwater level of the Obs.3 well will be close to 19.78 m.a.s.l. for the second scenario, while it could reach 21.1 m.a.s.l. for the third scenario. Consequently, the groundwater level in the area near the KWWTP can be increased to more than 1.3 m as a result of doubling the pumping rates (3rd scenario) during the next eight years. It is important to note that the models were calibrated based on the regional interpolation parameters due to the availability of data, while the effect of artificial recharge was evident in a specific area near the treatment plant where the sites of the proposed pumping wells (for economic requirements) were located. Therefore, these factors may cause predictive uncertainty in the expected results. As in many previous studies [5,6,8], the effect of hydrogeological properties was very significant in the rates of change of groundwater levels as a result of artificial recharge.

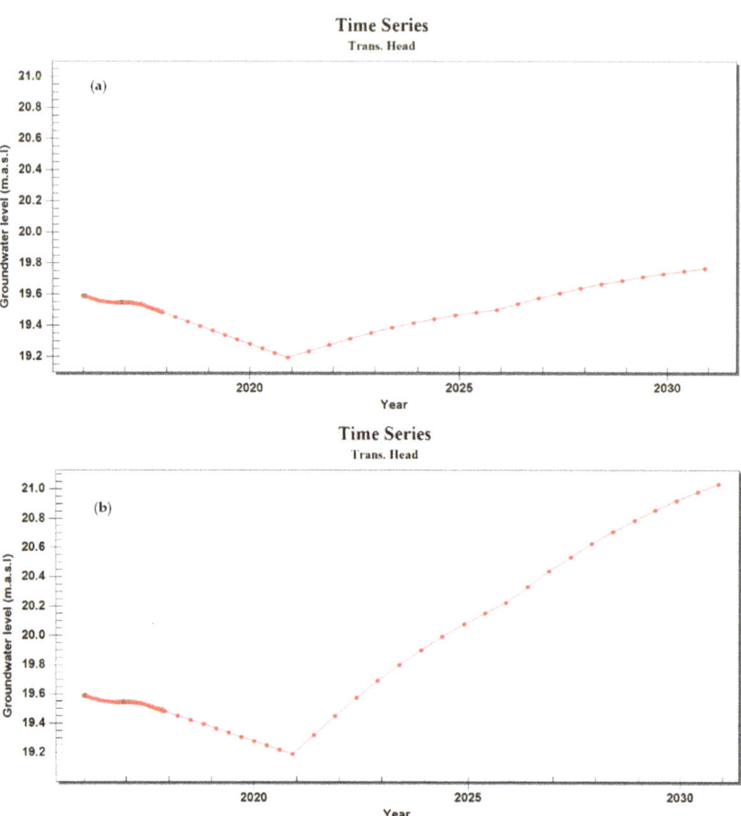

Figure 11. Groundwater level variations in observation well No.3 (Obs.3). (**a**) MOD2 with 5000 m^3/day and (**b**) MOD3 with 10,000 m^3/day, artificial recharge rate for period (January 2022 to December 2030).

The estimation of water demand in the study area depends fundamentally on two significant factors. The first one is the climate status, which involves rainfall, temperature, sunshine duration, wind speed, and humidity. The second factor is the type of cultivation, which determines the demand for irrigation water and reflects the plant's modulus [35]. According to the Ministry of Agriculture and Ministry of Water Resources (Iraq), the maximum irrigation water demand for a proposed crop plan has been determined to be equal to 3 m^3/donum/day, (one donum = 2500 m^2). Considering the simulation results for the next ten years, a new agricultural area could be added to the region, with an area of more than 5800 dunams (14.6 km^2) if 5% of the treated water production from WWTP was used for the artificial recharge process. The additional areas could be increased to 25,000 dunums if 10% of the treated water production was used. This expected increase in cultivated land will be very useful in facing the phenomenon of desertification, global warming, and improving the environment in the study area.

3.4. Sensitivity Analysis

Sensitivity analysis is a measure of uncertainty in the calibrated model caused by uncertainty in the aquifer parameters and boundary conditions. The main objective of the sensitivity analysis is to understand the influence of various model parameters and hydrogeological stresses on the aquifer system and to identify the most sensitive parameter(s) that will need spatial attention in the future studies. Sensitivity analysis was performed at the end of PEST iterations of each of the parameters used. For steady state, the hydraulic conductivity and natural recharge values for each zone within the study area (Figure 4) were examined as illustrated in Figure 12. It is clear to note that the steady-state model is sensitive to both recharge and hydraulic conductivity to a different degree. The model is highly sensitive to changes in natural recharge compared to its sensitivity to changes in hydraulic conductivity. The natural recharge of the largest zone (RECH-1) has the most significant impact in terms of predicting groundwater levels compared to other input parameters. The hydraulic conductivity parameters for regions 1, 5, and 7 had the least influence on the simulation model results. This behavior may be due to the large area represented by this parameter (RECH-1) relative to the other areas of parameters. The relatively small areas (HK-1, HK-2, HK-6, HK-7, and RECH-3) were the least sensitive compared to the other areas. It can therefore be noted that the spatial variability of parameters could influence predictions for steady-state calibration. However, the transient model shows sensitivity due to the increase and decrease in the specific yield but with a slow response. This analysis is useful in identifying the parameters that have the greatest impact on the model as well as the parameters that have the least impact on the model. Thus, non-sensitive parameters can be kept constant or removed in future studies, while it is necessary to give attention to the parameters that have a high sensitivity in the simulation model.

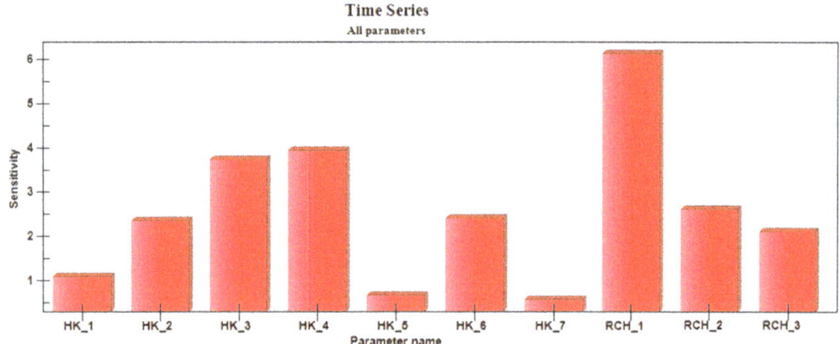

Figure 12. Sensitivity analysis of modeling parameters.

4. Conclusions

One of the main issues concerning the evolution of groundwater resource management and its sustainability is the efficient storage and control of its recharge and consumption rates. Artificial recharge is considered to be one of the most important tools that can help the sustainability of groundwater aquifers. In this study, a percentage of treated water (tertiary treatment) from the wastewater treatment plant (WWTP) in Kerbala was utilized to investigate the impact of artificial recharge on the groundwater levels in the Dibdibba unconfined aquifer. A three-dimensional numerical model was applied to simulate the flow system of the unconfined aquifer using MODFLOW with GMS 10.4. The developed models were calibrated automatically using the PEST tool. The modeled groundwater levels matched the observed groundwater levels for steady-state and transient simulations for the period 2016–2017.

The calibrated models were used to simulate three different scenarios. One scenario included applying natural conditions without artificial recharge and two scenarios included artificial recharge through wells (5000 and 10,000 m^3/day) to show the response of the aquifer for the future period 2022–2030. The results of the simulations illustrated that during the study period, the artificial recharge of wells with a pumping rate of 5000 and 10,000 m^3/day distributed over 20 injection wells would induce an annual increase of 7 cm and 20 cm in groundwater levels, respectively. This increase would lead to recovery of groundwater levels of up to 36 and 81 km^2 around the recharge site for the 2nd and 3rd scenarios, respectively, and extend about 20 km along the river. Moreover, it was shown that the application of an artificial recharge project would reduce groundwater decline during long-term periods. In the first simulation using the artificial recharge scenario MOD2, the total increase in water storage volume was 6.4×10^6 m^3/year with a minimum increase in groundwater level of 0.2 m at the end of the simulation period in 2030. This means that the pumping rates could be raised to reclaim more than 14.6 km^2 of the new agricultural area. In the second artificial recharge scenario MOD3, the new area of reclaimed agricultural land could be increased by more than 62 km^2. These reclaimed areas could represent a significant addition to important agricultural production areas since the reclaimed areas consist of sandy soils that are suitable for cultivating multiple types of plants.

Author Contributions: Conceptualization, W.H.H., B.K.N. and K.M.; methodology, W.H.H., B.K.N., J.W. and C.R.; software, W.H.H. and J.W.; validation, B.K.N., K.M., J.W. and C.R.; formal analysis, W.H.H.; investigation, K.M., J.W. and C.R.; resources, W.H.H. and B.K.N.; data curation, W.H.H. and B.K.N.; writing—original draft preparation, W.H.H.; writing—review and editing, B.K.N., K.M., J.W. and C.R.; visualization, W.H.H.; supervision, W.H.H., B.K.N., K.M., J.W. and C.R.; project administration, B.K.N. and K.M.; funding acquisition, K.M. All authors have read and agreed to the published version of the manuscript.

Funding: This Research is fully funded by NUFFIC Orange Knowledge Program OKP-IRA-104278.

Data Availability Statement: All data are contained within the article.

Acknowledgments: This study has been funded by Nuffic, the Orange Knowledge Programme, through the OKP-IRA-104278 project titled "Efficient water management in Iraq switching to climate smart agriculture: capacity building and knowledge development" coordinated by Wageningen University & Research, The Netherlands.

Conflicts of Interest: The authors declare no conflict of interest.

References

1. Shahid, S.; Alamgir, M.; Wang, X.J.; Eslamian, S. Climate change impacts on and adaptation to groundwater. In *Handbook of Drought and Water Scarcity*; CRC Press: Boca Raton, FL, USA, 2017; pp. 107–124.
2. Hassan, W.H.; Nile, B.K. Climate change and predicting future temperature in Iraq using CanESM2 and HadCM3 modeling. *Modeling Earth Syst. Environ.* **2021**, *7*, 737–748. [CrossRef]
3. Green, T.R. Linking climate change and groundwater. In *Integrated Groundwater Management*; Springer: Cham, Switzerland, 2016; pp. 97–141.
4. Hassan, W.H. Climate change impact on groundwater recharge of Umm er Radhuma unconfined aquifer Western Desert, Iraq. *Int. J. Hydrol. Sci. Technol.* **2020**, *10*, 392–412. [CrossRef]
5. Horriche, F.J.; Benabdallah, S. Assessing Aquifer Water Level and Salinity for a Managed Artificial Recharge Site Using Reclaimed Water. *Water* **2020**, *12*, 341. [CrossRef]
6. Abraham, M.; Mathew, R.A.; Jayapriya, J. Numerical Modeling as an Effective tool for Artificial Groundwater Recharge Assessment. *J. Phys. Conf. Ser.* **2021**, *1770*, 012097. [CrossRef]
7. Chitsazan, M.; Movahedian, A. Evaluation of artificial recharge on groundwater using MODFLOW model (case study: Gotvand Plain-Iran). *J. Geosci. Environ. Prot.* **2015**, *3*, 122. [CrossRef]
8. Gale, I.; Neumann, I.; Calow, R.; Moench, D.M. *The Effectiveness of Artificial Recharge of Groundwater: A Review*; Institute of Social and Environmental Transition: Katmandu, Nepal, 2002.
9. Bouwer, H. Artificial recharge of groundwater: Hydrogeology and engineering. *Hydrogeol. J.* **2002**, *10*, 121–142. [CrossRef]
10. Sheng, Z. An aquifer storage and recovery system with reclaimed wastewater to preserve native groundwater resources in El Paso, Texas. *J. Environ. Manag.* **2005**, *75*, 367–377. [CrossRef]
11. Cazurra, T. Water reuse of south Barcelona's wastewater reclamation plant. *Desalination* **2008**, *218*, 43–51. [CrossRef]
12. Shammas, M.I. The effectiveness of artificial recharge in combating seawater intrusion in coastal aquifer Salalah, Oman. *Environ. Earth Sci.* **2008**, *55*, 191–204. [CrossRef]
13. Tamer, A.; Al-Assa, D.F.; Ahmad, A. Artificial groundwater recharge to a semi-arid basin: Case study of Mujib aquifer, Jordan. *Environ. Earth Sci.* **2010**, *60*, 845–859.
14. Salem, S.B.; Chkir, N.; Zouari, K.; Cognard-Plancq, A.L.; Valles, V.; Marc, V. Natural and artificial recharge investigation in the Zeroud Basin, Central Tunisia: Impact of Sidi Saad Dam storage. *Environ. Earth Sci.* **2012**, *66*, 1099–1110. [CrossRef]
15. Ketata, M.; Gueddari, M.; Bouhlila, R. Hydrodynamic and salinity evolution of groundwater during artificial recharge within semi-arid coastal aquifers: A case study of El Khairat aquifer system in Enfidha (Tunisian Sahel). *J. Afr. Earth Sci.* **2014**, *97*, 224–229. [CrossRef]
16. Bouri, S.; Ben Dhia, H. A thirty-year artificial recharge experiment in a coastal aquifer in an arid zone:The Teboulba aquifer system (Tunisian Sahel). *Comptes Rendus Geosci.* **2010**, *342*, 60–74. [CrossRef]

17. Hassan, W.H. Climate change projections of maximum temperatures for southwest Iraq using statistical downscaling. *Clim. Res.* **2021**, *83*, 187–200. [CrossRef]

18. Hassan, W.H.; Khalaf, R.M. Optimum Groundwater use Management Models by Genetic Algorithms in Karbala Desert, Iraq. *IOP Conf. Ser. Mater. Sci. Eng.* **2020**, *928*, 022141. [CrossRef]

19. Al-Aboodi, A.H.; Ibrahim, H.T.; Ibrahim, N. Estimation of groundwater recharge in Safwan-Zubair area, South of Iraq, using water balance and inverse modeling methods. *Int. J. Civ. Eng. Technol.* **2019**, *10*, 202–210.

20. Al-Mussawi, W.H. Kriging of groundwater level-a case study of Dibdiba Aquifer in area of Karballa-Najaf. *J. Kerbala Univ.* **2008**, *6*, 170–182.

21. Nile, B.K. Effectiveness of Hydraulic and Hydrologic Parameters in Assessing Storm System Flooding. *Adv. Civ. Eng.* **2018**, *2018*, 4639172. [CrossRef]

22. Chenini, I.; Ben Mamou, A. Groundwater Recharge Study in Arid Region. An Approach Using GIS Techniques and Numerical Modeling. *Comput. Geosci.* **2010**, *36*, 801–817. [CrossRef]

23. Katibeh, H.; Hafezi, S. Application of MODFLOW Model and Management by Utilizing Groundwater and Evaluation of the Performance of Artificial Recharge of Ab Barik Plain in Bam. *J. Water Wastewater* **2004**, *50*, 45–58.

24. Martinez-Santos, P.; Martinez-Alfaro, P.; Murillo, J.M. A method to estimate the artificial recharge capacity of the Crestatx aquifer (Majorca, Spain). *Environ. Earth Sci.* **2005**, *47*, 1155–1161. [CrossRef]

25. Izbicki, J.A.; Flint, A.L.; Stamos, C.L. Artificial recharge through a thick, heterogeneous unsaturated zone. *Ground Water* **2008**, *46*, 475–488. [CrossRef]

26. Vandenbohede, A.; Van Houte, E.; Lebbe, L. Groundwater flow in the vicinity of two artificial recharge ponds in the Belgian coastal dunes. *Hydrogeol. J.* **2008**, *16*, 1669–1681. [CrossRef]

27. Zhang, H.; Xu, Y.; Kanyerere, T. Site Assessment for MAR through GIS and Modeling in West Coast, South Africa. *Water* **2019**, *11*, 1646. [CrossRef]

28. General Commission of Groundwater. *Advanced Survey of Hydrogeologic Resources in Iraq, Phase II (ASHRI-2)*; Ministry of Water Resources of Iraq: Bagdad, Iraq, 2018; Unpublish report.

29. Al-Sudani, H.I.Z. Groundwater system of Dibdibba sandstone aquifer in south of Iraq. *Appl. Water Sci.* **2019**, *9*, 1–11. [CrossRef]

30. Abdulameer, A.; Thabit, J.M.; Kanoua, W.; Wiche, O.; Merkel, B. Possible Sources of Salinity in the Upper Dibdibba Aquifer, Basrah, Iraq. *Water* **2021**, *13*, 578. [CrossRef]

31. Al-Jiburi, H.J.; Al-Basrawi, N.H.; Ibrahim, S.A.R. *Hydrogeological and Hydrochemical Study of Karbala Quadrangle (NI-38-14), Scale 1: 250000*; GEOSURV: Sydney, Australia, 2002.

32. Hassan, W.H. Application of a genetic algorithm for the optimization of a location and inclination angle of a cut-off wall for anisotropic foundations under hydraulic structures. *Geotech. Geol. Eng.* **2019**, *37*, 883–895. [CrossRef]

33. Al-Ghanimy, M.A. Assessment of Hydrogeological Condition in Karbala—Najaf Plateau, Iraq. Ph.D. Thesis, (unpublished). University of Baghdad, Baghdad, Iraq, 2018.

34. Anderson, M.P.; Woessner, W.W.; Hunt, R.J. *Applied Groundwater Modeling: Simulation of Flow and Advective Transport*; Academic Press: London, UK, 2015.

35. Agriculture Organization of the United Nations. Fisheries Department. In *The State of World Fisheries and Aquaculture, 1998*; Food Agriculture Organization: Rome, Italy, 1999.

Article

Drought Severity and Frequency Analysis Aided by Spectral and Meteorological Indices in the Kurdistan Region of Iraq

Heman Abdulkhaleq A. Gaznayee [1,*], Ayad M. Fadhil Al-Quraishi [2,*], Karrar Mahdi [3], Joseph P. Messina [4], Sara H. Zaki [1], Hawar Abdulrzaq S. Razvanchy [5], Kawa Hakzi [5], Lorenz Huebner [6], Snoor H. Ababakr [5], Michel Riksen [3] and Coen Ritsema [3]

[1] Department of Forestry, College of Agriculture Engineering Science, Salahaddin University, Erbil 44003, Kurdistan Region, Iraq
[2] Petroleum and Mining Engineering Department, Faculty of Engineering, Tishk International University, Erbil 44001, Kurdistan Region, Iraq
[3] Soil Physics and Land Management Group, Wageningen University and Research, 6700 AA Wageningen, The Netherlands
[4] Department of Geography, The University of Alabama, Tuscaloosa, AL 35405, USA
[5] Department of Soil and Water, College of Agricultural Engineering Science, Salahaddin University-Erbil, Erbil 44003, Kurdistan Region, Iraq
[6] 24943 Flensburg, Germany
* Correspondence: heman.ahmed@su.edu.krd (H.A.A.G.); ayad.alquraishi@tiu.edu.iq (A.M.F.A.-Q.)

Citation: Gaznayee, H.A.A.; Al-Quraishi, A.M.F.; Mahdi, K.; Messina, J.P.; Zaki, S.H.; Razvanchy, H.A.S.; Hakzi, K.; Huebner, L.; Ababakr, S.H.; Riksen, M.; et al. Drought Severity and Frequency Analysis Aided by Spectral and Meteorological Indices in the Kurdistan Region of Iraq. *Water* **2022**, *14*, 3024. https://doi.org/10.3390/w14193024

Academic Editor: Luis Gimeno

Received: 21 August 2022
Accepted: 20 September 2022
Published: 26 September 2022

Publisher's Note: MDPI stays neutral with regard to jurisdictional claims in published maps and institutional affiliations.

Abstract: In the past two decades, severe drought has been a recurrent problem in Iraq due in part to climate change. Additionally, the catastrophic drop in the discharge of the Tigris and Euphrates rivers and their tributaries has aggravated the drought situation in Iraq, which was formerly one of the most water-rich nations in the Middle East. The Kurdistan Region of Iraq (KRI) also has catastrophic drought conditions. This study analyzed a Landsat time-series dataset from 1998 to 2021 to determine the drought severity status in the KRI. The Modified Soil-Adjusted Vegetation Index (MSAVI2) and Normalized Difference Water Index (NDWI) were used as spectral-based drought indices to evaluate the severity of the drought and study the changes in vegetative cover, water bodies, and precipitation. The Standardized Precipitation Index (SPI) and the Spatial Coefficient of Variation (CV) were used as meteorologically based drought indices. According to this study, the study area had precipitation deficits and severe droughts in 2000, 2008, 2012, and 2021. The MSAVI2 results indicated that the vegetative cover decreased by 36.4%, 39.8%, and 46.3% in 2000, 2008, and 2012, respectively. The SPI's results indicated that the KRI experienced droughts in 1999, 2000, 2008, 2009, 2012, and 2021, while the southeastern part of the KRI was most affected by drought in 2008. In 2012, the KRI's western and southern parts were also considerably affected by drought. Furthermore, Lake Dukan (LD), which lost 63.9% of its surface area in 1999, experienced the most remarkable shrinkage among water bodies. Analysis of the geographic distribution of the CV of annual precipitation indicated that the northeastern parts, which get much more precipitation, had less spatial rainfall variability and more uniform distribution throughout the year than other areas. Moreover, the southwest parts exhibited a higher fluctuation in annual spatial variation. There was a statistically significant positive correlation between MSAVI2, SPI, NDWI, and agricultural yield-based vegetation cover. The results also revealed that low precipitation rates are always associated with declining crop yields and LD shrinkage. These findings may be concluded to provide policymakers in the KRI with a scientific foundation for agricultural preservation and drought mitigation.

Keywords: drought; Iraqi Kurdistan Region; normalized difference water index; standardized precipitation index

1. Introduction

Drought is a complicated natural disaster that is difficult to diagnose (including its onset, duration, intensity, and scope), forecast, and manage in a broader context; it has a very negative effect on the social, environmental, and economic status of the affected region [1]. In general, drought results in water scarcity and is caused by low precipitation averages, high evapotranspiration rates, a lack of natural water resources, over-exploitation of water resources, or a combination of these factors [1,2]. Several additional climatic elements have an essential role in the incidence of drought [3], including high temperature, strong winds, relatively low air humidity, timing and rain patterns (particularly during agricultural growth seasons), severity, and length [4,5]. Drought and climate variability, as well as their associated impacts on water resources, have gained increased attention in recent decades as nations seek to enhance mitigation and adaptation mechanisms [2]. Besides precipitation, the most crucial component of the hydrologic budget is water stress, which can result from excessive evapotranspiration rates [6,7], overexploitation of water resources, or a combination of these variables [8]. Drought poses significant hazards to individuals and the environment; hence, it is crucial to understand the spatiotemporal pattern of drought [9]. Various parts of the world are predicted to experience increasingly frequent and severe droughts as a result of climate change [10]. When there is an extended lack of precipitation, meteorologists talk of a meteorological drought [11]. We refer to an agricultural drought when a lack of precipitation results in depleted soil moisture and inadequate plant cover [12].

The periods of drought substantially harmed the agriculture sector and vulnerable populations in the Kurdistan Region [3,13,14]. The Kurdistan Region of Iraq (KRI) has sufficient water resources; however, these supplies are restricted and unpredictable in time and area. According to the Ministry of Agriculture and Water Resources in the KRI, nearly 40% of the KRI's springs dried up during prior droughts. In addition, the water resources in Turkey and Iran [13] mostly depend on the amount of precipitation and seasonal snowfall, as well as the policy of running dams and reservoirs in rivers with shared watersheds. Without international water-sharing agreements between these nations, Iraq's water supplies change from year to year. Water shortage and water quality will be anticipated to deteriorate, especially once Turkey completes its dam projects and Iran builds its planned irrigation projects. In addition, the area anticipates that population expansion, rising water consumption, and climate change will significantly impact water supplies. According to the 2011 Regional Development Strategy for KRI, the Tigris faced a 40% water shortfall in 2016 [15,16].

However, further research is required to comprehend drought events' historical frequency, length, and spatial extent and identify the most susceptible water-using sectors. The studies aid academics, decision-makers, and drought planners in mitigating the negative effects of crisis-based management measures [17–19]. The estimations of surface and groundwater are the primary sources of irrigation water required for agricultural sustainability [20]. Given the limited study on assessing LD in terms of climate change, evaluating how the climate has changed and fluctuated historically in connection to this and other lakes is essential. Moreover, using satellite pictures and remote sensing [21], we investigate the fluctuations in LD's water area extent.

Utilizing remote sensing (RS) techniques for drought monitoring is an efficient and effective method, especially for developing drought indices as well as related spatial data analysis tools, while models and databases also significantly contribute nowadays in predicting, preventing, researching, addressing, rehabilitating, and managing these phenomena of drought [1,22]. This is partly because remote sensing techniques enable more data collection over a larger geographical area and with fewer resources than ground-based observations [22]. Whether the purpose is agricultural, meteorological, or hydrological, satellite data can be exploited for drought monitoring. This data enables one to comprehend the manifestations of drought in a greater region more directly and in less time than previous techniques [23]. Numerous studies utilize meteorological drought indices for drought evaluation, monitoring, and decision-making. The Standardized Precipitation Index (SPI) [24–26] is a frequently employed drought characterization index. This precipitation-based indicator is practical and straightforward. In addition, SPI might be measured at various intervals during meteorological drought monitoring [27,28]. The Modified Soil-Adjusted Vegetation Index (MSAVI2) is also considered an excellent predictor of dry and semi-arid vegetation cover [29,30], and was designed for low-cover areas to map vegetation in arid mountainous environments [31]. In mountainous areas, primarily topographic gradients govern species distributions; thus, they must be incorporated into the mapping process [32].

The primary objectives of this study are to provide an insight into the historical frequency, duration, and spatial extent of drought episodes and agricultural drought by: (1) analyzing temporal trends in annual total precipitation, vegetation cover, and water body area over the period 1998–2021; (2) calculating the frequency, degree, and variation of drought and drought intensity over the past two decades; and (3) identifying spatial variations in drought and drought rates based on climate variables. Agriculture and water resources in the KRI require such an evaluation and information on vegetation cover to launch vegetation conservation and restoration activities. This study may help decision-makers design better strategies to enhance the KRI's land and water management sector to achieve the second sustainability development goal (SDG 2) adopted by the United Nations (UN) and Nuffic program goals in Iraq for agricultural strategy planners and regional authorities.

2. Materials and Methods

2.1. Study Area

This study was conducted in the KRI, which is situated between latitudes $33°57'58.5''$–$37°20'33.55''$ N and longitudes $42°20'25.36''$–$46°19'16.475''$ E (Figure 1A), with its elevation ranges between 88 and 3600 m (Figure 1C). It encompassed the governorates of Duhok, Erbil, and Sulaimaniyah. The KRI has a Mediterranean climate that is cold and wet in the winter and hot and dry in the summer [33,34]. Generally, the climate is determined by high precipitation rates in the north and a dryer climate in the plains [35,36]. From October to May, precipitation ranges from 350 mm in the southern regions to more than 1200 mm in the northern and northeastern regions (Figure 1D). The rainfall distribution is unimodal and concentrated from December to April [37]. The average daily temperature ranges from 5 °C in the winter to 30 °C in the summer, but in the south, it can reach 50 °C [37]. Physiographically, the KRI can be divided into the Zagros Mountains and the foothills. The precipitation pattern is influenced by the Mediterranean climate. On the other hand, the KRI is split into three categories based on average annual precipitation: assured rainfall area (above 500 mm), semi-assured rainfall area (350–500 mm), and unassured rainfall area (below 350 mm) [33,34]. Furthermore, the total area of rainfed arable land is 10,682 km^2, which accounts for 87.6% of all agricultural land (Figure A1). Approximately 7202 km^2 of the KRI's agricultural area is devoted to the production of field crops, constituting a significant share of the KRI's agricultural acreage. Two field crops comprise most of the total land area dedicated to field crops [38].

2.2. Datasets

2.2.1. Satellite Images Data

For this study, 144 Landsat images have been downloaded from the Landsat databases on the U.S. Geological Survey website (glovis.usgs.gov (accessed on 28 May 2022)). MSAVI2 was calculated using the Google Earth Engine (GEE). Table A4 displays the JavaScript code used to construct MSAVI2. The images were obtained between 1998 and 2021, during April and May, when yearly vegetation growth was at its highest in the study area. The datasets were gathered from three different Landsat satellites: L5 Thematic Mapper (TM), L7 Enhanced Thematic Mapper Plus (ETM+), and Landsat 8 OLI, which represents the data of (Path/row: 170/34, 170/35, 169/35, 169/34, 168/35, 168/36). Landsat images offer a 30 m spatial resolution (Table A2).

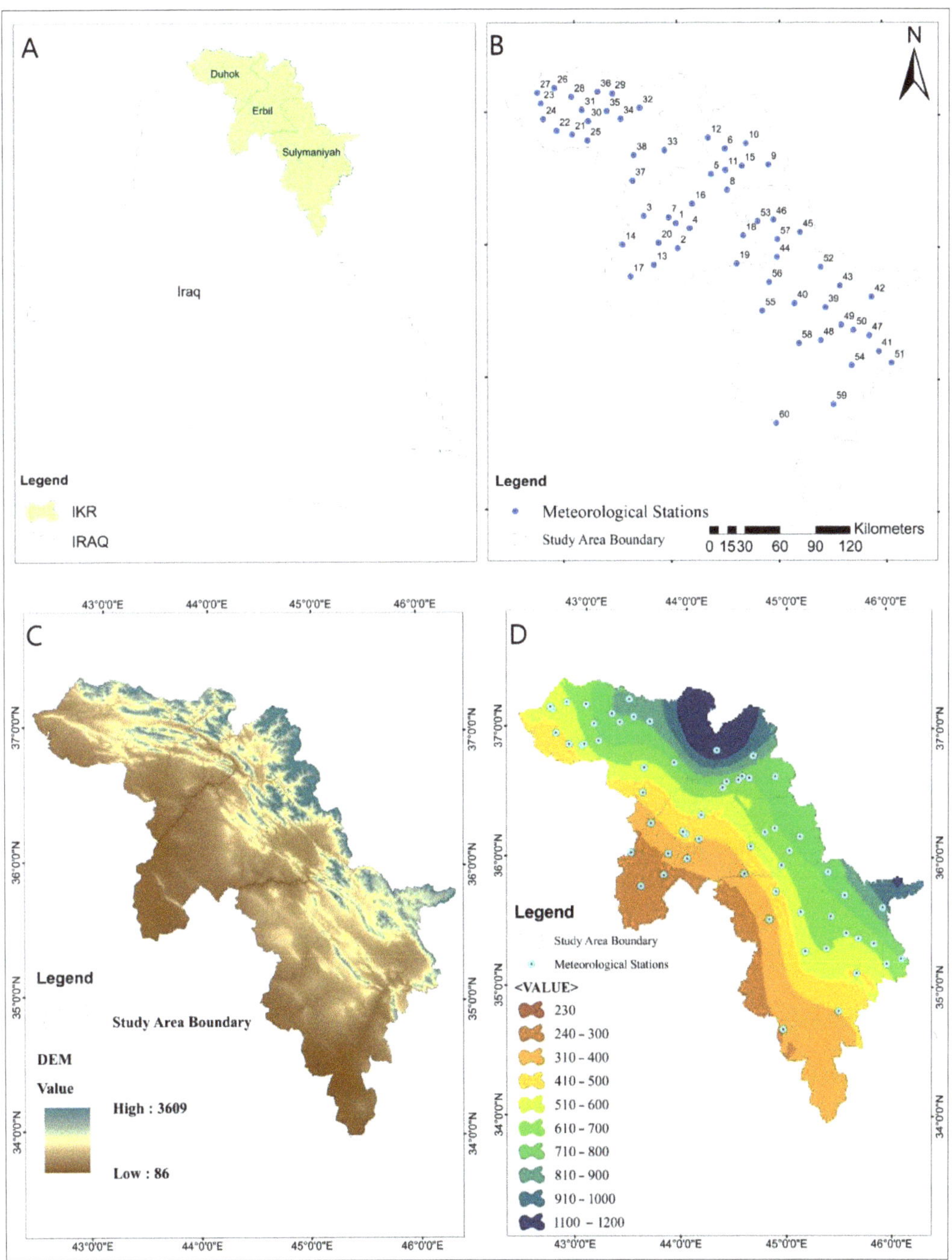

Figure 1. (**A**) Site map of the study area in the KRI, (**B**) the weather stations' locations and their codes, (**C**) the elevation map of the study area, and (**D**) the average of 24 years' precipitation (mm) map.

2.2.2. Meteorological Data

Data on annual precipitation (AP), and geographical coordinates (longitude, lati-tude, and elevation) for 60 stations were obtained from the Ministry of Agriculture and Water Resources of KRI for the period from 1998 to 2021 (Table A1). Additionally, Figure 1B shows the spatial distribution of these stations. Moreover, Figure 2 shows the overall methodological flowchart utilized in this work, illustrating the whole drought trend analysis procedure. These data were used to estimate the SPI and CV indices.

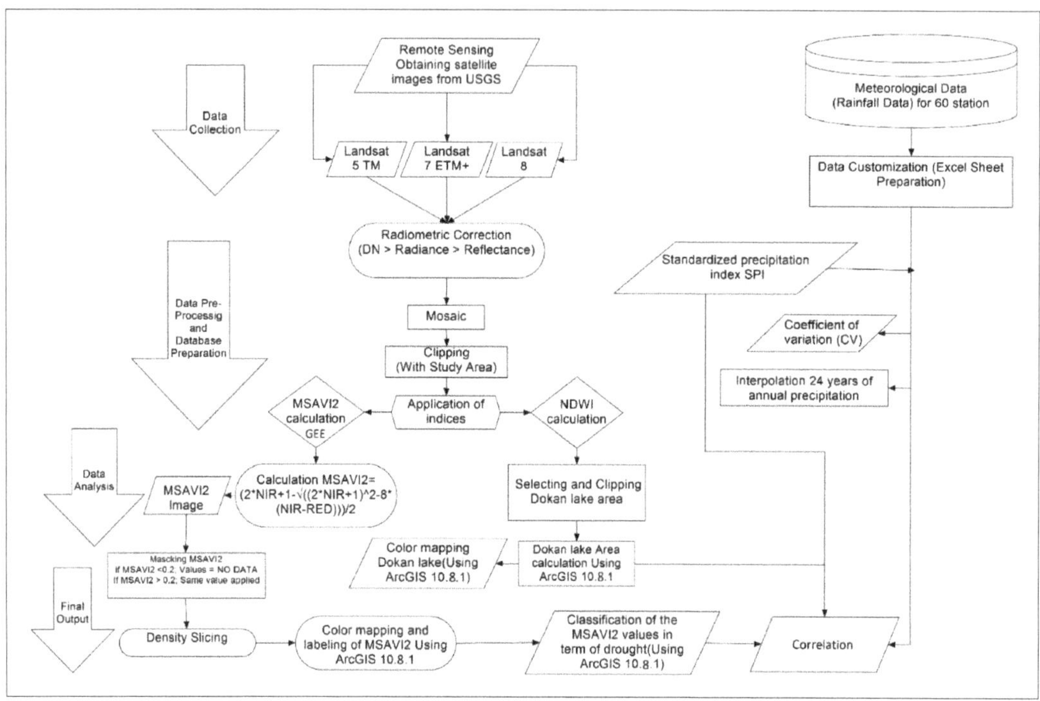

Figure 2. Flowchart of the methodology.

2.3. Spectral Drought Indices

The spectral datasets were used to calculate the MSAVI2 and NDWI [39] to identify vegetation and drought trends in time and space from a long-term sequence between 1998 and 2021 [40]. Furthermore, using ArcGIS software, 30 m high resolution satellite imagery was used to calculate LD area in km^2.

2.3.1. The Modified Soil-Adjusted Vegetation Index (MSAVI2)

MSAVI2 is an upgrade to MSAVI; although it is comparable to the SAVI index, it is more accurate for high-exposure soil locations and simply calculates a correction factor for soil brightness [41,42]. MSAVI2 values vary from −1 to +1, with values between −1 and 0 signifying non-plant features such as bare surface, built-up area, and water body, and values greater than 0 representing vegetation cover. The primary objective of this step is to mask non-vegetated areas, such as meadows, residential sites, and roadways, so that only vegetated regions remain. Using the following formula [41,42], MSAVI2 is calculated per pixel:

$$\text{MSAVI2} = \frac{2*NIR + 1 - \sqrt{(2*NIR+1)^2 - 8*(NIR-RED)}}{2} \tag{1}$$

2.3.2. The Normalized Difference Water Index (NDWI)

LD is located in Sulaimaniyah (SU), between latitudes 35°:30″ and 36°:40″ N and longitudes 44°:30″ and 46°:20″ E. It is considered the largest lake in the KRI and is a reservoir created on the Little Zab River by the Dukan Dam, which was built to provide water storage, irrigation, and hydroelectricity [15,16]. NDWI was employed to map the surface area of LD [15,16,43].

According to McFeeters [44], water bodies can be mapped using a threshold value to separate surfaces with detectable water from those without (NDWI values less than 0.3 vs. NDWI values higher than or equal to 0.3). The NIR and green bands were used to calculate the NDWI according to the equation below.

$$\text{NDWI} = \frac{Green - NIR}{Green + NIR} \tag{2}$$

where *Green* refers to the green wavelengths and *NIR* refers to the near-infrared wavelengths.

2.4. Meteorological Drought Indices
2.4.1. Standardized Precipitation Index (SPI)

McKee [24] developed the SPI, which has grown in favor over the past two decades due to it is substantial theoretical development, robustness, and applicability in drought analyses. This study relies on spectral and meteorological indices; therefore, selecting an appropriate index for comparing values across varied climatic regions is crucial. Consequently, the SPI index was used for various analyses [45,46], including frequency and temporal-spatial studies [47]. The SPI is the number of standard deviations from the long-term mean of a normally distributed random variable, which is the observed value in this case [48,49]. The drought severity varied from region to region during the stated drought years. Moreover, the SPI index provides trend analysis for the specified regions. Using DrinC software and the hydrological year (October–September), the default calculation period begins in October with an annual first calculation step. The anomalous strength was categorized after normalized SPI readings, as shown in (Table 1).

Table 1. SPI drought severity classes for wet and dry periods [26].

SPI	Class
2.0 or more	Extremely wet
1.5 to 1.99	Very wet
1.0 to 1.49	Moderately wet
0.99 to −0.99	Near normal
−1.0 to −1.49	Moderate drought
−1.5 to −1.99	Severe drought
−2.0 or less	Extreme drought

The SPI is computed by dividing the difference between the normalized seasonal precipitation and its long-term seasonal mean by the standard deviation. It can be calculated using the formula:

$$\text{SPI} = \frac{X_{ij} - X_{im}}{\sigma} \tag{3}$$

where X_{ij} is the seasonal precipitation at the rain gauge station and the observation, X_{im} is the long-term seasonal mean, and σ is its standard deviation.

2.4.2. Spatial Distribution of Rainfall across the Study Area

The coefficient of variation (CV) is a statistical measure of the deviation of individual data points from the mean. The higher the CV value, the greater the spatial variability, and vice versa [50]. CV is used to determine the spatial distribution of annual precipitation variability depending on data obtained from 60 locations in the KRI, using ArcGIS and the Kriging spatial interpolation technique. The CV applied to precipitation is especially relevant when comparing the results of two separate surveys or tests with different measures or values. Multiplying the coefficient by 100 is an optional step to calculate a percentage [50]. For example, we compare the results of two tests with varying scoring mechanisms. If sample A has a CV of 12% and sample B has a CV of 25%, then sample B has more variation relative to its mean. The coefficient of variation is expressed as:

$$CV = \frac{\sigma}{\mu} * 100 \tag{4}$$

where: σ = standard deviation and μ = mean.

2.5. The Statistical Analyses
The Correlation Coefficient (r)

Bivariate correlations (Pearson correlation coefficient) were adopted in order to find if the variables Crop yield (ton)/year, Crop area (km^2), Average SPI (60 stations), LD area (km^2), MSAVI2 (Mean Values), and Vegetative cover based on MSAVI2 (km^2) are related to one another.

3. Results
3.1. Modified Soil-Adjusted Vegetation Index (MSAVI2)

The MSAVI2 calculated for the study area from 1998 to 2021 is presented in Table 2 for each year. The lowest mean values of MSAVI2 (0.02, 0.23, and 0.25) were recorded in 2000, 2008, and 2021, respectively. These low values occurred due to the decrease in yearly precipitation, a crucial factor in determining the vegetation cover and MSAVI2 score in those years. The years 2015 and 2016 produced the highest MSAVI2 rating (0.46), indicating greater vegetation cover, as illustrated by Figures 3 and 4. The drought's effects in 2000, 2008, and 2021 suggest that nearly all regions were affected. According to MSAVI2 results, the most substantial loss in vegetation cover in 2000 occurred during the growing season (April and May). Severe drought affected 7865.6 km^2 (42.9%), particularly in the KRI's southern, central, and southeastern portions. In 2008, the percentage of land covered by vegetation was 0.2, or 10,018.0 km^2. The low vegetation percentage may have resulted from a mismatch between seasonal precipitation and plant needs during the evaluation of the critical growth stage.

Three key factors explained the loss and worsening of the vegetation cover in 2000. Firstly, 1999 was also a drought year, and it may have played a significant role in the return of drought for two consecutive years. Secondly, overgrazing; due to the severe drought in 1999 and 2000, many livestock breeders in central and southern Iraq sought to feed and pasture in the KRI [51,52]. During 1999 and 2000, grasses, bushes, and forests experienced a drastic reduction in vegetation coverage. Thirdly, a physiological explanation is that drought, in most circumstances, results in an incomplete seed production physiological cycle. In addition, it may fail to produce a sufficient number of viable seeds for the bush, pasture, and grass, which substantially impacts the germination of seeds and the growth of vegetation in subsequent years [11,51,53]. Figures 3–5 illustrate the spatial and temporal distribution of MSAVI2 in the KRI from 1998 to 2021. The vegetation cover showed significant spatial variation at the spatial scale, particularly in the middle of the KRI, whereas the northeastern and southern regions remained the most and most minor vegetative areas, respectively. There was an essential relationship between MSAVI2 and precipitation averages across the KRI from 1998 to 2021.

Table 2. The max, min, mean, std. dev. of MSAVI2 values and the area of vegetative cover and the MSAVI2—based vegetation density classes in the KRI from 1998 to 2021.

Years	Max	Min.	Mean	Std. Dev.	Class 1 Values <0.2		Class 2 Values 0.2–<0.6		Class 3 Values 0.6–1		Sparse and Non-Vegetation	Total Vegetative Cover			Total Study Area
					Very Low MSAVI2 (km²)	(%)	Low to Moderately Low MSAVI2 (km²)	(%)	Moderately High to High MSAVI2 (km²)	(%)	(km²)	(km²)	(%)	(+ − %)	(km²)
1998	1.00	0.20	0.42	0.15	0.0	0.0	21,347.0	86.2	3411.3	13.7	25,506.1	24,758.3	49.2	−5.8	50,350.6
1999	0.99	0.22	0.39	0.12	0.0	0.0	23,223.8	94.6	1336.6	5.4	25,695.5	24,560.5	48.8	−6.2	50,350.6
2000	0.99	0.03	0.02	0.19	7865.6	42.9	9199.60	50.2	1274.5	6.9	31,917.9	18,339.6	36.4	−18.5	50,350.6
2001	0.84	0.19	0.41	0.14	764.9	3.3	19,843.3	86.4	2362.8	10.2	27,289.9	22,971.0	45.6	−9.3	50,350.6
2002	0.84	0.16	0.38	0.14	2906.1	10.5	22,677.0	81.9	2111.8	7.6	22,563.3	27,694.9	55.0	0.0	50,350.6
2003	0.84	0.13	0.38	0.15	3769.2	13.8	21,276.2	77.7	2352.8	8.6	22,861.1	27,398.1	54.4	−0.6	50,350.6
2004	0.84	0.10	0.35	0.15	5542.4	19.2	22,003.7	76.2	1337.6	4.6	21,371.6	28,883.7	57.4	2.4	50,350.6
2005	0.84	0.14	0.34	0.13	3813.5	15.7	19,858.4	81.7	647.9	2.7	25,933.4	24,319.8	48.3	−6.7	50,350.6
2006	0.88	0.09	0.36	0.17	5834.1	22.6	17,734.6	68.8	2190.5	8.5	24,499.9	25,759.2	51.2	−3.8	50,350.6
2007	0.84	0.21	0.44	0.13	0.00	0.0	26,028.6	88.5	3388.6	11.5	20,844.9	29,417.2	58.4	3.5	50,350.6
2008	0.78	0.05	0.23	0.13	10,018	50.0	9856.50	49.2	154.60	0.8	30,222.3	20,029.1	39.8	−15.2	50,350.6
2009	0.92	0.15	0.39	0.14	2348.8	9.4	20,656.6	82.5	2030.0	8.1	25,223.3	25,035.4	49.7	−5.2	50,350.6
2010	0.84	0.23	0.43	0.12	0.00	0.0	25,131.1	89.2	3034.1	10.7	22,096.2	28,165.1	55.9	1.0	50,350.6
2011	0.86	0.15	0.36	0.15	3540.8	14.7	18,352.7	76.1	2217.4	9.2	26,148.9	24,110.9	47.9	−7.1	50,350.6
2012	0.84	0.10	0.35	0.15	4391.9	18.9	17,575.1	75.5	1324.5	5.7	26,964.8	23,291.5	46.3	−8.7	50,350.6
2013	0.77	0.28	0.44	0.10	0.0	0.0	26,300.6	93.3	1880.4	6.7	22,076.3	28,181.0	56.0	1.0	50,350.6
2014	0.77	0.30	0.45	0.09	0.0	0.0	29,578.3	93.2	2161.0	6.8	18,518.1	31,739.3	63.0	8.1	50,350.6
2015	0.78	0.29	0.46	0.10	0.0	0.0	30,243.0	91.6	2787.5	8.4	17,228.6	33,030.4	65.6	10.6	50,350.6
2016	0.84	0.30	0.46	0.09	0.0	0.0	29,637.6	92.2	2498.5	7.8	18,122.3	32,136.1	63.8	8.9	50,350.6
2017	0.78	0.30	0.44	0.09	0.0	0.0	26,111.7	96.7	896.8	3.3	23,245.4	27,008.5	53.6	−1.3	50,350.6
2018	0.90	0.20	0.30	0.28	10,529.2	32.8	15,936.6	49.7	5593.3	17.4	18,291.5	32,059.1	63.7	8.7	50,350.6
2019	0.93	0.20	0.36	0.16	9501.9	22.6	20,926.6	49.9	11,547.8	27.5	8374.3	41,976.3	83.4	28.4	50,350.6
2020	0.99	0.10	0.30	0.14	10,998.5	28.0	20,420.2	51.9	7920.5	20.1	11,011.4	39,339.2	78.1	23.2	50,350.6
2021	0.90	0.10	0.25	0.12	10,772.9	44.9	10,707.3	44.6	2535.9	10.6	26,334.5	24,016.1	47.7	−7.3	50,350.6

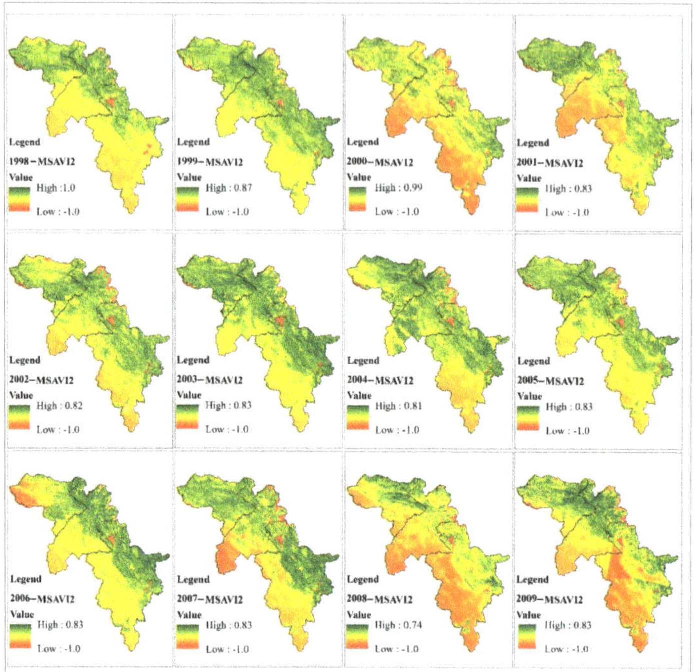

Figure 3. Spatial variation of the MSAVI2−based vegetation from 1998 to 2009.

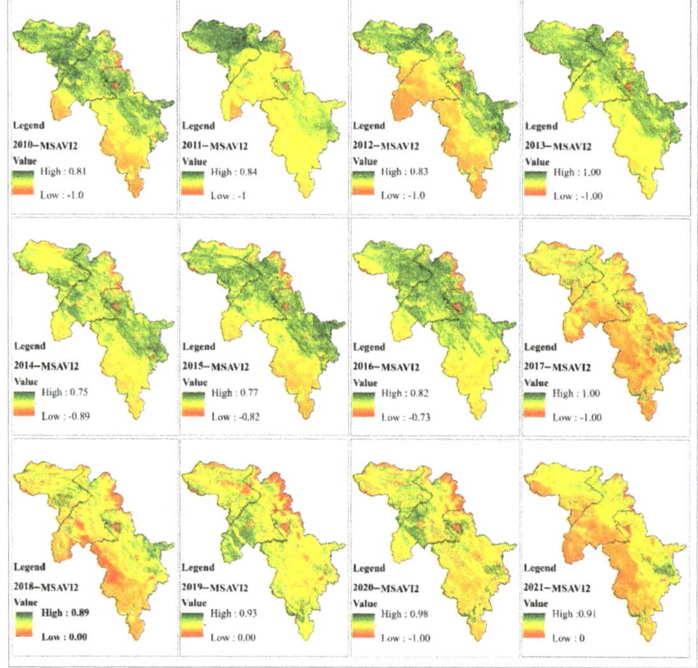

Figure 4. Spatial variation of the MSAVI2−based vegetation from 2010 to 2021.

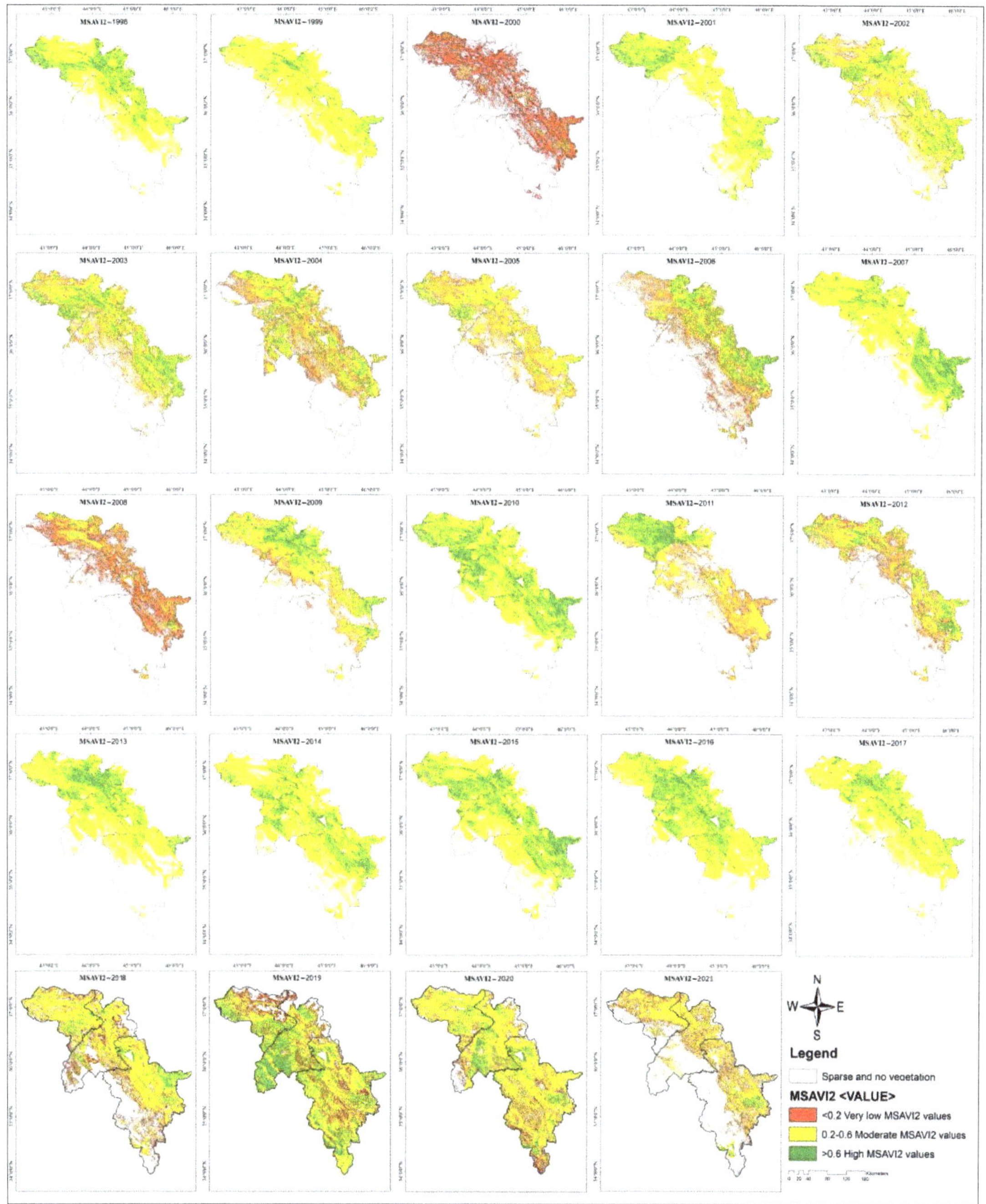

Figure 5. Spatial variation of the MSAVI2−based vegetation density classes from 1998 to 2021.

This trend is consistent with the meteorological features of the study area, namely the average rainfall and temperature. In general, precipitation was highest (about 1000 mm) in the northeast and gradually decreased in the southwest (to around 150 mm). Additionally, elevation followed the same pattern of decline and indirectly influenced temperature and precipitation. The association between vegetation cover area and MSAVI2 values and elevation data was statistically significant [14]. The results reported in Table 2 and Figures 3–5 make this very obvious. Based on the results presented in Table 2 and Figures 3–5, the years 2000, 2008, 2012, and 2021 were the most vulnerable to drought, as detected by the vegetation growth in the region.

In comparison to earlier years, the vegetation cover was drastically reduced throughout these years. During the most severe drought in 2000, the vegetative cover was reduced to 18,339.6 km^2 (representing 36.4% of the overall study area). During 1998–2021, the average vegetation coverage was 55%, although the vegetation coverage in 2000 varied by 36.4% from the average, and the vegetative cover area declined to 24016.1 km^2 in 2021 (representing 47.7% of the overall study area).

3.2. NDWI (Waterbody Area of LD)

The spatiotemporal analysis found that LD reached its greatest extent of 282 km^2 in 2019 and its smallest extent of 125 km^2 in 2009 (Table 3 and Figures 6 and 7). In addition, Figures 6 and 7 and Table 3 show that the most severe droughts in the LD area occurred during the hydrologic years 1999, 2000, 2008, and 2009, by 140 km^2, 137 km^2, 135 km^2, and 125 km^2, respectively. Numerous causes, such as bordering countries prohibiting water imports and territorial laws, decreasing yearly precipitation, constructing various dams in all riparian countries, and rising water demand for agricultural activities, have been attributed to the LD level decline [13]. Low water levels have resulted from drought years in Iraq's river basins, particularly the Tigris, which contributes 70% of the country's water resources [54].

Table 3. Area of water body in (LD) for 1998–2021.

Time, Year	(LD) Area (km^2)	Area Ave.	% (+ −)
1998	258	195	62
1999	140	195	−55
2000	137	195	−58
2001	185	195	−10
2002	225	195	30
2003	267	195	72
2004	254	195	59
2005	238	195	43
2006	216	195	21
2007	189	195	−6
2008	135	195	−60
2009	125	195	−70
2010	159	195	−37
2011	137	195	−59
2012	170	195	−26
2013	200	195	5
2014	158	195	−37
2015	149	195	−46
2016	229	195	33
2017	224	195	28
2018	207	195	12
2019	282	195	87
2020	220	195	25
2021	185	195	−10

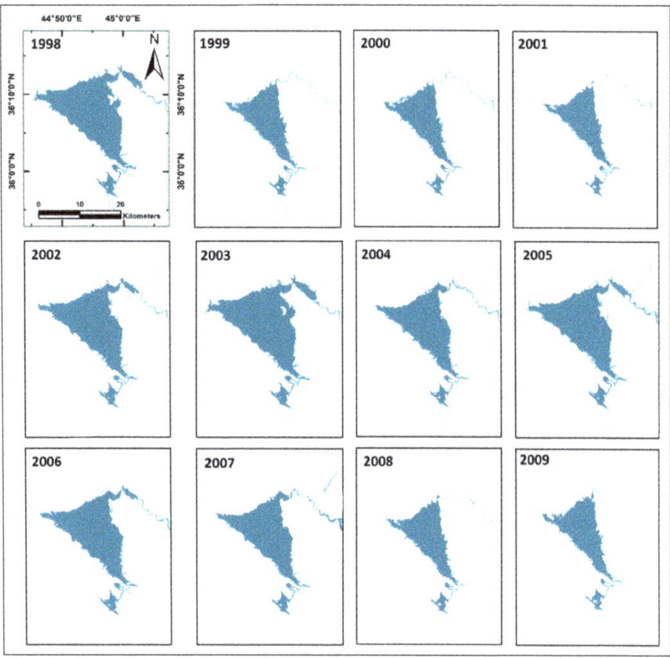

Figure 6. Dukan Lake area change from 1998–2009.

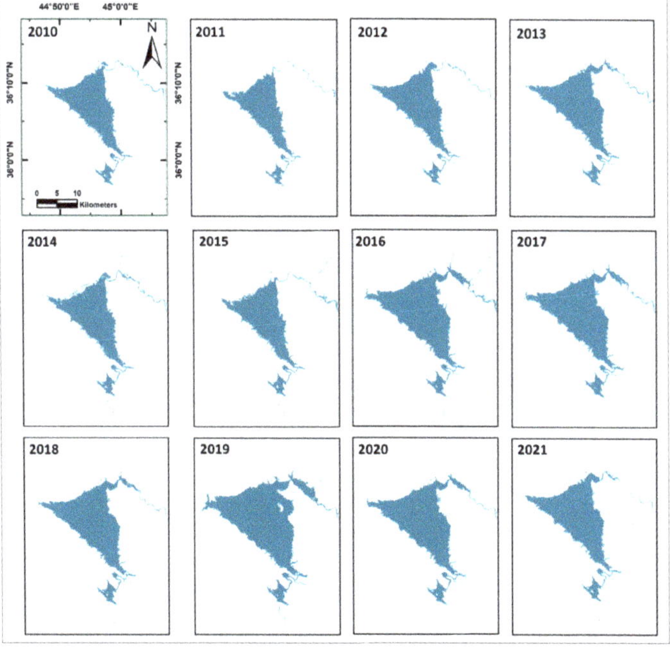

Figure 7. Dukan Lake area change from 2010–2021.

From 1997 to 2002, the average annual discharge of the Tigris River into Iraq fell below 43,000 million cubic meters, and from 1997 to 2001, it dropped precipitously to less than 19,000 million cubic meters, or about 40% less than the average annual discharge. Some of the low discharges are attributed to decreased precipitation in the Tigris River watersheds, which is consistent with the expected drop in precipitation in the country due to climate change [55–57]. According to [58], precipitation in the Turkish highlands is anticipated to decline by 10–60% by the end of the century, resulting in a 29% reduction in Tigris flow. According to a study undertaken at the University of California, Irvine, the total water storage in the Tigris and Euphrates rivers, which flow through Turkey, Syria, Iraq, and Iran, is diminishing at an alarming rate. Between 2003 and 2009, the researchers discovered that the river basin lost around 144 km^3 of fresh water [16]. Approximately 60% of this loss is related to groundwater extraction from aquifers, which is frequently used to meet demand when surface water resources are insufficient [36].

3.3. Standardized Precipitation Index SPI

Figure 8 depicts the spatiotemporal trends of SPI for 60 meteorological stations in the KRI. During the drought years, the severity varied from area to area. According to McKee et al. [26], drought arises when the SPI value is negative and disappears when the SPI value is positive. Four years in the studied historical record, specifically 1999, 2000, 2008, and 2021, saw severe drought, measured by the SPI values. Two years, 2009 and 2012, experienced moderate drought (Figures 8 and 9). Stations 11, 16, 19, 22, 23, 24, 40, 48, 49, 55, 56, 57, and 58 had the most severe drought in 2008, with average SPI values of -2.28, -2.26, -2.27, -2.25, -2.19, -2.54, -2.38, -2.92, -2.24, -2.17, -2.56, -2.35, and -2.28, respectively (Table A3).

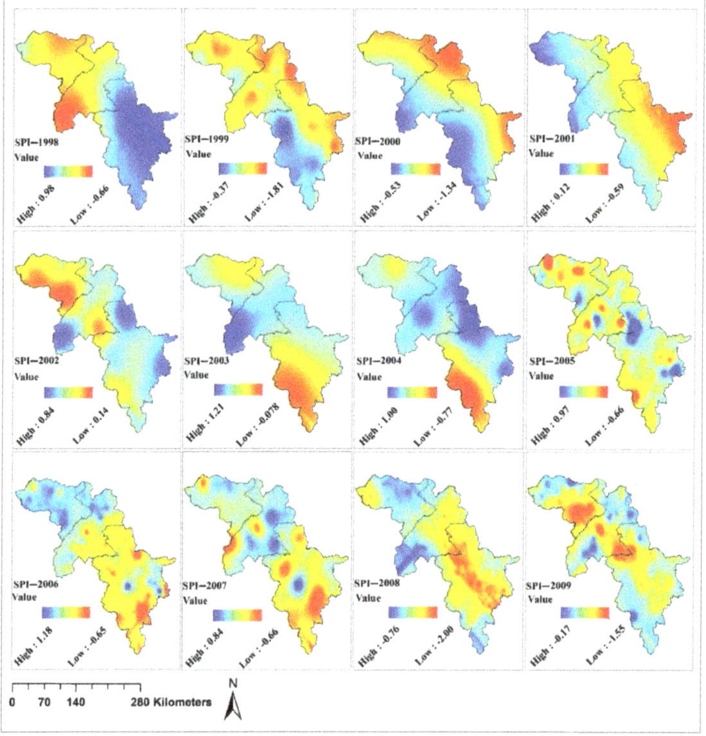

Figure 8. Temporal pattern of SPI drought and wet periods for 60 meteorological stations from 1998 to 2009.

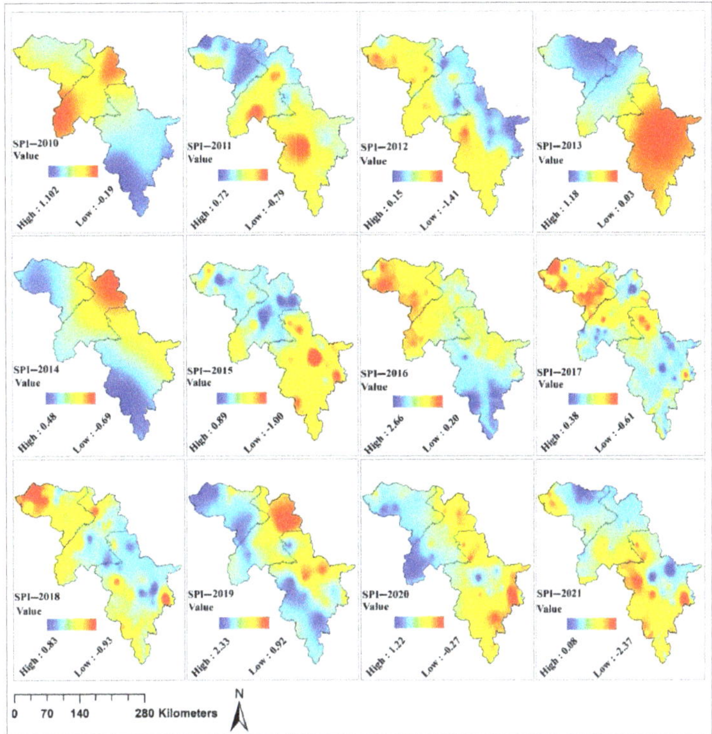

Figure 9. Temporal pattern of SPI drought and wet periods for 60 meteorological stations from 2009 to 2021.

In addition, approximately 60% of the studied years fell under the near-normal drought class. The range of the normalized precipitation index for the near-normal class is between −1.0 and 1.0. Overall, there is no discernible trend in the SPI values across the study years (Table 4), with negative and positive SPI values fluctuating over the study period (Figures 8 and 9). However, a comprehensive review of this graph revealed three instances of more severe drought, notably 1999, 2000, 2008, 2012, and 2021 (Figures 8 and 9). The precipitation deficits continued for at least three years, making the drought throughout these three eras long-term. Indeed, dry years were marked by poor river flow, low groundwater and reservoir levels, extremely dry soil, and decreased crop yields or crop failure [59]. Regardless of the severity of the drought, the entire study area in 2008 and 2012 suffered exceptional dryness.

The SPI values computed for each site revealed that the frequency of severe drought to extreme drought has risen in the KRI by more than three to four times during the previous 24 years. This study demonstrates that severe and intense drought occurred intermittently over the study area, resulting in varying implications on agricultural practices and water supplies in the KRI. The spatiotemporal patterns of SPI distribution for 60 meteorological stations in the KRI's sub-districts indicated drought had varying severity in most studied areas between 1999 and 2021. The severity of these drought years varied from area to area. Figures 8 and 9 demonstrate the trend of drought severity for each of the 60 KRI stations.

Table 4. The frequency of drought SPI index in 60 weather stations during the 24 years.

Station No.	Station Name	Extremely Wet 2.00 or More	Very Wet 1.50 to 1.99	Moderately Wet 1.00 to 1.49	Near Normal 0.99 to −0.99	Moderate Drought −1.00 to −1.49	Severe Drought −1.50 to −1.99	Extreme Drought −2 or Less
				Erbil				
1	Erbil	0	2	1	16	2	1	2
2	Qushtapa	0	1	1	18	1	1	2
3	Khabat	1	0	3	15	3	1	1
4	Bnaslawa	0	1	3	17	1	1	1
5	Harir	0	1	4	16	1	2	0
6	Soran	0	0	7	14	1	2	0
7	Shaqlawa	0	2	1	17	2	2	0
8	Khalifan	0	1	4	16	1	0	2
9	Choman	0	1	3	17	1	1	1
10	Sidakan	0	1	3	16	1	2	1
11	Rwanduz	0	0	6	13	4	0	1
12	Mergasur	0	1	3	17	1	0	2
13	Dibaga	1	2	4	12	3	2	0
14	Gwer	1	2	1	14	5	1	0
15	Barzewa	1	0	0	20	2	1	0
16	Bastora	0	1	3	18	0	0	2
17	Makhmor	0	2	3	15	3	0	1
18	Koya	0	2	2	17	1	1	1
19	Taqtaq	0	2	1	16	3	0	2
20	Shamamk	2	0	3	15	2	1	1
				Duhok				
21	Duhok	2	2	9	7	4	0	0
22	Semel	1	1	13	6	1	2	0
23	Zakho	2	1	11	7	1	2	0
24	Batel	1	3	9	8	2	1	0
25	Dam-DU	2	1	9	9	3	0	0
26	Darkar.H	1	4	7	10	2	0	0
27	Zaxo-A.S	2	0	12	7	2	1	0
28	Batifa	1	2	11	8	0	2	0
29	Kani Masi	1	2	10	8	3	0	0
30	Zaweta	2	2	10	7	2	1	0
31	Mangish	1	3	9	10	0	1	0
32	Deraluke	0	4	8	10	0	2	0
33	Akre	1	3	10	7	3	0	0
34	Amadia	1	3	8	11	0	1	0
35	Sarsink	1	2	12	8	0	1	0
36	Bamarni	0	5	8	9	2	0	0
37	Bardarash	2	3	7	8	4	0	0
38	Qasrok	1	2	11	8	2	0	0
				Sulaimaniyah				
39	SU	0	2	3	15	3	0	1
40	Bazian	0	0	5	16	1	1	1
41	Halabja	0	1	4	15	1	2	1
42	Penjwen	0	1	2	18	1	0	2
43	Chwarta	0	0	6	14	2	2	0
44	Dukan	0	2	3	15	2	1	1
45	Qaladiza	0	2	3	16	0	2	1
46	Rania	0	1	4	15	2	2	0
47	Said Sadiq	1	2	1	15	4	1	0
48	Qaradagh	0	2	0	18	3	0	1
49	Arbat	1	1	3	16	1	1	1
50	K-Panka	0	1	4	15	2	2	0
51	Byara	0	1	3	17	1	2	0
52	Mawat	0	2	2	15	3	1	1
53	Dar-Dikhan	1	1	3	14	3	2	0
54	Chamchamal	0	2	2	15	3	1	1
55	Kalar	2	1	2	17	0	1	1
56	Agjalar	0	1	4	16	3	0	0
57	Bngrd	0	1	4	14	3	1	1
58	Sangaw	1	0	4	15	2	1	1
59	Bawanor	2	0	1	17	3	0	1
60	Kifri	1	1	2	17	3	0	0

Figures 8 and 9 show that the drought zone was determined by interpolating SPI data using Kriging. The KRI's SPI values from 1998 through 2021 are displayed in Tables 4 and A3. The data suggest an irregular cyclical pattern of dry/wet spells during the past 24 years. The initial decline in SPI values began in 1999 and continued until 2001. This drop closely parallels the precipitation decrease seen in DU, ER, and SU provinces during the same year. The SPI index findings were determined to be comparable to the NDWI index results (Table 3). In 1999 and 2000, the drought was extremely severe, but in 2008, the part of the KRI worst hit was the southeast. In 2008 and 2012, the western and southern portions of the study area suffered moderate drought. Figures 8 and 9 depicted the SPI values when drought conditions were found in 1999, 2000, 2008, 2012, and 2021.

In comparison, the wettest years were 2003, 2016, and 2019, respectively. According to McKee et al. (1993), drought occurs when the SPI value is negative and dissipates when it is positive. Table 4 demonstrates that around 57% of the studied years fell into the near-normal drought class, with an SPI range of -1.0 to 1.0 for the near-normal class. Table 2 displays that negative and positive SPI values alternate over the study period. The SPI values show no clear trend throughout the studied periods (Tables 4 and A3). However, a closer look (Figures 8 and 9) revealed that the drought was more severe in 1999, 2000, and 2008 than in any other studied year. Drought can occur despite average precipitation in hydrological and vegetative realms [26]. During the growing season, the absence of a relationship between vegetative and each hydrological drought and SPI is most evident.

3.4. Spatial Pattern Variation of Precipitation

The Zagros Mountains receive the most precipitation from October through May. To examine the spatial pattern of precipitation variability over the study area, which encompasses the whole KRI, the CV was calculated for each of the 60 study stations. Figure 10 and Table 5 depict the average (24−year) precipitation (mm), maximum precipitation (mm), lowest precipitation (mm), standard deviation (%), and coefficient of variation (%). The annual precipitation variability indicates that station #60, with a CV of 56.7%, displayed the most temporal variability, while station #10 exhibited the least, with a CV of 23.0%. Similarly, stations #10 and #17 had the highest and lowest annual precipitation averages, with 1370.3 mm and 244.3 mm, respectively (Table 5).

Generally, the highest CV values are seen in the study area's southern parts, which receive the least precipitation. The statistical results in Table 5 reveal that annual precipitation varies significantly over time. The CV ranged from a low of around 23.1% at station #10 to a high of approximately 56.7% at station #60. (Figure 10). In addition, the lowest annual precipitation averages, less than 244 mm, 297.4 mm, and 293.1 mm at stations #17, #20, and #60, respectively, occurred in low-latitude and low-elevation portions of the KRI. In contrast, higher than 1370.3 mm of precipitation was reported at station #12 in the northern area of the KRI. As the temperature falls with increasing height, Figure 8 depicts a rise from all directions toward high-elevation parts. The spatial variation study reveals, in Figure 10, that the northeast area, which received much more precipitation than other parts, had less regional rainfall variability and a more uniform rainfall distribution than other parts.

In the southwest area, station numbers 2, 13, 14, 15, and 20 (ER), 29, 31,32, 35, and 38 (DU), and 58, 59, and 60 (SU) exhibited a large range of annual spatial variation, with CVs of 43.9, 46.7, 51.5, and 42.8%, and 44.0, 44.3, and 41.5%, respectively. The CV was utilized for the analysis of variability. The study findings showed a downward trend in the KRI's annual and seasonal rainfall series. At station #12 (1370.3 mm) and station #17 (244.3 mm), the maximum and minimum annual precipitation averages, respectively, were recorded. In the southern parts of the KRI, the CV% exhibited significant interannual fluctuation. Figure A2 in Appendix A depicts the CV annual precipitation at 60 selected meteorological stations throughout the KRI. The highest CV values are recorded in the southern parts, characterized by low rainfall.

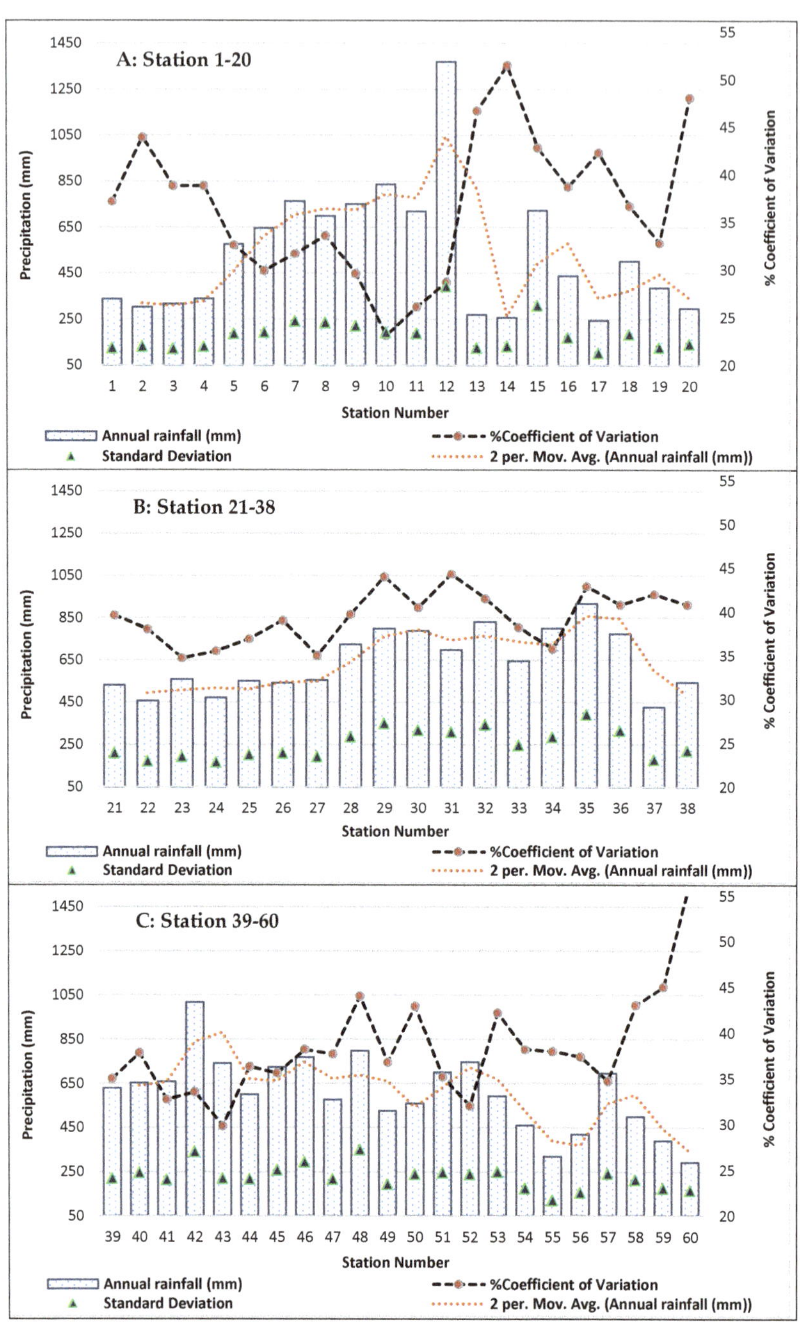

Figure 10. Descriptive statistics of the average annual precipitation series data recorded at each of the 60 weather stations.

Table 5. Descriptive statistics of the annual precipitation series data recorded at each of the 60 weather stations.

Station No.	Geographical Coordinates		Record (Years)	Maximum Rainfall (mm)	Minimum Rainfall (mm)	Average (Annual Rainfall) (mm)	Standard Deviation	Coefficient of Variation CV
	Longit	Latitude						
				Erbil				
1	44.009	36.191	24	645.6	114.2	337.3	125.4	37.2
2	44.028	36.001	24	681.5	106.1	301.3	132.2	43.9
3	43.674	36.273	24	733.0	125.7	317.0	122.9	38.8
4	44.140	36.154	24	694.1	118.0	338.9	131.6	38.8
5	44.365	36.551	24	1042.1	264.5	576.8	188.0	32.6
6	44.561	36.638	24	963.3	290.5	647.2	193.6	29.9
7	43.985	36.209	24	1295.5	360.5	762.9	241.9	31.7
8	44.404	36.599	24	1241.3	263.6	699.3	235.3	33.6
9	44.889	36.637	24	1131.0	271.3	750.8	221.9	29.6
10	44.671	36.797	24	1173.0	463.7	835.3	192.6	23.1
11	44.525	36.612	24	1012.4	342.4	719.6	188.0	26.1
12	44.306	36.838	24	2111.1	624.7	1370.3	392.9	28.7
13	43.805	35.873	24	663.9	94.0	267.5	125.0	46.7
14	43.481	36.045	24	601.6	93.0	256.6	132.2	51.5
15	44.633	36.627	24	1889.0	284.2	722.9	309.3	42.8
16	44.160	36.339	24	870.4	139.7	436.8	169.1	38.7
17	43.583	35.783	24	530.3	92.0	244.3	103.4	42.3
18	44.648	36.099	24	1047.6	216.8	501.8	184.1	36.7
19	44.586	35.887	24	677.6	154.9	386.2	126.7	32.8
20	43.847	36.040	24	746.4	91.0	297.4	142.9	48.1
				Duhok				
21	42.979	36.868	24	1120.0	217.2	531.0	210.0	39.6
22	42.854	36.873	24	995.0	142.7	455.5	172.9	38.0
23	42.682	37.144	24	1165.4	232.5	557.9	193.8	34.7
24	42.722	36.959	24	1004.0	157.4	472.2	167.4	35.5
25	43.003	36.876	24	1135.0	233.1	550.0	202.9	36.9
26	42.823	37.199	24	1187.0	242.0	540.4	210.7	39.0
27	42.659	37.160	24	1165.4	247.8	554.0	194.1	35.0
28	43.013	37.184	24	1705.5	257.2	724.8	288.1	39.7
29	43.441	37.229	24	1688.0	269.5	798.2	350.8	44.0
30	43.143	36.906	24	1768.6	280.1	788.7	319.2	40.5
31	43.093	37.035	24	1657.0	175.4	699.3	309.5	44.3
32	43.649	37.059	24	1867.0	286.8	830.1	344.3	41.5
33	43.893	36.741	24	1425.8	274.9	644.7	246.4	38.2
34	43.487	37.093	24	1650.0	349.4	800.4	286.6	35.8
35	43.350	37.050	24	2015.0	219.2	918.0	393.4	42.9
36	43.269	37.115	24	1677.5	316.4	774.6	316.2	40.8
37	43.589	36.508	24	1014.6	187.1	427.2	179.4	42.0
38	43.598	36.701	24	1262.5	201.8	543.9	222.2	40.8
				Sulaimaniyah				
39	45.436	35.557	24	1147.5	230.2	627.7	219.5	35.0
40	45.140	35.589	24	1209.8	201.6	652.9	246.5	37.8
41	45.974	35.186	24	1081.4	295.4	658.1	215.2	32.7
42	45.941	35.620	24	1873.4	384.0	1017.6	341.3	33.5
43	45.575	35.720	24	1212.5	355.4	741.4	220.8	29.8
44	44.953	35.954	24	1058.2	224.6	599.5	217.7	36.3
45	45.133	36.176	24	1374.5	271.2	723.1	257.6	35.6
46	44.886	36.239	24	1618.4	307.4	768.6	293.4	38.2
47	45.853	35.344	24	1159.9	265.0	575.8	217.3	37.7
48	45.390	35.309	24	1727.5	103.6	798.0	350.7	44.0
49	45.587	35.425	24	1029.7	184.3	525.0	193.4	36.8
50	45.705	35.385	24	1275.0	205.4	558.3	239.4	42.9
51	46.116	35.225	24	1300.7	285.5	700.6	246.4	35.2
52	45.410	35.901	24	1296.6	326.2	746.6	238.9	32.0
53	44.787	36.210	24	1338.6	218.1	592.1	249.8	42.2
54	45.686	35.116	24	914.3	148.9	459.8	175.7	38.2
55	44.833	35.533	24	681.8	106.3	320.4	121.7	38.0
56	44.897	35.748	24	805.0	125.0	418.6	156.5	37.4
57	45.030	36.066	24	1213.5	241.4	695.9	241.5	34.7
58	45.182	35.286	24	1089.0	144.4	499.1	214.5	43.0
59	45.509	34.823	24	900.0	139.1	389.9	175.3	45.0
60	44.966	34.683	24	868.8	134.3	293.1	166.2	56.7

3.5. The Correlation Coefficient

The significant spatiotemporal variability of precipitation in the KRI indicates and forecasts an increase in drought frequency and duration. The correlation coefficients between precipitation, SPI, MSAVI2 mean and vegetation area, crop area, and crop production from 1998 to 2021 are shown in Table 6 (average of 24 years). The analysis of variance for drought indices indicated statistically significant differences between the studied years at p of 0.01 and p of 0.05. The results demonstrated a substantial positive correlation between MSAVI2 and precipitation (Table 6). There was a statistically significant correlation between remote sensing-derived spectral indices and precipitation.

Table 6. Correlation coefficients between spectral indices, meteorological indices crop area, crop yield, and annual average precipitation.

	Crop Area (km²)	(LD) Area (km²)	MSAVI2 (km²)	SPI	MSAVI2 (Mean)	Precipitation (mm)	Crop Yield (Ton)/Year
Crop Area (km²)	1	−0.05	0.37	0.28	0.35	0.28	0.71 **
(LD) Area (km²)	−0.05	1	0.33	0.68 **	0.22	0.69 **	0.05
MSAVI2 Area(km²)	0.37	0.33	1	0.69 **	0.78 **	0.68 **	0.73 **
SPI	0.281	0.68 **	0.69 **	1	0.53 *	0.995 **	0.42
MSAVI2 (Mean Value)	0.35	0.22	0.77 **	0.53 *	1	0.51*	0.61 **
Precipitation (mm)	0.28	0.69 **	0.68 **	0.995 **	0.51 *	1	0.39
Crop Yield (Ton)/Year	0.71 **	0.05	0.73 **	0.42	0.61 **	0.39	1

* Correlation is significant at the 0.05 level (2-tailed). ** Correlation is significant at the 0.01 level (2-tailed).

Table 6 illustrates the relationship between the mean values of vegetation cover based on MSAVI2 characteristics, elevation, latitude, and longitude (precipitation). The graph indicates that as terrain elevation rises, precipitation and elevation increase, but event duration increases. Consequently, mountain regions receive relatively heavy, strong, and long-lasting precipitation. MSAVI2 and elevation are significantly correlated with the event features of the study region. Surface relief greatly influences land characteristics and productivity [60]. The lowland parts receive less precipitation than the mountainous parts. Nevertheless, MSAVI2 measurements are related to precipitation quantity and elevation [61].

4. Discussion

LD's surface area has witnessed major expansions and contractions over the years. Drought estimates are crucial and are required to evaluate how the climates of these lakes and their environs have altered [16]. It is similar to the reports of UNESCO [36] and Fadhil [62]. In addition, the studies show that LD experienced severe droughts in 2008 and 2009. These results are equivalent to those of prior research [15,16]. The years 1999 and 2008 experienced the most severe drought conditions, followed by 2009 and 2012. In 2008, the southeast of the study area, comprising three stations, was the most affected region, with this finding also supported by [63,64]. In addition, the western and southern parts of the study area experienced mild drought conditions in 1999. A previous study [7,65] indicated that the SPI was an effective index; for instance, the SPI at station #9 at the Choman site for the hydrological year 2007–2008 was −2.52, while at station #13 at the Bastora site for the same year it was −1.94. The discrepancy indicates that the precipitation at station #9 (Choman site) in 2007–2008 was less than that at station #13 (Bastora site) in the same time period.

These CV results accord with the findings of [33], indicating a considerable climatic gradient from the south's semi-arid climate to the north's semi-wet climate. This also supports past rainfall patterns and our region's understanding [66]. Elevation is the most influential factor in the regional variation of rainfall. On the other hand, the CV exhibited the reverse tendency [67]. In addition, the KRI's mountainous areas receive an abundance of seasonal precipitation. We discovered that the high seasonal precipitation in mountainous areas is mostly the result of frequent and prolonged rainstorm episodes. However, seasonal precipitation in certain portions of the border area is characterized by low-intensity, short-duration occurrences [64,65,68]. Nearly everywhere to the south-southwest of the KRI, where precipitation and altitude are frequently limited (Figure 1C,D), the MSAVI2 reported much lower values. At all sites, MSAVI2 levels declined concurrently with lower elevations [69–71].

The majority of KRI regions had a severe drought between 1999 and 2008. However, the drought intensity dropped to moderate in 2000, 2009, and 2012, as confirmed by 90% of KRI weather stations. There have been five major droughts in the previous two decades Other than years of severe and moderate drought, the remaining years experienced drought conditions that were near average. The lowest water levels in LD were recorded in 1999, 2000, 2008, and 2009, which is consistent with the decline in SPI values. The MSAVI2 data, on the other hand, indicated droughts in 2000, 2008, 2012, and 2021. Therefore, SPI is a better indicator of drought in the region than MSAVI2, in which the SPI is dependent entirely on precipitation, whereas vegetation cover (MSAVI2) is affected by a more significant number of factors, such as precipitation, temperature, DEM, and soil qualities [72].

5. Conclusions

Between 1999 and 2008, most KRI faced a severe drought, but 90% of KRI weather stations indicated that the drought severity decreased to moderate in 2000, 2009, and 2012. According to the findings, the lowest water levels in LD were recorded in 1999, 2000, 2008, and 2009. This study shed light on historical and agricultural drought events' frequency, length, and spatial extent. Based on the study of rainfall data collected in the KRI from 1998 to 2021, the following may be determined: 1. The yearly precipitation is highest in the northern portion of KRI and lowest in the southern part. 2. The yearly rainfall is quite irregular, with a coefficient of variation of 30%. In the south and southwest of the KRI, the precipitation's CV was reported to vary the most spatially by 56.7%. 3. The SPI data indicated that 2007–2008 was the driest hydrological year between 1998 and 2021. 4. The annual precipitation series exhibits a significant correlation coefficient at most stations. The correlations between the SPI series and the area of LD, vegetation cover, crop area, and crop yield were significant and positive. 5. Between 1999 and 2008, spatial patterns of drought frequency based on the SPI revealed substantial increasing trends of drought severity at stations in the northeast, mid-latitude, and southwest parts of the KRI.

According to the spatiotemporal drought map pattern, the top and middle regions of the KRI had moderate droughts in 1999 and 2008. SPI, NDWI, and MSAVI2 all showed identical drought patterns, consistent with the fall in SPI values. The remainder of the region had acute drought conditions. In contrast, the MSAVI2 data suggested droughts in 2000, 2008, 2012, and 2021. SPI relies solely on precipitation, whereas vegetation cover (MSAVI2) is controlled by a greater variety of parameters, including precipitation, temperature, elevation, latitude, and soil quality [72]. The results of the past 24 years indicate that the drought's consequences were more pronounced in the southern and southeastern regions. In addition, these parts are characterized by expansive grain-growing plains, and the absence of methods to mitigate the consequences of frequent droughts has led to the desertification of these regions. Using MSAVI2 and NDWI, the present work seeks to determine the spatiotemporal extent of drought across KRG and evaluates the performance of the indices by comparing the estimations to the meteorological drought indicator SPI.

In general, we may infer that the drought indicators included in this study demonstrated comparable patterns. Between indices for all analyzed meteorological stations, robust coefficients of determination (R2) were determined. However, it is difficult to infer from this study the precise driving mechanism underlying MSAVI2, as global warming, climate change, temperature fluctuations, and variation in geopotential height may have all had a substantial effect. Continuous observation of rainfall levels and comparisons with current consumption levels can prevent human caused drought and aid in developing an intense drought management program [59]. The findings give better insight into the importance of remote sensing applications to better understand the agricultural and water situations in data-scarce regions such as the KRI.

Author Contributions: Conceptualization, A.M.F.A.-Q., H.A.A.G. and K.M.; Data curation, H.A.A.G., A.M.F.A.-Q., H.A.S.R., S.H.A., K.H. and S.H.Z.; Formal analysis, H.A.A.G. and A.M.F.A.-Q.; Investigation, A.M.F.A.-Q., C.R., K.M., J.P.M., M.R. and H.A.A.G.; Methodology, H.A.A.G., K.H. and A.M.F.A.-Q.; Resources, A.M.F.A.-Q., C.R., K.M., L.H., H.A.S.R., S.H.Z. and H.A.A.G.; Supervision, A.M.F.A.-Q.; Validation, H.A.A.G., S.H.A. and A.M.F.A.-Q.; Visualization, A.M.F.A.-Q., C.R., M.R., K.H. and H.A.A.G.; Writing—original draft, A.M.F.A.-Q. and H.A.A.G.; Writing—review and editing, A.M.F.A.-Q., M.R., C.R., J.P.M. and K.M. All authors have read and agreed to the published version of the manuscript.

Funding: This study has received partial funding from Nuffic, the Orange Knowledge Programme, through the OKP-IRA-104278 project titled "Efficient water management in Iraq switching to climate smart agriculture: capacity building and knowledge development", Coordinated by Wageningen University and Research, The Netherlands and Salahaddin University, Erbil, Kurdistan Region, Iraq.

Institutional Review Board Statement: Not applicable.

Informed Consent Statement: Not applicable.

Data Availability Statement: Some data in this manuscript were obtained from the Ministry of Agriculture and Water Resources, Kurdistan Region, Iraq, while other data was provided by the United States Geological Service (USGS), along with Landsat images being freely available on its website and statistical analysis of the parameters.

Acknowledgments: The authors would like to thank the United States Geological Service (USGS) for providing the Landsat images freely on its website. The authors would like to thank Tariq H. Kakahama and Fuad M. Ahmad at the College of Agricultural Engineering Sciences, Salahaddin University, for their sincere assistance. We are extremely grateful to the anonymous reviewers for their insightful comments and suggestions that significantly enhanced the quality of our paper. We are also thankful to Nuffic, the Orange Knowledge Programme, through the OKP-IRA-104278, Wageningen University and Research, The Netherlands; the Ministry of Agriculture and Water Resources, Water Resources Department, Salahaddin University; and the Tishk International University, Erbil, Kurdistan Region, Iraq, for their valuable support.

Conflicts of Interest: The authors declare no conflict of interest.

Appendix A

Table A1. The Annual Precipitation (AP) (mm) (average of 24 years), DEM, and coordinates (latitude and longitude) of the 60 meteorological stations in the IKR used in this study.

Station No.	Station Name	Lat-	Long-	DEM (m)	AP (mm)	Station No.	Station Name	Lat-	Long-	DEM (m)	AP (mm)
1	Erbil	36.1911	44.0092	412.7	337.3	31	Mangish	37.0351	43.0925	1030.2	689.0
2	Qushtapa	36.0009	44.0285	390.8	301.3	32	Deraluke	37.0586	43.6493	706.8	819.5
3	Khabat	36.2728	43.6739	285.9	317.0	33	Akre	36.7414	43.8933	683.1	633.7
4	Bnaslawa	36.1538	44.1400	540.7	338.9	34	Amadia	37.0925	43.4872	1148.5	790.7
5	Harir	36.5511	44.3648	837.3	576.8	35	Sarsink	37.0503	43.3503	957.1	905.9
6	Soran	36.6385	44.5614	701.6	647.2	36	Bamarni	37.1151	43.2693	1203.0	763.4
7	Shaqlawa	43.9851	36.2094	966.5	762.9	37	Bardarash	36.5082	43.5894	363.6	418.4
8	Khalifan	36.5986	44.4038	697.1	699.3	38	Qasrok	36.7009	43.5980	414.8	533.7
9	Choman	36.6374	44.8893	1178.4	750.8	39	SU	35.5572	45.4356	870.8	617.3
10	Sidakan	36.7974	44.6714	1011.3	835.3	40	Bazian	35.5890	45.1395	943.7	652.9

Table A1. *Cont.*

Station No.	Station Name	Lat-	Long-	DEM (m)	AP (mm)	Station No.	Station Name	Lat-	Long-	DEM (m)	AP (mm)
11	Rwanduz	36.6119	44.5247	801.6	719.6	41	Halabja	35.1864	45.9739	716.6	641.4
12	Mergasur	36.8382	44.3062	1108.9	1370.3	42	Penjwen	35.6197	45.9414	1442.9	1004.2
13	Dibaga	35.8730	43.8050	328.3	267.5	43	Chwarta	35.7197	45.5747	1011.6	741.1
14	Gwer	36.0449	43.4808	309.7	256.6	44	Dukan	35.9542	44.9528	700.4	586.4
15	Barzewa	36.6268	44.6333	798.3	722.9	45	Qaladiza	36.1755	45.1333	628.2	711.7
16	Bastora	36.3389	44.1605	630.0	436.8	46	Rania	36.2391	44.8855	607.8	753.5
17	Makhmoor	35.7833	43.5833	287.7	244.3	47	Said Sadiq	35.3437	45.8534	544.1	564.6
18	Koya	36.0994	44.6481	724.5	501.8	48	Qaradagh	35.3093	45.3896	887.9	784.9
19	Taqtaq	35.8874	44.5856	397.5	386.2	49	Arbat	35.4246	45.5868	701.6	515.2
20	Shamamk	36.0400	43.8467	310.6	297.4	50	KaniPanka	35.3850	45.7046	685.8	549.6
21	Duhok	36.8679	42.9790	588.3	520.0	51	Byara	35.2251	46.1163	1333.5	693.3
22	Semel	36.8733	42.8540	491.6	445.2	52	Mawat	35.9007	45.4105	1063.8	735.4
23	Zakho	37.1436	42.6819	501.4	547.0	53	D-dikhan	35.1163	45.6863	534.6	577.4
24	Batel	36.9595	42.7217	531.0	461.1	54	Chamchamal	35.5333	44.8333	726.6	452.5
25	Dam-DU	36.8758	43.0029	605.6	538.3	55	Kalar	34.6411	45.3293	243.2	313.9
26	Dar. hajam	37.1988	42.8227	649.8	533.7	56	Agjalar	35.7483	44.8974	702.3	410.6
27	Zaxo-farh	37.1599	42.6587	447.1	542.6	57	Bngrd	36.0660	45.0299	841.2	683.5
28	Batifa	37.1840	37.1840	930.2	713.6	58	Sangaw	35.2862	45.1825	704.4	484.9
29	Kani Masi	37.2291	37.2291	1332.3	795.6	59	Bawanor	34.8233	45.5087	358.4	379.9
30	Zaweta	36.9058	36.9058	1006.4	775.6	60	Kifri	34.6833	44.9664	238.7	279.2

Table A2. Landsat data chosen for analysis were a mixture of Landsat TM5, ETM7, and Landsat OLI8.

Date Years	Sensor	Target_WRS_Path Target_WRS_Row Path/Row	Date_Acquired	Resolutions
1998	Landsat 5 TM	170/34,170/35, 169/35, 169/34, 168/35, 168/36	10/04, 10/04, 21/05, 21/05, 30/05, 30/05	30 m
1999	Landsat 5 TM	170/34,170/35, 169/35, 169/34, 168/35, 168/36	13/04, 13/04, 22/04, 22/04,01/05, 01/05	30 m
2000	Landsat 5 TM Landsat 7 ETM+	170/34, 170/35, 169/35, 169/34, 168/35, 168/36	15/05, 15/05, 16/04, 16/04, 25/04, 25/04	30 m
2001	Landsat 7 ETM+	170/34,170/35, 169/35, 169/34, 168/35, 168/36	26/04, 26/04, 21/05, 21/05, 28/04, 28/04	30 m
2002	Landsat 7 ETM+	170/34,170/35, 169/35, 169/34, 168/35, 168/36	13/04, 13/04, 08/05, 08/05, 01/05, 01/05	30 m
2003	Landsat 7 ETM+	170/34,170/35, 169/35, 169/34, 168/35, 168/36	02/05, 02/05, 11/05, 11/05, 20/05, 20/05.	30 m
2004	Landsat 7 ETM+	170/34,170/35, 169/35, 169/34, 168/35, 168/36	06/05, 06/05,11/04, 27/04, 06/05, 06/05	30 m
2005	Landsat 7 ETM+	170/34,170/35, 169/35, 169/34, 168/35, 168/36	23/04, 23/04, 30/04, 30/04, 23/04, 23/04	30 m
2006	Landsat 7 ETM+	170/34,170/35, 169/35, 169/34, 168/35, 168/36	26/05, 26/05, 19/05, 19/05, 12/05, 28/05	30 m
2007	Landsat 5 TM Landsat 7 ETM+	170/34,170/35, 169/35, 169/34, 168/35, 168/36	05/05,05/05, 20/04, 13/04, 07/05, 07/05	30 m
2008	Landsat 7 ETM+	170/34,170/35, 169/35, 169/34, 168/35, 168/36	15/05, 15/05, 22/04, 24/05, 15/04, 15/04	30 m
2009	Landsat 5 TM Landsat 7 ETM+	169/35, 169/34, 170/34,170/35, 168/35, 168/36	03/05, 03/05, 02/05, 02/05, 20/05, 20/05	30 m
2010	Landsat 5 TM Landsat 7 ETM+	170/34,170/35, 169/35, 169/34, 168/35, 168/36	26/05, 29/05, 22/05, 04/04, 05/04, 19/04	30 m
2011	Landsat 5 TM Landsat 7 ETM+	170/34,170/35, 169/34, 168/35, 168/36169/35,	16/05, 16/05, 08/05, 16/04, 16/04, 15/04	30 m
2012	Landsat 7 ETM+	170/34,170/35, 169/35, 169/34, 168/35, 168/36	26/04, 26/04, 19/05, 19/05, 26/04, 26/04	30 m
2013	Landsat 8 OLI	170/34,170/35, 169/35, 169/34, 168/35, 168/36	05/05, 05/05, 28/04, 28/04, 23/05, 23/05,	30 m
2014	Landsat 8 OLI	170/34,170/35, 169/35, 169/34, 168/35, 168/36	06/04, 06/04, 15/04, 01/05, 24/04, 24/04	30 m
2015	Landsat 8 OLI	170/34,170/35, 169/35, 169/34, 168/35, 168/36	09/04, 25/04,18/04, 01/04, 27/04, 27/04	30 m
2016	Landsat 8 OLI	170/34,170/35, 169/35, 169/34, 168/35, 168/36	13/05, 13/05, 20/04, 20/04, 15/05, 15/05	30 m
2017	Landsat 8 OLI	170/34,170/35, 169/35, 169/34, 168/35, 168/36	30/04, 30/04, 09/05, 09/05, 18/05, 18/05	30 m
2018	Landsat 8 OLI	170/34,170/35, 169/35, 169/34, 168/35, 168/36	04,10/04, 10/04, 26/04, 19/04, 19/04	30 m
2019	Landsat 8 OLI	170/34,170/35, 169/35, 169/34, 168/35, 168/36	4/04, 4/04, 13/04, 13/04, 24/05, 24/05	30 m
2020	Landsat 8 OLI	170/34,170/35, 169/35, 169/34, 168/35, 168/36	08/05, 08/05, 15/04, 15/04, 23/03, 23/03	30 m
2021	Landsat 8 OLI	170/34,170/35, 169/35, 169/34, 168/35, 168/36	25/4, 10/05, 20/04, 20/04, 26/03, 26/03	30 m

Table A3. The duration, frequency, and severity of droughts based on the SPI index in 60 weather stations in the KRI from 1998 to 2021.

Station No.	Long-	Lat-	1997–1998	1998–1999	1999–2000	2000–2001	2001–2002	2002–2003	2003–2004	2004–2005	2005–2006	2006–2007	2007–2008	2008–2009
1	44.009	36.191	−0.68	−1.94	−0.76	−0.15	0.61	1.31	1.17	0.71	0.69	0.45	−1.16	−0.46
2	44.028	36.001	−1.11	−1.44	−1.24	0.08	0.68	0.79	0.66	0.49	0.11	0.63	−0.57	−0.37
3	43.674	36.273	0.05	−1.01	−0.84	0.32	0.25	0.74	0.65	0.25	0.56	−0.01	−1.52	−0.9
4	44.14	36.154	−0.64	−1.62	−0.6	−0.08	0.16	0.98	1.06	0.49	0.53	0.48	−1.38	−0.95
5	44.365	36.551	0.26	−1.44	−1.02	−0.72	0.69	0.71	0.77	0.32	0.38	0.58	−1.44	−0.57
6	44.561	36.638	−0.32	−0.75	−1.57	−1.01	0.76	0.9	0.79	0.34	0.8	0.64	−1.37	−0.52
7	43.985	36.209	0.35	−1.58	−1.25	−0.39	0.64	1.05	0.8	0.47	0.54	0.81	−1.71	−0.65
8	44.404	36.599	−0.27	−1.68	−1.68	0.07	0.87	0.8	0.54	0.01	0.64	0.53	−1.01	−0.45
9	44.889	36.637	−0.17	−2.09	−1.29	−0.59	0.65	0.27	1.03	0.04	0.31	0.56	−1.13	−0.39
10	44.671	36.797	0.6	−1.27	−1.24	−0.49	0.67	0.45	0.82	0.34	0.86	0.32	−1.68	−0.73
11	44.525	36.612	1.01	−1.3	−0.53	0.24	−0.06	0.25	0.99	0.42	0.87	0.94	−2.03	−0.81
12	44.306	36.838	−0.91	−1.94	−1.86	0	0.71	0.16	0.54	0.27	0.94	0.22	−1.29	−0.56
13	43.805	35.873	−1.12	−1.4	−0.78	−0.2	0.71	1.02	0.38	0.14	0.77	0.42	−0.94	−0.42
14	43.481	36.045	−0.82	−1.22	−0.47	0.13	1.05	1.75	0.22	0.07	0.41	−0.91	−0.82	−1.07
15	44.633	36.627	0.53	−0.99	−1.34	0.55	0.3	2.89	0.39	0.23	0.23	0.61	−1.81	−0.91
16	44.16	36.339	0.52	−0.74	−0.53	−0.08	0.69	0.61	0.57	−0.13	−0.14	−0.53	−1.72	−1.57
17	43.583	35.783	−0.52	−1.45	−0.46	0	0.93	1.28	0.9	0.3	0.63	0.29	−0.97	−0.82
18	44.648	36.099	0.23	−1.01	−0.62	−0.61	0.04	0.52	−0.22	−0.32	0.1	0.88	−1.43	−1.15
19	44.586	35.887	0.56	−0.89	−1	−0.41	0.04	0.35	0.47	0.18	0.27	0.51	−1.72	−1.56
20	43.847	36.04	−0.53	−1.62	−0.45	0.04	0.81	1.56	0.8	−0.17	0.21	0.17	−0.84	−0.68
21	42.979	36.868	−0.11	−1.26	−1.39	0.4	0.24	0.9	0.31	0.3	0.77	0.11	−1.4	−0.92
22	42.854	36.873	0.08	−0.99	−0.63	0.65	0.19	0.47	0.55	0.19	0.62	0.4	−1.83	−0.96
23	42.682	37.144	0.58	−1.58	−0.74	0.14	0.49	0.74	0.25	0.17	0.53	0.32	−1.71	−0.9
24	42.722	36.959	0.71	−0.9	−1.02	0.3	0.25	0.73	0.4	0.48	0.87	0.33	−1.89	−0.53
25	43.003	36.876	−0.08	−1.47	−0.46	−0.28	0.23	0.72	0.46	0.2	0.69	0.53	−1.43	−1.01
26	42.823	37.199	0.04	−1.32	−1.43	0.26	0.66	1.02	0.61	−0.67	0.18	−0.63	−0.96	−0.5
27	42.659	37.16	0.06	−1.17	−1.38	−0.1	0.28	0.5	0.48	0.26	0.44	0.27	−1.61	−0.96
28	43.013	37.184	−0.26	−1.57	−1.6	−0.45	0.3	0.7	0.23	0.32	0.75	0.53	−0.91	−0.65
29	43.441	37.229	−0.61	−1.28	−1.34	−1.01	0.4	0.16	0.24	0.38	0.59	0.55	−1.18	−0.15
30	43.143	36.906	−0.38	−1.53	−0.28	0.07	0.25	0.49	0.37	−0.03	0.76	0.18	−0.91	−1.13
31	43.093	37.035	−0.3	−1.87	−1.08	−0.24	0.2	0.54	0.3	0.01	0.73	0.33	−1.11	−0.58
32	43.649	37.059	−0.71	−1.47	−1.44	0	0.55	0.4	0.64	−0.11	0.71	0.32	−0.69	−0.75
33	43.893	36.741	0.72	−1.26	−0.74	−0.15	0.23	0.52	0.36	0.25	0.5	0.25	−1.03	−1.39
34	43.487	37.093	0.06	−1.4	−0.8	−0.45	0.5	0.23	−0.07	−0.15	0.32	0.67	−0.99	−1.03
35	43.35	37.05	−0.73	−1.83	−1.14	0.25	0.54	0.28	0.09	0.13	0.57	0.19	−0.96	−0.89
36	43.269	37.115	−0.64	−1.34	−1.29	−0.21	0.78	0.25	0.1	0.06	0.93	0.51	−1.1	−0.93
37	43.589	36.508	0.25	−0.71	−0.67	−0.49	−0.31	0.79	0.75	0.67	1.0	0.33	−1.23	−1.23
38	43.598	36.701	−0.06	−1.11	−0.93	0.04	0.3	0.57	0.55	0.44	0.89	0.19	−1.4	−1.46
39	45.436	35.557	1.28	−1.78	−0.83	−0.21	0.71	1.0	0.92	0.28	0.6	0.11	−0.92	−0.66
40	45.14	35.589	0.7	−1.28	−0.64	0.05	0.4	0.69	0.5	0.35	0.41	0.17	−1.59	−0.91
41	45.974	35.186	1.62	−2.16	−1.38	−1.01	1.08	0.76	1.46	0.96	1.17	0.32	−2.14	−0.77
42	45.941	35.62	−0.13	−1.68	−1.74	−0.65	0.72	1.02	0.64	0.3	0.69	0.41	−1.19	−0.76
43	45.575	35.72	0.81	−1.28	−1.1	−0.4	0.35	0.46	0.58	0.22	0.42	−0.03	−1.15	−0.78
44	44.953	35.954	1.71	−1.28	−0.83	−0.41	0.65	0.76	1.17	0.98	0.41	0.22	−1.85	−1.38
45	45.133	36.176	0.01	−1.68	−1.37	−0.48	0.91	1.23	1.05	0.15	0.13	−0.43	−1.19	−0.47
46	44.886	36.239	0.99	−1.35	−1.05	−0.24	0.72	0.78	0.87	0.49	0.15	0.48	−1.44	−1.06
47	45.853	35.344	1.59	−1.26	−1.27	−0.83	0.81	0.47	0.48	−0.07	0.81	0.12	−1.47	−1.0
48	45.39	35.309	0.59	−1.15	−0.86	−0.33	0.43	0.48	0.37	0.28	0.46	0.1	−2.25	−0.93
49	45.587	35.425	1.55	−1.49	−0.5	−0.46	0.74	0.42	0.34	0.02	0.32	0.02	−1.74	−0.92
50	45.705	35.385	0.79	−1.29	−0.9	−0.68	0.5	0.22	0.19	0.09	0.8	0.08	−1.27	−0.81
51	46.116	35.225	0.95	−1.42	−1.46	−0.61	0.65	0.58	0.57	0.38	−0.69	0.06	−1.1	−0.64
52	45.411	35.901	1.28	−1.23	−0.86	−0.86	0.72	0.69	0.9	0.38	−0.49	0.23	−1.61	−1.14
53	44.787	36.21	0.62	−1.45	−1.15	−1.0	1.13	0.84	0.59	0.56	0.42	−0.2	−1.62	−0.78
54	45.686	35.116	0.25	−0.86	−1.12	0.01	0.54	0.72	0.77	0.6	−0.03	−0.55	−1.66	−0.88
55	44.833	35.533	0.78	−0.06	0.1	0.16	0.96	−0.16	−0.17	0.2	−0.03	−0.53	−2.09	−0.73
56	44.897	35.748	0.45	−0.73	−0.92	−0.28	0.6	0.99	1.01	0.76	0.45	−0.22	−1.83	−1.09
57	45.03	36.066	1.29	−1.24	−1.04	−0.29	0.86	0.64	0.94	0.84	0.43	0.27	−1.98	−1.02
58	45.183	35.286	0.62	−0.81	−0.85	−0.28	0.61	0.48	0.57	0.18	1.24	1.09	−1.97	−1.12
59	45.509	34.823	0.66	−0.48	−0.54	0.35	0.7	0.22	−0.04	0.2	−0.67	−0.51	−1.67	−1.07
60	44.966	34.683	0.91	−0.68	−0.56	0.15	0.17	−0.75	−1.15	−0.38	−0.1	−0.22	−0.56	−0.19

Table A3. *Cont.*

Station No.	Long-	Lat-	2009–2010	2010–2011	2011–2012	2012–2013	2013–2014	2014–2015	2015–2016	2016–2017	2017–2018	2018–2019	2019–2020	2020–2021
1	44.009	36.191	0.25	−0.09	−1.05	0.65	−0.35	0.02	0.56	−0.28	0.25	1.8	0.59	−1.3
2	44.028	36.001	0.22	−0.73	−1.12	0.62	0.12	0.4	0.85	0.06	0.38	1.9	1.03	−1.06
3	43.674	36.273	−0.3	−0.03	−0.5	0.82	−0.1	0.36	0.64	−0.38	0.03	2.19	1.0	−0.76
4	44.14	36.154	−0.2	−0.25	−0.51	0.88	0.02	0.56	0.64	−0.31	0.46	1.79	0.64	−0.66
5	44.365	36.551	0.3	−0.4	−0.43	0.8	−0.64	0.5	0.88	−0.19	0.29	1.65	0.76	−0.65
6	44.561	36.638	−0.01	−0.56	−0.48	0.53	−0.54	0.79	0.95	0.09	0.39	1.17	0.69	−0.32
7	43.985	36.209	0.31	−0.21	−0.85	1.42	−0.34	0.07	0.77	−0.56	−0.01	1.71	0.4	−1.24
8	44.404	36.599	0.23	−0.57	−0.61	0.92	−0.22	0.37	0.95	−0.01	0.46	1.54	0.5	−0.52
9	44.889	36.637	0.08	0.26	−0.27	1.09	−0.37	0.79	1.2	−0.38	0.45	1.25	0.43	−0.61
10	44.671	36.797	−0.12	0.01	−0.31	0.57	−1.18	0.31	1.29	0.44	0.39	1.3	0.2	−0.68
11	44.525	36.612	−0.27	−0.5	−0.99	0.97	−0.93	0.55	1.26	−0.27	0.17	1.27	0.4	−1.08
12	44.306	36.838	0.59	0.21	0.05	1.46	−0.32	0.27	1.33	−0.04	−0.25	1.49	0.23	−0.7
13	43.805	35.873	0.03	−0.5	−0.8	0.98	0.0	0.28	0.41	0.03	0.21	2.09	1.05	−0.8
14	43.481	36.045	−0.21	0.08	−0.78	0.37	0.41	0.06	0.56	0.18	0.2	1.82	1.07	−0.38
15	44.633	36.627	−0.41	−0.36	−0.8	0.47	−0.73	0.72	0.72	−0.4	0.16	0.9	0.24	−0.84
16	44.16	36.339	0.02	−0.3	−0.38	1.06	0.17	0.85	0.98	−0.11	0.64	1.7	0.68	−0.79
17	43.583	35.783	−0.26	−0.15	−1.05	0.6	−0.23	−0.03	0.32	−0.24	0.13	1.89	1.14	−0.8
18	44.648	36.099	0.76	0.09	0.05	0.53	−0.09	0.41	1.18	−0.23	0.5	1.95	0.82	−0.94
19	44.586	35.887	0.51	0.01	−0.33	0.72	0.14	0.47	1.26	−0.16	0.65	1.6	0.8	−1.35
20	43.847	36.04	0.12	−0.28	−1.17	0.36	−0.22	0.21	0.68	0.08	0.36	2.16	0.76	−1.06
21	42.979	36.868	0.43	−0.12	−1.03	1.21	0.68	0.27	0.39	−0.44	−0.04	1.96	0.87	−1.06
22	42.854	36.873	0.39	−0.1.0	−1.17	0.77	0.38	0.36	0.2	−0.35	−0.04	2.09	1.13	−1.23
23	42.682	37.144	0.44	0.35	−0.88	0.56	−0.38	0.42	1.18	−0.52	−0.3	2.26	0.51	−1.22
24	42.722	36.959	0.42	−0.23	−1.36	0.36	−0.11	0.11	0.53	−0.11	0.2	2.15	0.7	−1.45
25	43.003	36.876	0.6	−0.08	−1.07	1.28	0.63	0.19	0.25	−0.57	−0.03	2.03	0.88	−1.22
26	42.823	37.199	0.48	0.71	−0.2	0.99	0.68	−0.17	0.17	−0.65	−0.94	2.23	0.91	−0.57
27	42.659	37.16	0.45	0.41	−0.79	0.44	0.48	0.46	1.47	−0.52	−0.34	2.33	0.5	−1.36
28	43.013	37.184	0.51	0.14	−0.68	0.63	0.15	0.33	0.58	0.14	0.02	2.23	0.64	−0.68
29	43.441	37.229	0.68	0.27	−1.0	1.26	0.08	0.29	0.84	−0.05	0.52	1.64	0.58	0.07
30	43.143	36.906	0.43	−0.14	−1.21	1.14	0.29	0.47	0.58	−0.28	0.03	2.04	1.05	−0.72
31	43.093	37.035	0.61	0.06	−1.02	1.02	0.43	0.9	0.47	−0.17	−0.23	2.06	0.87	−0.54
32	43.649	37.059	0.28	0.47	−0.75	0.9	−0.15	0.23	0.82	−0.18	0.26	1.94	0.9	−0.44
33	43.893	36.741	0.7	0.44	−1.26	1.01	0.13	0.12	0.44	−0.59	0.04	2.13	0.67	−0.84
34	43.487	37.093	0.5	0.3	−0.68	1.28	−0.01	0.46	0.76	−0.34	0.18	1.97	0.79	−0.53
35	43.35	37.05	0.37	0.15	−0.71	1.2	0.15	0.53	0.93	0.03	0.45	1.8	0.74	−0.42
36	43.269	37.115	0.66	0.4	−0.97	1.02	0.13	0.32	0.92	−0.33	0.19	1.95	0.55	−0.59
37	43.589	36.508	0.22	0.44	−1.2	0.57	−0.38	0.38	0.3	−0.38	0.17	2.22	0.79	−1.0
38	43.598	36.701	0.47	0.46	−0.9	0.71	0.09	0.47	0.37	−0.52	0.22	2.12	0.7	−0.86
39	45.436	35.557	0.76	−0.04	−0.12	−0.62	−0.58	−1.01	0.65	−0.1	0.24	1.72	0.59	−0.88
40	45.14	35.589	0.58	−0.32	−0.5	−0.23	−0.06	0.07	0.69	−0.14	0.38	1.33	1.42	0.19
41	45.974	35.186	1.2	0.03	−0.16	0.26	−0.78	−0.37	0.84	−0.65	−0.48	1.98	−0.43	−2.51
42	45.941	35.62	0.65	−0.03	0.17	0.26	−0.07	−0.07	1.02	−0.09	0.42	1.73	0.2	−0.64
43	45.575	35.72	0.75	−0.08	−0.47	−0.06	−0.08	0.23	0.68	0.01	0.44	1.32	1.07	0.13
44	44.953	35.954	0.05	−0.35	−0.61	0	−0.47	0.17	0.94	−0.15	0.42	1.66	0.03	−1.39
45	45.133	36.176	0.34	0.03	−0.09	0.47	0.11	−0.04	1.01	−0.31	0.59	1.78	0.29	−0.79
46	44.886	36.239	0.41	−0.09	−0.49	0.37	−0.35	−0.06	0.64	−0.48	0.41	1.96	0.61	−0.99
47	45.853	35.344	0.7	−0.03	−0.46	0.08	−0.24	−0.04	1.3	−0.03	0.15	2.1	−0.09	−1.19
48	45.39	35.309	0.4	−0.19	−0.14	0.02	0.24	0.1	1.32	0.25	0.66	1.67	0.88	−0.51
49	45.587	35.425	0.74	0.03	−0.39	0.09	−0.1	−0.04	0.97	−0.24	0.42	1.84	0.54	−0.99
50	45.705	35.385	0.76	0.13	−0.39	0.06	−0.08	−0.08	0.82	−0.07	0.85	1.9	0.91	−0.54
51	46.116	35.225	0.8	−0.02	−0.15	0.23	−0.04	0.12	0.99	0.06	0.19	1.63	0.93	−0.4
52	45.411	35.901	0.69	−0.11	−0.13	0.26	−0.34	0.19	0.9	0.01	0.43	1.66	0.37	−0.84
53	44.787	36.21	0.86	0.3	−0.53	0.35	−0.22	−0.15	1.22	−0.25	0.1	2.23	0.1	−1.48
54	45.686	35.116	0.5	−0.25	−1.2	0.56	0.37	0.29	1.12	0.07	0.52	1.72	0.69	−0.66
55	44.833	35.533	0.66	−0.43	−1.46	0.62	0.34	−0.1	1.68	−0.21	−0.18	2.29	0.53	−1.91
56	44.897	35.748	0.49	−0.08	−1.11	0.04	−0.13	0.19	0.86	−0.11	0.53	1.69	0.71	−0.93
57	45.03	36.066	0.64	−0.18	−0.47	−0.03	−0.24	−0.25	0.96	−0.51	0.41	1.65	0.34	−1.29
58	45.183	35.286	0.88	−0.78	−1.12	−0.1	0.06	0.01	1.15	−0.26	0.12	2.13	0.12	−1.64
59	45.509	34.823	0.8	−0.06	−1.04	0.51	0.33	0.02	1.98	−0.32	0.22	2.29	−0.07	−1.08
60	44.966	34.683	1.24	−0.31	−1.05	−0.04	0.65	−0.28	2.82	−0.23	−0.04	1.85	0.47	−1.39

Table A4. Google Earth Engine JavaScript for estimating MSAVI2.

```javascript
/** Kawa Hakzi 2022 kawahakzy@gmail.com MSAVI2 */
// Assign a common name to the sensor-specific bands.
var LC9_BANDS = ['B2', 'B3', 'B4', 'B5', 'B6', 'B7', 'B10']; //Landsat 8
var LC8_BANDS = ['B2', 'B3', 'B4', 'B5', 'B6', 'B7', 'B10']; //Landsat 8
var LC7_BANDS = ['B1', 'B2', 'B3', 'B4', 'B5', 'B7', 'B6_VCID_2']; //Landsat 7
var LC5_BANDS = ['B1', 'B2', 'B3', 'B4', 'B5', 'B7', 'B6']; //Llandsat 5
var STD_NAMES = ['blue', 'green', 'red', 'nir', 'swir1', 'swir2', 'temp'];
var l9 = ee.ImageCollection('LANDSAT/LC09/C02/T1_TOA').select(LC9_BANDS,
STD_NAMES)// Landsat 8
//Bands are not arranged yet
var l8 = ee.ImageCollection('LANDSAT/LC08/C01/T1_TOA').select(LC8_BANDS,
STD_NAMES)// Landsat 8
//print(l8, 'Landsat 8')
var l7 = ee.ImageCollection('LANDSAT/LE07/C01/T1_TOA').select(LC7_BANDS, STD_NAMES)
//Landsat 7
//print(l7, 'Landsat 7')
var l5 = ee.ImageCollection('LANDSAT/LT05/C01/T1_TOA').select(LC5_BANDS, STD_NAMES)
//Landsat 5
//print(l5, 'Landsat 5')
var images = ee.ImageCollection(l5.merge(l7).merge(l8));//.merge(l9)
var table = ee.FeatureCollection("projects/ee-kawa/assets/kurdistan"),
Map.addLayer(table);
//var images = ee.ImageCollection('LANDSAT/LC08/C01/T1_TOA')
.filterBounds(table)
.filterDate('2019-04-01', '2019-05-01')
.select('B4', 'B5', 'B2', 'B3');
print(images.size());
var nir = images.select('B5');
var red = images.select('B4');
var ndvi = nir;
var clipnir = nir.filterBounds(table).mosaic().clip(table);
var clipred = red.filterBounds(table).mosaic().clip(table);
var msavi2imgmosaic = clipnir.multiply(2).add(1)
.subtract(clipnir.multiply(2).add(1).pow(2)
.subtract(clipnir.subtract(clipred).multiply(8)).sqrt()
).divide(2).rename("MSAVI2");
Map.addLayer(msavi2imgmosaic);
Map.centerObject(table, 7);
Export.image.toDrive({
image: msavi2imgmosaic,
description: 'imageToDrive_year()',
crs: 'EPSG:4326',
scale: 30,
maxPixels:200000000,
region: table )};
```

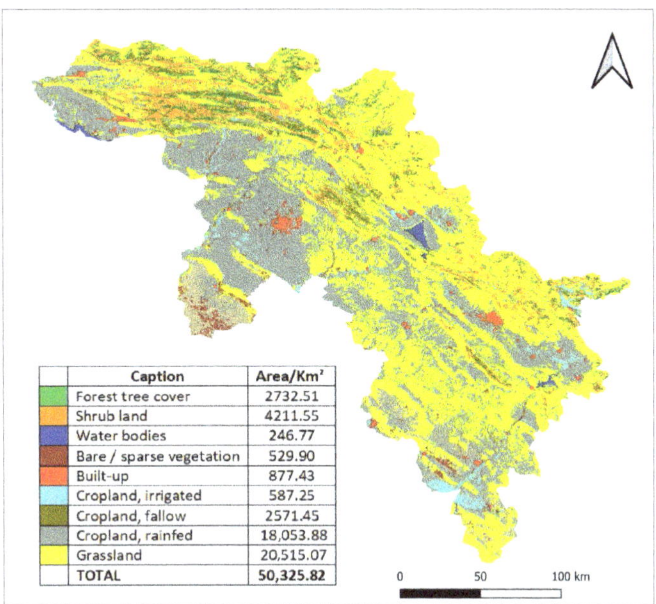

Figure A1. Land use and land cover classes in the KRI.

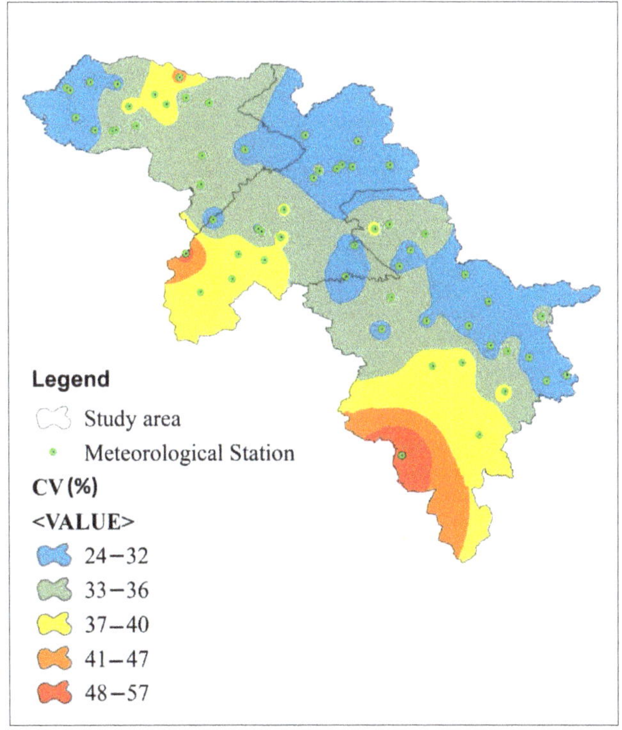

Figure A2. The coefficient of variation (CV%) map of annual precipitation for 60 selected meteorological stations across the KRI.

References

1. Tsatsaris, A.; Kalogeropoulos, K.; Stathopoulos, N.; Louka, P.; Tsanakas, K.; Tsesmelis, D.E.; Krassanakis, V.; Petropoulos, G.P.; Pappas, V.; Chalkias, C. Geoinformation Technologies in Support of Environmental Hazards Monitoring under Climate Change: An Extensive Review. *ISPRS Int. J. Geo-Inf.* **2021**, *10*, 94. [CrossRef]
2. Bhaga, T.D.; Dube, T.; Shekede, M.D.; Shoko, C. Impacts of Climate Variability and Drought on Surface Water Resources in Sub-Saharan Africa Using Remote Sensing: A Review. *Remote Sens.* **2020**, *12*, 4184. [CrossRef]
3. Gaznayee, H.A.A.; Al-Quraishi, A.M.F.; Al-Sulttani, A.H.A. Drought Spatiotemporal Characteristics Based on a Vegetation Condition Index in Erbil, Kurdistan Region, Iraq. *Iraqi J. Sci.* **2021**, *62*, 4545–4556. [CrossRef]
4. Yang, F.; Wang, G.F.; Long, J.M.; Wang, B.L. Influence of Surface Energy on the Pull-in Instability of Electrostatic Nano-Switches. *J. Comput. Theor. Nanosci.* **2013**, *10*, 1273–1277. [CrossRef]
5. Das, S.; Angadi, D.P. Land Use Land Cover Change Detection and Monitoring of Urban Growth Using Remote Sensing and GIS Techniques: A Micro-Level Study. *GeoJournal* **2022**, *87*, 2101–2123. [CrossRef]
6. Martínez-Vilalta, J.; Lloret, F. Drought-Induced Vegetation Shifts in Terrestrial Ecosystems: The Key Role of Regeneration Dynamics. *Glob. Planet. Change* **2016**, *144*, 94–108. [CrossRef]
7. Gaznayee, H.A.A.; Al-Quraishi, A.M.F.; Mahdi, K.; Ritsema, C. A Geospatial Approach for Analysis of Drought Impacts on Vegetation Cover and Land Surface Temperature in the Kurdistan Region of Iraq. *Water* **2022**, *14*, 927. [CrossRef]
8. Al-Quraishi, A.M.F.; Negm, A.M. *Environmental Remote Sensing and GIS in Iraq*; Springer: Berlin/Heidelberg, Germany, 2020; ISBN 9783030213435.
9. Zhao, S.; Cong, D.; He, K.; Yang, H.; Qin, Z. Spatial-Temporal Variation of Drought in China from 1982 to 2010 Based on a Modified Temperature Vegetation Drought Index (MTVDI). *Sci. Rep.* **2017**, *7*, 1–12. [CrossRef]
10. Sharma, A. Spatial Data Mining for Drought Monitoring: An Approach Using Temporal NDVI and Rainfall Spatial Data Mining for Drought Monitoring: An Approach Using Temporal NDVI and Rainfall. MSc Thesis, University of Twente, Faculty of Geo-Information Sci, Enschede, The Netherlands, 2006. Unpublished.
11. Habibi, M.; Babaeian, I.; Schöner, W. Changing Causes of Drought in the Urmia Lake Basin—Increasing Influence of Evaporation and Disappearing Snow Cover. *Water* **2021**, *13*, 3273. [CrossRef]
12. Fan, Y.; Wang, L.; Su, T.; Lan, Q. Spring Drought as a Possible Cause for Disappearance of Native Metasequoia in Yunnan Province, China: Evidence from Seed Germination and Seedling Growth. *Glob. Ecol. Conserv.* **2020**, *22*, e00912. [CrossRef]
13. Al-Quraishi, A.M.F.; Gaznayee, H.A.A.; Crespi, M. Drought Trend Analysis in a Semi-Arid Area of Iraq Based on Normalized Difference Vegetation Index, Normalized Difference Water Index and Standardized Precipitation Index. *J. Arid Land* **2021**, *13*, 413–430. [CrossRef]
14. Gaznayee, H.A.A.; Al-Quraishi, A.M.F. Analysis of Agricultural Drought, Rainfall, and Crop Yield Relationships in Erbil Province, the Kurdistan Region of Iraq Based on Landsat Time-Series Msavi2. *J. Adv. Res. Dyn. Control Syst.* **2019**, *11*, 536–545. [CrossRef]
15. Husain, Y. Monitoring and Calculating the Surface Area of Lakes in Northern Iraq. *Appl. Res. J.* **2016**, *2*, 54–62.
16. Yaseen, A.; Mahmood, M.I.; Yaseen, G.; Ali, A.A. Area Change Monitoring of Dokan & Darbandikhan Iraqi Lakes Using Satellite Data. *Sustain. Resour. Manag. J.* **2018**, *3*, 1–16. [CrossRef]
17. Anderson, M.C.; Zolin, C.A.; Sentelhas, P.C.; Hain, C.R.; Semmens, K.; Tugrul Yilmaz, M.; Gao, F.; Otkin, J.A.; Tetrault, R. The Evaporative Stress Index as an Indicator of Agricultural Drought in Brazil: An Assessment Based on Crop Yield Impacts. *Remote Sens. Environ.* **2016**, *174*, 82–99. [CrossRef]
18. Samui, P.; Sitharam, T.G. Machine Learning Modelling for Predicting Soil Liquefaction Susceptibility. *Nat. Hazards Earth Syst. Sci.* **2011**, *11*, 1–9. [CrossRef]
19. Wilhite, D. Breaking the Hydro-Illogical Cycle: Progress or Status Quo for Drought Management in the United States. *Eur. Water* **2011**, *34*, 5–18.
20. Toma, J.J. Limnological Study of Dokan, Derbendikhan and Duhok Lakes, Kurdistan Region of Iraq. *Open J. Ecol.* **2013**, *03*, 23–29. [CrossRef]
21. Abdullah, S.M.A. Parasitic Fauna of Some Freshwater Fishes from Darbandikhan Lake, North of Iraq. *J. Dohuk Univ.* **2016**, *8*, 29–35.
22. Ahamed, A.; Bolten, J.; Doyle, C.; Fayne, J. Near Real-Time Flood Monitoring and Impact Assessment Systems Case Study: 2011 Flooding in Southeast Asia. *Remote Sens. Hydrol. Extrem.* **2017**, 105–118. [CrossRef]
23. Menon, D.K.; Bhavana, V. An Overview of Drought Evaluation and Monitoring Using Remote Sensing and GIS. *Pdfs.Semant.org* **2016**, *3*, 32–37.
24. McKee, T.B. Drought Monitoring with Multiple Time Scales. In Proceedings of the Conference on Applied Climatology, Boston, MA, USA, 15–20 January 1995.
25. Indexed, S.; Kadhim, K.N. Estimating of Consumptive Use of Water in Babylon Governorate-Iraq. *Int. J. Civ. Eng. Technol.* **2018**, *9*, 798–807. [CrossRef]
26. Thomas, B.; McKee, N.J.D. Analysis of Standardized Precipitation Index (SPI) Data for Drought Assessment. In Proceedings of the Eighth Conference on Applied Climatology, Anaheim, CA, USA, 17–22 January 1993; Volume 26, pp. 1–72. [CrossRef]
27. Arshad, S.; Morid, S.; Mobasheri, M.R.; Alikhani, M.A. Development of Agricultural Drought Risk Assessment Model for Kermanshah Province (Iran), Using Satellite Data and Intelligent Methods. In Proceedings of the: The First International Ionference on Drought Management, Zaragoza, Spain, 12–14 June 2008; 12, pp. 303–310.

28. Salih, S.; Alzwainy, F. Microstructure Analysis and Residual Strength of Fiber Reinforced Eco-Friendly Self- Consolidating Concrete Subjected To. *Glob. Planet. Change* **2018**, *144*, 94–108.

29. Gessner, U.; Machwitz, M.; Conrad, C.; Dech, S. Estimating the Fractional Cover of Growth Forms and Bare Surface in Savannas. A Multi-Resolution Approach Based on Regression Tree Ensembles. *Remote Sens. Environ.* **2013**, *129*, 90–102. [CrossRef]

30. Baugh, W.M.; Groeneveld, D.P. Broadband Vegetation Index Performance Evaluated for a Low-Cover Environment. *Int. J. Remote Sens.* **2006**, *27*, 4715–4730. [CrossRef]

31. Vanselow, K.A.; Samimi, C. Predictive Mapping of Dwarf Shrub Vegetation in an Arid High Mountain Ecosystem Using Remote Sensing and Random Forests. *Remote Sens.* **2014**, *6*, 6709–6726. [CrossRef]

32. Saleh, A.M. Relationship Betweenvegetation Indicesof Landsat-7 ETM+, MSS Data and Some Soil Properties: Case Study of Baqubah, Diyala, Iraq. *IOSR J. Agric. Vet. Sci. Ver. II* **2015**, *8*, 2319–2372. [CrossRef]

33. Keya, D.R. Building Models to Estimate Rainfall Erosivity Factor from Rainfall Depth in Iraqi Kurdistan Region. Ph.D. Thesis, Salahaddin University, Erbil, Iraq, 2020; pp. 1–3.

34. Gaznayee, H.A.A. Modeling Spatio-Temporal Pattern of Drought Severity Using Meteorological Data and Geoinformatics Techniques for the Kurdistan Region of Iraq. Ph.D. Thesis, Salahaddin University, Erbil, Iraq, 2020; pp. 1–11.

35. UNESCO Survey of Infiltration Karez in Northern Iraq. In *History and Current Status of Underground Aqueducts A Report Prepared for UNESCO*; A report prepared for UNESCO; Department of Geography, Oklahoma State University: Stillwater, OK, USA, 2009; p. 56.

36. UNESCO Integrated Drought Risk Management–DRM Executive. In *National Framework for Iraq, an Analysis Report, Technical Report*, 2nd ed.; UNESCO Office: Jordan, Iraq, 2014. Available online: http://www.unesco.org/new/fileadmin/MULTIMEDIA/FIELD/Iraq/pdf/Publications/DRM.pdf (accessed on 28 May 2022).

37. Al-Quraishi, A.M.F.; Mustafa, Y.T.; Negm, A.M. *Environmental Degradation in Asia: Land Degradation, Environmental Contamination, and Human Activities*; Springer: Cham, Switzerland, 2022; ISBN 978-3-031-12111-1. (In production)

38. Eklund, L.; Persson, A.; Pilesjö, P. Cropland Changes in Times of Conflict, Reconstruction, and Economic Development in Iraqi Kurdistan. *Ambio* **2016**, *45*, 78–88. [CrossRef]

39. Chander, G.; Markham, B.L.; Helder, D.L. Summary of Current Radiometric Calibration Coefficients for Landsat MSS, TM, ETM+, and EO-1 ALI Sensors. *Remote Sens. Environ.* **2009**, *113*, 893–903. [CrossRef]

40. Wang, H.; Lin, H.; Liu, D. Remotely Sensed Drought Index and Its Responses to Meteorological Drought in Southwest China. *Remote Sens. Lett.* **2014**, *5*, 413–422. [CrossRef]

41. Qi, J.; Kerr, Y.; Chehbouni, A. External Factor Consideration in Vegetation Index Development. In Proceedings of the 6th International Symposium on Physical Measurements and Signatures in Remote Sensing, ISPRS, Val d'Isère, France, 17–21 January 1994; pp. 723–730. [CrossRef]

42. Ahmad, F. Spectral Vegetation Indices Performance Evaluated for Cholistan Desert. *J. Geogr. Reg. Plan.* **2012**, *5*, 165–172. [CrossRef]

43. Jassim, M.A. TIN Model Extraction for Dukan Lake Bed Using HYPACK System. *ZANCO J. Pure Appl. Sci.* **2016**, *27*, 113–120.

44. McFeeters, S.K. The Use of the Normalized Difference Water Index (NDWI) in the Delineation of Open Water Features. *Int. J. Remote Sens.* **1996**, *17*, 1425–1432. [CrossRef]

45. Rajsekhar, D.; Singh, V.P.; Mishra, A.K. Multivariate Drought Index: An Information Theory Based Approach for Integrated Drought Assessment. *J. Hydrol.* **2015**, *526*, 164–182. [CrossRef]

46. Mishra, A.K.; Desai, V.R.; Singh, V.P. Drought Forecasting Using a Hybrid Stochastic and Neural Network Model. *J. Hydrol. Eng.* **2007**, *12*, 626–638. [CrossRef]

47. Mishra, A.K.; Desai, V.R. Drought Forecasting Using Stochastic Models. *Stoch. Environ. Res. Risk Assess.* **2005**, *19*, 326–339. [CrossRef]

48. Feilhauer, H.; Schmid, T.; Faude, U.; Sánchez-Carrillo, S.; Cirujano, S. Are Remotely Sensed Traits Suitable for Ecological Analysis? A Case Study of Long-Term Drought Effects on Leaf Mass per Area of Wetland Vegetation. *Ecol. Indic.* **2018**, *88*, 232–240. [CrossRef]

49. Saavedra, C. *Estimating Spatial Patterns of Soil Erosion and Deposition in the Andean Region Using Geo-Information Techniques: A Case Study in Cochabamba, Bolivia*; Wageningen University and Research: Wageningen, The Netherlands, 2005; ISBN 9798516031052.

50. Curtis, A.; Byron, I.; MacKay, J. Integrating Socio-Economic and Biophysical Data To Underpin Collaborative Watershed Management 1. *JAWRA J. Am. Water Resour. Assoc.* **2005**, *41*, 549–563. [CrossRef]

51. Thiébault, S.; Moatti, J.-P. The Mediterranean region under climate change: The 22nd Conference of the Parties to the United Nations Framework Convention on Climate Change. In Proceedings of the 22nd Conference of the Parties to the United Nations Framework Convention on Climate Change COP22, Morocco, Marrakech, 7–18 November 2016; Chapter 2. pp. 71–72; ISBN 978-2-7099-2219-7.

52. Jaradat, A. Agriculture in Iraq: Resources, Potentials, Constraints, Research Needs and Priorities. *Agriculture* **2003**, *1*, 83.

53. Båld, M. Water Scarcity & Migration: A Comparative Case Study of Egypt and Iraq. . Iraq's Drought Crisis and the Damaging Effects on Communities "Our Source of Living Has Dried Up". Master Thesis, Uppsala University, Uppsala, Sweden, 2022.

54. Lucani, P.; Saade, M. Iraq Agriculture Sector Note. FAO Investment Centre. Food and Agriculture Organisation of the United Nations and the World Bank. *Near East North Afr. Rep.* **2012**, 1–75. Available online: http://www.fao.org/3/a-i2877e (accessed on 28 May 2022).

55. Abdullah, T.O.; Ali, S.S.; Al-Ansari, N.A.; Knutsson, S. Hydrogeochemical Evaluation of Groundwater and Its Suitability for Domestic Uses in Halabja Saidsadiq Basin, Iraq. *Water* **2019**, *11*, 11. [CrossRef]

56. Al-Saady, Y.; Merkel, B.; Al-Tawash, B.; Al Suhail, Q. Land Use and Land Cover (LULC) Mapping and Change Detection in the Little Zab River Basin (LZRB), Kurdistan Region, NE Iraq and NW Iran. *FOG-Freib. Online Geosci.* **2015**, *43*, 1–32.

57. Talab, A.A. *Evaluation of Some Irrigation Projects in Dukan Watershed as Controlling and Conservation of Water Resources*; Iraqi Ministry of Water Resources: Baghdad, Iraqi, 2007.

58. Gitz, V.; Meybeck, A.; Lipper, L.; Young, C.D.; Braatz, S. Climate Change and Food Security: Risks and Responses; Risks and Responses. *Food Agric. Organ. United Nations Rep* **2016**, *110*, 2–4; ISBN 978925108998.

59. Mzuri, R.T.; Mustafa, Y.T.; Omar, A.A. Land Degradation Assessment Using AHP and GIS-Based Modelling in Duhok District, Kurdistan Region, Iraq. *Geocarto Int.* **2021**, *36*, 1–19. [CrossRef]

60. Rahimzadeh-Bajgiran, P.; Omasa, K.; Shimizu, Y. Comparative Evaluation of the Vegetation Dryness Index (VDI), the Temperature Vegetation Dryness Index (TVDI) and the Improved TVDI (ITVDI) for Water Stress Detection in Semi-Arid Regions of Iran. *ISPRS J. Photogramm. Remote Sens.* **2012**, *68*, 1–12. [CrossRef]

61. Hameed, M.; Ahmadalipour, A.; Moradkhani, H. Apprehensive Drought Characteristics over Iraq: Results of a Multidecadal Spatiotemporal Assessment. *Geosciences* **2018**, *8*, 58. [CrossRef]

62. Fadhil, A.M. Land Degradation Detection Using Geo-Information technology for Some Sites in Iraq. *J. Al-Nahrain Univ. Sci.* **2009**, *12*, 94–108. [CrossRef]

63. Fadhil, A.M. Drought Mapping Using Geoinformation Technology for Some Sites in the Iraqi Kurdistan Region. *Int. J. Digit. Earth* **2011**, *4*, 239–257. [CrossRef]

64. Seeyan, S.; Merkel, B.; Abo, R. Investigation of the Relationship between Groundwater Level Fluctuation and Vegetation Cover by Using NDVI for Shaqlawa Basin, Kurdistan Region–Iraq. *J. Geogr. Geol.* **2014**, *6*, 187–202. [CrossRef]

65. Saeed, M.A. Analysis of Climate and Drought Conditions in the Fedral. *Int. Sci. J. Environ. Sci.* **2012**, *2*, 953.

66. Chakraborty, S.; Pandey, R.P.; Chaube, U.C.; Mishra, S.K. Trend and Variability Analysis of Rainfall Series at Seonath River Basin, Chhattisgarh (India). *Int. J. Appl. Sci. Eng. Res.* **2013**, *2*, 425–434. [CrossRef]

67. Karim, T.H.; Talab, A.A.; Yaseen, A.; Mahmood, M.I.; Yaseen, G.; Ali, A.A.; Karim, T.H.; Keya, D.R.; Amin, Z.A.; Profile, S.E.E. Temporal and Spatial Variations in Annual Rainfall Distribution in Erbil Province. *Outlook Agric.* **2018**, *3*, 1–16. [CrossRef]

68. Zakaria, S.; Mustafa, Y.T.; Mohammed, D.A.; Ali, S.S.; Al-Ansari, N.; Knutsson, S. Estimation of Annual Harvested Runoff at Sulaymaniyah Governorate, Kurdistan Region of Iraq. *Nat. Sci.* **2013**, *5*, 1272–1283. [CrossRef]

69. Al-Shwani, F.M.A. Land Cover Change Detection in Erbil Governorate Using Remote Sensing Techniques. Master's Thesis, Salahaddin University, Erbil, Iraq, 2009.

70. Razvanchy, H.A.S. Modelling Some of the Soil Propertiesin the Iraqi Kurdistan Region Using Landsat Datasets and Spectroradiometer. Master's Thesis, Salahaddin University, Erbil, Iraq, 2008.

71. Bazzaz, A.O.H. Drought Monitoring Using Geoinformatics Techniques in Several Districts of Iraqi Kurdistan Region. Master's Thesis, Salahadin Universty, Erbil, Iraq, 2016; pp. 1–68.

72. Karavitis, C.A.; Alexandris, S.; Tsesmelis, D.E.; Athanasopoulos, G. Application of the Standardized Precipitation Index (SPI) in Greece. *Water* **2011**, *3*, 787–805. [CrossRef]

Article

A Geospatial Approach for Analysis of Drought Impacts on Vegetation Cover and Land Surface Temperature in the Kurdistan Region of Iraq

Heman Abdulkhaleq A. Gaznayee [1,*], Ayad M. Fadhil Al-Quraishi [2,*], Karrar Mahdi [3] and Coen Ritsema [3]

[1] Department of Forestry, College of Agriculture Engineering Science, Salahaddin University, Erbil 44003, Kurdistan Region, Iraq
[2] Petroleum and Mining Engineering Department, Faculty of Engineering, Tishk International University, Erbil 44001, Kurdistan Region, Iraq
[3] Soil Physics and Land Management Group, Wageningen University & Research, 6700 AA Wageningen, The Netherlands; karrar.mahdi@wur.nl (K.M.); coen.ritsema@wur.nl (C.R.)
* Correspondence: heman.ahmed@su.edu.krd (H.A.A.G.); ayad.alquraishi@tiu.edu.iq (A.M.F.A.-Q.)

Abstract: Drought is a common event in Iraq's climate, and the country has severely suffered from drought episodes in the last two decades. The Kurdistan Region of Iraq (KRI) is geographically situated in the semi-arid zone in Iraq, whose water resources have been limited in the last decades and mostly shared with other neighboring countries. To analyze drought impacts on the vegetation cover and the land surface temperature in the KRI for a span of 20 years from 1998 to 2017, remote sensing (RS) and Geographical Information Systems (GIS) have been adopted in this study. For this study, 120 Landsat satellite images were downloaded and utilized, whereas six images covering the entire study area were used for each year of the study period. The Normalized Difference Vegetation Index (NDVI) and Land Surfaces Temperature Index (LST) were applied to produce multi-temporal classified drought maps. Changes in the area and values of the classified NDVI and LST were calculated and mapped. Mann–Kendall and Sen's Slope statistical tests were used to assess the variability of drought indices variation in 60 locations in the study area. The results revealed increases in severity and frequency of drought over the study period, particularly in the years 2000 and 2008, which were characterized by an increase in land surface temperatures, a decrease in vegetation area cover, and a lack of precipitation averages. Climate conditions affect the increase/decrease of the vegetated cover area, and geographical variability is also one factor that significantly influences the distribution of vegetation. It can be concluded that the southeast and southwestern parts of the KRI were subjected to the most severe droughts over the past 20 years.

Keywords: drought; KRI; NDVI; LST

Citation: Gaznayee, H.A.A.; Al-Quraishi, A.M.F.; Mahdi, K.; Ritsema, C. A Geospatial Approach for Analysis of Drought Impacts on Vegetation Cover and Land Surface Temperature in the Kurdistan Region of Iraq. *Water* **2022**, *14*, 927. https://doi.org/10.3390/w14060927

Academic Editor: Ana Iglesias

Received: 9 February 2022
Accepted: 10 March 2022
Published: 16 March 2022

Publisher's Note: MDPI stays neutral with regard to jurisdictional claims in published maps and institutional affiliations.

1. Introduction

Among all natural disasters, drought can be considered the most complex due to the difficulties in identifying its start, end, intensity, and extent [1]. Droughts cause enormous sufferings for the society and the environment. Consequently, it is important to learn drought's spatial-temporal pattern [2]. Several environmental factors play significant roles in the occurrence of droughts, high temperature and winds, relatively low humidity, timing, characteristics, and patterns of rains—especially during crop growing seasons, intensity and duration of rainfall, and onset and termination [3]. Although drought has no universal definition, it can be simply defined as the deficit in precipitation and terrestrial water storage (the sum of surface and subsurface water), which adversely impacts agriculture, the environment, and the economy [4,5]. Drought has a significant adverse impact on the socio-economic, agricultural, and environmental sectors [2]. During drought periods, severe water stress can occur in a region due to lack of precipitation, water resources overexploitation, high rates of evapotranspiration, and/or an amalgamation of those factors [6,7].

Remote sensing plays a vital role in detecting, mapping, assessing, and monitoring the earth's resources and natural hazards at spatiotemporal scales [2]. Various techniques and indices have been developed to address and manage drought status. The leading cause of drought is the lack of rainfall averages below normal levels; however, human and social activities also lead to drought [7,8]. The occurrence of high temperatures and low moisture levels is often related to drought events, which have become quite frequent in recent years; thus, it is predominantly associated with climate change [9]. The influence of drought might also vary geographically due to variability in precipitation patterns and human resilience [7]. The National Oceanic and Atmospheric Administration (NOAA) has defined the drought background and its effect on Iraqi lands as the decline in rainfall averages for long periods, for a season or more, which leads to water stress-causing negative effects on the water resources and consequently adverse impacts on the plants, animals, and people [10]. In the series of drought development, there are two phases. The first is the meteorological drought that occurs when there is an extended decrease in rainfall rates compared to the normal rates. Secondly, the lack of rainfall is one reason that leads to a decrease in soil moisture. Thus, the lack of suitable conditions for plant growth and the dwindling of vegetation cover is called agricultural drought [11]. The drought situation in Iraq has been stated by several researchers [12]. In recent years, the annual precipitation averages have been declining due to global warming [13]. The Iraqi report 2009 issued by the Coordination of Humanitarian Affairs, UNAMI, and IAU office, considered the most important reasons for the successive droughts events in Iraq are the decrease in rainfall rates and the water discharge rates decline of the main rivers in Iraq. Consequently, these lead to reduced groundwater levels, the river flows, and draining water sources (springs, deep, and shallow wells) [14]. On the other side, in a span of ten years, from 2003 to 2012, Iraq has suffered several severe droughts, which were results from different reasons, such as low average precipitations, higher temperatures rates, lower water income from the upstream countries, and low efficiency in water utilization [15,16].

Iraq's location in arid and semi-arid regions led to a high frequency of droughts, especially during the last two decades [17]. Low precipitation and its fluctuation during the season are normal in most North African and West Asian countries. This puts Iraq, among other countries, in a place where serious actions on drought management must be adopted [18]. Moreover, in 1999 a severe drought occurred in Erbil and Dohuk, where it also suffered from moderate drought in 1986–1987, 1989–1991, 1999, and 2008 [11,19]. The annual precipitation average in the KRI ranges is from less than 100 mm in the south to 1200 mm in the northeastern mountainous region [14]. From 1999 to 2002, Erbil suffered from a decrease in rainfall averages and drought suffering. It also went through another drought period in 2007–2011, indicating that Erbil is an area prone to drought [20]. Moreover, Sulaymaniyah was subjected to severe droughts from 1994 to 1998 [20,21]. On the other hand, precipitations were significantly decreased in 2008, then a drought took place in the governorate, and similar observations are also noted in the Duhok governorate [12].

Although a few methods were developed in remote sensing for drought monitoring, some others further considered the influence of drought on vegetation. The Normalized Difference Vegetation Index (NDVI) is one of the earliest vegetation indices used to monitor drought; it has been used since the 1980s [18,22,23]. Different studies have been conducted to explore spatiotemporal patterns of drought; however, most of those studies focused on the methods of drought detecting and evaluating the agricultural drought's relationships with each rainfall average and the crop yield using the Landsat time-series dataset [18].

In Iraq, including the Kurdistan region, drought is a common event causing significant agro-economic losses, but there is a significant lack of detailed information on the spatiotemporal patterns of drought severity in the KRI, for which it can be employed to take extra precautions for mitigating its negative impacts [20,21]. Therefore, a detailed analysis of seasonal drought dynamics is required to identify spatiotemporal drought patterns at a meteorological scale and vegetative spheres [24]. Time-series patterns of droughts in the KRI have been mapped using remote sensing (RS) and Geographic Information Systems (GIS) using various drought indices. Since aquifer recharge, agricultural activities, and ecological changes are affected by rainfall, the focus was on drought during the agriculture growing season [25].

The NDVI, the land surface temperature (LST), and the LST/NDVI slope can have an essential role in monitoring drought, low rainfall, and tracking crop growth, crop yields, weather impact, and the environmental and economic effects [26]. For vegetated regions, the fluctuation in weather-related NDVI cannot be detected easily, as the integrated area of the weather component is smaller than the ecosystem component [7,27]. Hence, it is advised to separate weather components from an ecosystem component when using NDVI to analyze weather's impact on vegetation [28]. Drought analysis requires both drought-causative and responsive parameters, such as rainfall, soil moisture, potential evapotranspiration, vegetation condition, groundwater, and surface water levels. Since drought measuring parameters are not linearly correlated, the correlation among drought indices is usually weak, and typically, they are not predicting similar patterns [29,30].

Using NDVI data, the changes in vegetation cover in the study area were presented, and the trend in drought occurrences can be studied. The NDVI performance is not without errors, such as errors during the growing season and saturation effect on dense vegetation [31]. Therefore, the results have to be validated using other parameters to increase the accuracy [32]. The LST is a good index of the earth's surface's energy balance, providing important information about the surface's physical properties and climate [28]. It was found that there is a negative correlation between LST and NDVI, reported by [31], as an increase of LST was observed at several scales due to changes in vegetation cover and soil moisture, which indicates that the surface temperature can rise rapidly with water stress. Thus, the ratio of LST/NDVI increases during times of drought [31].

This study aims to analyze a spatial pattern for drought severity in the KRI to investigate the spatiotemporal drought characteristics to focus on the agricultural drought assessment by analyzing vegetation stress caused by the lower precipitation. Overall, there are two reasons for selecting the KRI as the research area. First, KRI is prone to drought because of its geographical location and climate. Thus, mastering the mapping and classification of drought characteristics is conducive to forecasting drought in the future. Second, the abovementioned three areas in KRI vary significantly in terms of their topography, NDVI, LST, and precipitation distribution, and thus, the drought characteristics differ considerably among those three areas using Landsat time-series image-based NDVI and LST indices for a span of twenty years from 1998 to 2017.

2. Materials and Methods

2.1. Study Area

The KRI territories were selected as the study area in this research, particularly in Erbil, Sulaimaniyah, and Duhok governorates. The study area is located in the northern part of Iraq. Syria borders the study area from the west, Iran from the east, and Turkey from the north [25]. The KRI is characterized by a Mediterranean climate, which is cold and rainy in winter and hot and dry in summer [33]. It is situated between latitudes 34° and 37° and longitudes 41° and 46°, covering an extent of about 53,000 km², which constitutes a large portion of the entire Iraq territory [34]. It has a diverse physical environment, whereas the elevation ranges from 88 m in its southern parts to more than 3603 m in the north and northeast parts (Figures 1 and 2).

KRI's climate is characterized by high precipitation rates in the northern and mountainous parts, while dry weather is governed in the plains in the southern parts [35]. In general, the precipitation starts from October to May, with 350 mm in the southwestern parts to more than 1200 mm in northern and northeastern parts [36]. Figure 1B and Table 1 explain the data collected from 60 meteorological stations for three different zones in the KRI: assured rainfall zone (>500 mm), semi-assured rainfall zone (350–500 mm), and unassured rainfall zone (<350 mm) [20]. The rain-fed lands represent approximately 37.2% of the total agricultural lands in the KRI [19]. The mean daily temperature varies from 5 °C in winter to 30 °C in summer; however, this rises to 50 °C in the region's southern parts [35]. The total area of forests and pastures in the KRI is 6486.9 and 8397.2 km^2, respectively, distributed as follows: Erbil 29%, Duhok 28.7%, and Sulaimaniyah 42.3% [25,37].

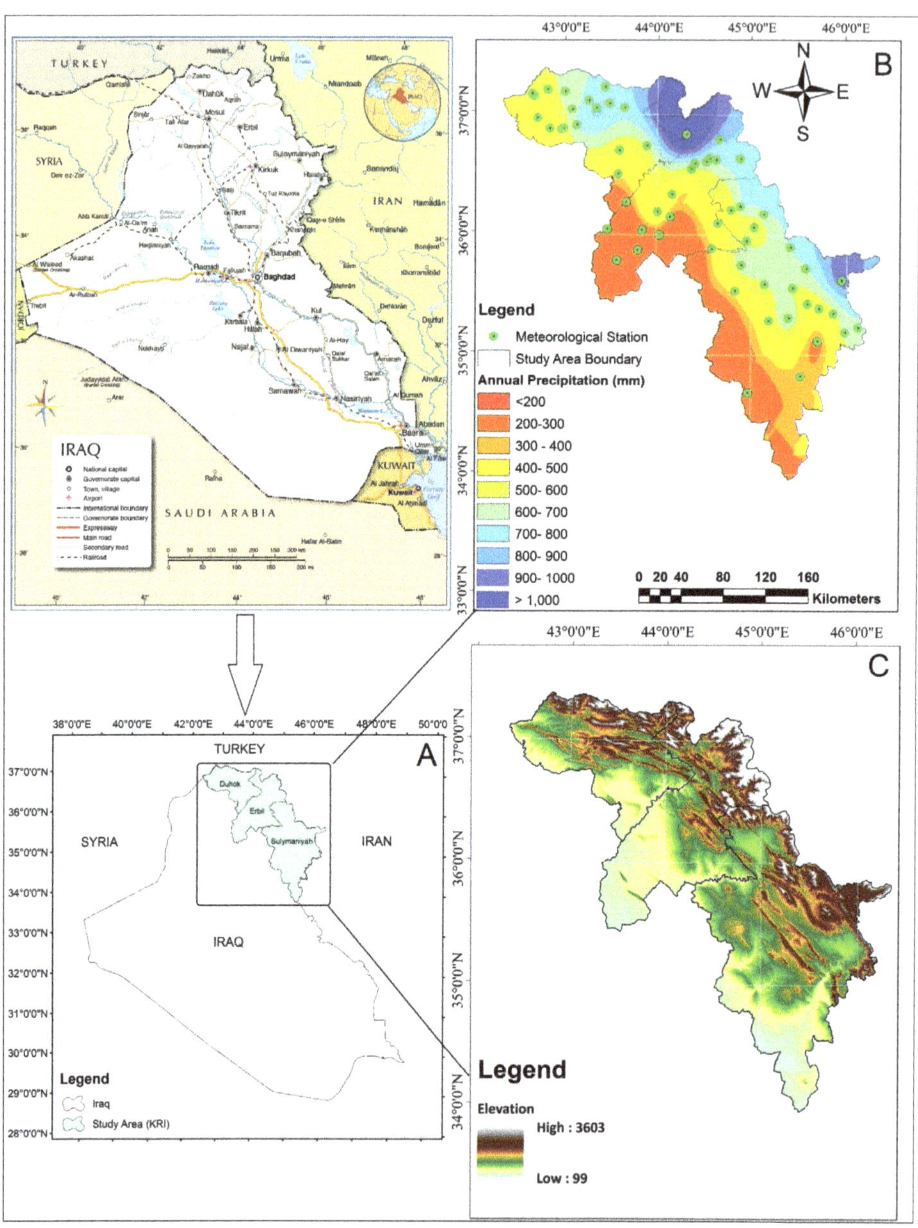

Figure 1. (**A**) Location map of the study area (**B**); The meteorological stations map and the spatial distribution of annual rainfall (mm/year) in the KRI in 1998–2017 (**C**); Digital Elevation Model (DEM).

Table 1. The (AP) annual precipitation (mm), elevation, and coordinates of the 60 (MT) meteorological stations in the KRI used in this study.

MT No.	Station Name	Lat	Long	DEM (m)	AP (mm)	MT No.	Station Name	Lat	Long	DEM (m)	AP (mm)
1	(ER)	36.19111	44.00917	412.7	326.2	31	Mangish	37.03513	43.09252	1030.2	645.0
2	Qushtapa	36.00085	44.02848	390.8	280.6	32	Deraluke	37.05859	43.64925	706.8	759.5
3	Khabat	36.27278	43.67389	285.9	290.9	33	Akre	36.74139	43.89333	683.1	600.0
4	Bnaslawa	36.1538	44.13999	540.7	320.2	34	Amadia	37.0925	43.48722	1148.5	745.7
5	Harir	36.5511	44.3648	837.3	552.2	35	Sarsink	37.05028	43.35028	957.1	841.6
6	Soran	36.63846	44.56136	701.6	625.7	36	Bamarni	37.11512	43.2693	1203.0	722.3
7	Shaqlawa	36.19111	44.00917	966.5	750.0	37	Barda	36.50822	43.58941	363.6	391.4
8	Khalifan	36.5986	44.4038	697.1	670.8	38	Qasrok	36.7009	43.59795	414.8	500.5
9	Choman	36.6374	44.8893	1178.4	732.2	39	(SU)	35.55722	45.43556	870.8	595.0
10	Sidakan	36.79736	44.6714	1011.3	822.5	40	Bazian	35.58902	45.13952	943.7	596.1
11	Rwanduz	36.61194	44.52472	801.6	712.3	41	Halabja	35.18639	45.97389	716.6	648.8
12	Mergasur	36.8382	44.3062	1108.9	1356.0	42	Penjwen	35.61972	45.94139	1442.9	968.7
13	Dibaga	35.87303	43.80496	328.3	246.2	43	Chwarta	35.71972	45.57472	1011.6	694.8
14	Gwer	36.04486	43.4808	309.7	235.3	44	Dukan	35.95417	44.95278	700.4	576.4
15	Barzewa	36.6268	44.6333	798.3	721.1	45	Qaladiza	36.1755	45.1333	628.2	681.9
16	Bastora	36.33888	44.16049	630.0	412.4	46	Rania	36.2391	44.8855	607.8	713.9
17	Makhmoor	35.7833	43.5833	287.7	228.2	47	S-sadiq	35.34369	45.85344	544.1	550.2
18	Koya	36.09944	44.64806	724.5	472.2	48	Qaradagh	35.30933	45.38961	887.9	721.7
19	Taqtaq	35.88737	44.58561	397.5	371.1	49	Arbat	35.42462	45.58683	701.6	492.5
20	Shamamk	36.0400	43.84669	310.6	276.2	50	Kani	35.38498	45.70458	685.8	498.7
21	(DU)	36.8679	42.97900	588.3	495.1	51	Byara	35.22507	46.11625	1333.5	656.3
22	Semel	36.87333	42.85400	491.6	414.4	52	Mawat	35.90074	45.4105	1063.8	712.0
23	Zakho	37.14361	42.68191	501.4	528.7	53	Darband	35.11626	45.68625	534.6	557.9
24	Batel	36.95946	42.72165	531.0	435.5	54	Chamcha	35.53333	44.83333	726.6	427.0
25	Dam-DU	36.87576	43.0029	605.6	514.2	55	Kalar	34.6411	45.32927	243.2	304.7
26	Dar. Hajam	37.19878	42.82273	649.8	509.5	56	Agjalar	35.74827	44.89741	702.3	390.0
27	zaxo-farh	37.15991	42.65873	447.1	525.2	57	Bngrd	36.06601	45.02989	841.2	666.7
28	Batifa	37.18404	37.18404	930.2	670.3	58	Sangaw	35.28623	45.1825	704.4	470.8
29	kanimasi	37.22906	37.22906	1332.3	736.2	59	Bawanor	34.82332	45.5087	358.4	364.3
30	Zaweta	36.90583	36.90583	1006.4	723.4	60	Kifri	34.68333	44.96639	238.7	279.2

The study area included Duhok (DU), Erbil (ER), and Sulaimaniyah (SU) governorates of the KRI. It is characterized by significant seasonal variations in precipitation, temperature, potential evaporation, wet winters, and dry summers (Figure 2). Most of the 586 mm precipitation amounts fall from October to May. During the study period between 1998 and 2017, the highest average monthly rainfall was 134.3 mm, in January. The highest average monthly evaporation rate was in July, with 250 mm in ER. The highest average monthly temperature recorded in July was 41.21 °C in Erbil, while the lowest monthly temperature was in January that reached 2.13 °C in SU.

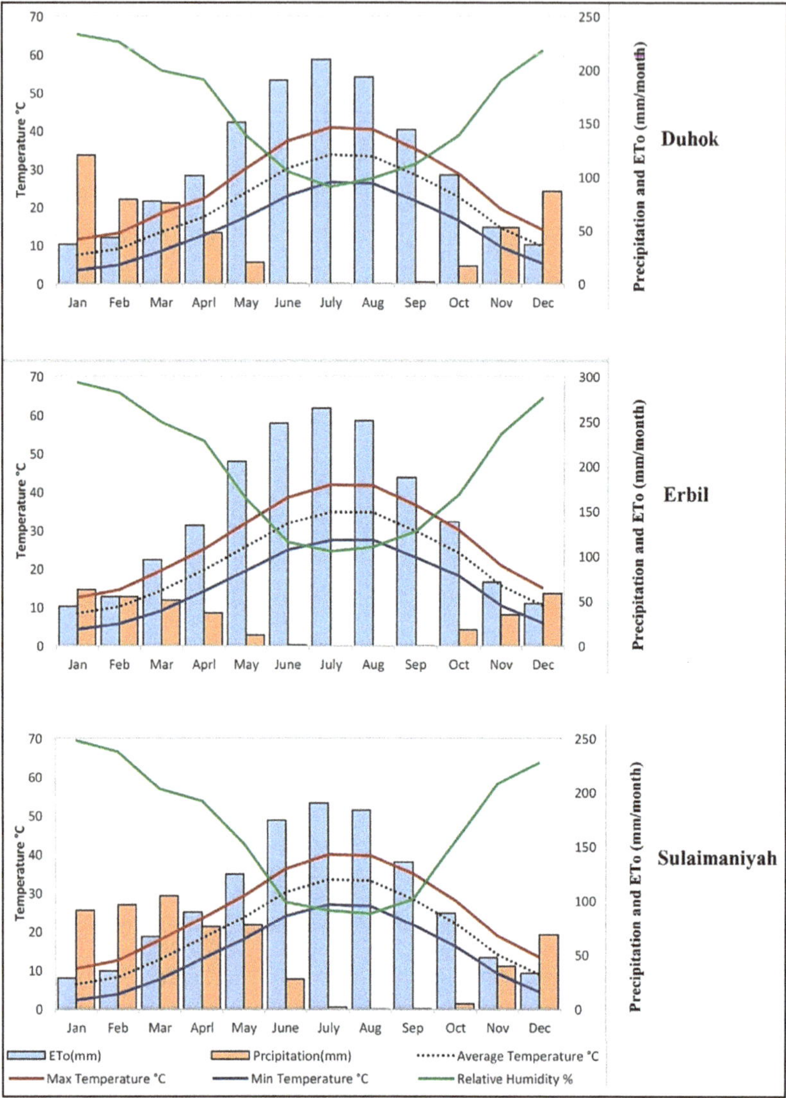

Figure 2. Monthly precipitation, relative humidity, potential evaporation, maximum, minimum, and mean temperature at Duhok (DU), Erbil (ER), Sulaimaniyah (SU), and surrounding areas recorded between 1998 and 2017.

2.2. Data

2.2.1. Landsat Datasets

For this study, 120 Landsat images were downloaded from the U.S. Geological Survey website (https://glovis.usgs.gov/, accessed on 5 January 2022). The images were acquired in April and May of 1997 to 2017, as the highest level of vegetation growth occurs every year in the two months in the study area. The remotely sensed datasets were a collection of three different sensors: L5 Thematic Mapper (TM), L7 Enhanced Thematic Mapper Plus (ETM+), and L8 Operational Land Imager (OLI) with a spatial resolution of 30 m. They were provided in geo-referenced format, cloudless, and free images type with (Path/row

170/34, 170/35, 169/35, 169/34, 168/35, 168/36). The characteristics of the images used in this study are provided in Supplementary Materials Table S1.

2.2.2. Landsat Images Preprocessing

The downloaded images were corrected by calibrating Digital Number (DN) into radiance by using the information from their metadata files. Then, the resultant images were converted into surface reflectance using Envi ver. 5.3. The images were then geo-referenced to the Universal Transverse Mercator (UTM), Zone 38 North with a World Geodetic System (WGS) 84 datum. To get good alignment of pixels in the respective images, an image-to-image registration was performed with a Root Mean Square Error (RMSE) of 0.4 pixels [38].

Six scenes of Landsat images were combined to create a mosaic covering the entire study area for each of the twenty years. The produced mosaic represents and covers the entire land in the KRI and the surrounding areas. The infrared thermal band (6th) of TM/ETM+ and Band 10 of OLI images were utilized for retrieving the LST images, while near-infrared (NIR) and red bands were also applied to calculate the NDVI images [39].

2.2.3. Image Processing
NDVI

The near-infrared (NIR) and red bands were also applied to calculate the NDVI images [39]. The NDVI index is calculated with the aid of the red (Red) and the near-infrared (NIR) bands of the Landsat images, using Formula (1), as follows:

$$NDVI = (NIR - Red)/(NIR + Red) \tag{1}$$

Theoretically, the NDVI values ranged between -1.0 and $+1.0$. However, the typical range of NDVI gauged from vegetation and other earth surface materials is between approximately -0.1 (NIR less than Red) for non-vegetated surfaces and as high as 0.9 for dense vegetative cover. The NDVI values increase with increasing green biomass, positive seasonal changes, and favorable factors (e.g., abundant precipitation) [40,41]. The NDVI-based vegetation density can be classified into three classes based on NDVI values, as shown in Table 2 The USGS remote sensing phenology states the following: Areas of barren rock, sand, or snow usually show very low NDVI values (for example, 0.1 or less) [42]. Sparse vegetation, such as shrubs and grasslands, or senescing crops may result in moderate NDVI values (approximately 0.2 to 0.5). High NDVI values (approximately 0.6 to 0.9) correspond to dense vegetation, such as that found in temperate and tropical forests or crops at their peak growth stage [41–43].

Table 2. Class Classification Standards for Description of NDVI Vegetation Cover.

Class	Class Classification Criterion
Bare soil and/or water (no vegetation)	$NDVI \leq 0$
Very Low NDVI	≤ 0.2
Low to Moderately Low NDVI	$0.2 < NDVI \leq 0.6$
Moderately High to High NDVI	$0.6 < NDVI \leq 1$

LST

The LST fraction images were produced using the Landsat thermal bands, the sixth bands of the L5 TM, L7 ETM+, and the 10–11 of L8 TIRS. Brightness temperature can be calculated using Plank's law using Top of the Atmosphere radiances obtained from TIR sensors [44]. Firstly, we calculated the changes in the five classes of droughts for the study area within 20 years (Figure 3). We then compared the changes among the five drought categories and selected the one which shows the most significant change than the other four categories as the dominating one. The fraction of lands dominated by each drought category is then counted for each period to show the temporal evolutions.

Equations used for converting digital numbers into land surface temperature are presented as follows:

Conversion of thermal DN values into satellite brightness temperature

$$TB = K2/\ln((K1/L\lambda) + 1) \tag{2}$$

One shows the largest change compared to the other four categories. The fraction of lands dominated by each drought category is then counted for each period to show the temporal evolution.

K1 = Band-specific thermal conversion constant (in watts/m^2 × srad × μm)

K2 = Band-specific thermal conversion constant (in kelvin)

Lλ is the spectral radiance at the sensor's aperture, measured in watts/(m^2 × star × μm).

Calculation of the Land Surface Temperature in Kelvin

$$T = TB/[1 + (\lambda \times TB/\varrho) \ln\varepsilon] \tag{3}$$

where λ = wavelength of emitted radiance; ϱ = h × c/σ (1.438 × 10^{-2} m·K); h = Planck's constant (6.626 × 10^{-34} J·s); σ = Boltzmann constant (1.38 × 10^{-23} J/K); c = velocity of light (2.998 × 10^{-8} m/s); ε = emissivity, which is given by the following [45]; ε = 1.009 + 0.047 ln(NDVI).

Conversion from Kelvin to Celsius

$$Tc = T - 273 \tag{4}$$

T = land surface temperature in Kelvin

Tc = land surface temperature in Celsius [44].

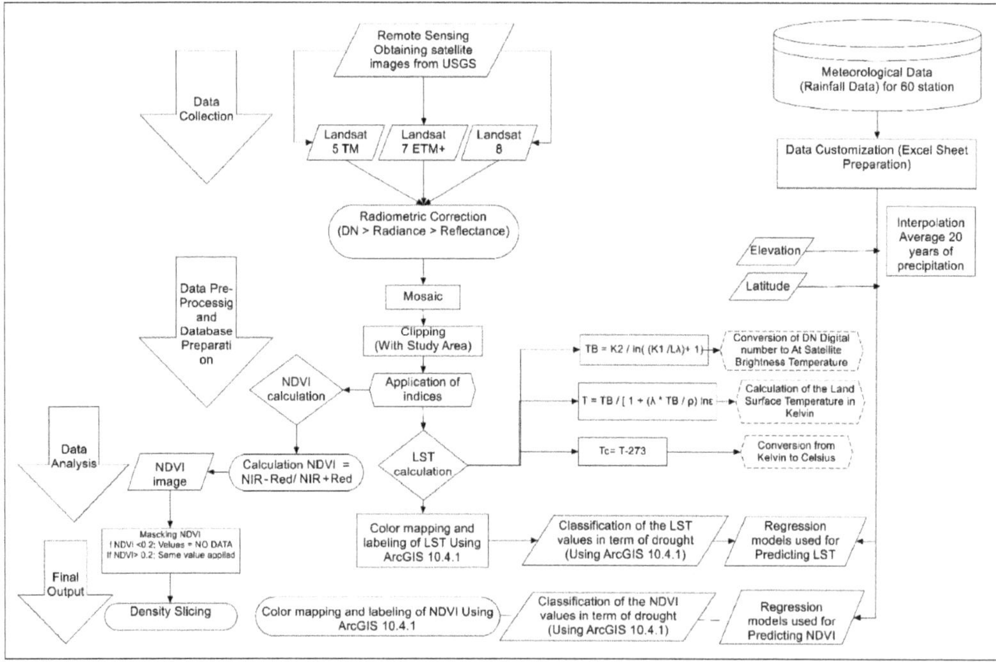

Figure 3. Flowchart of the methodology adopted in this study.

2.3. Statistical Analysis for Time Series

2.3.1. Trend Detection (Mann–Kendall Test)

The nonparametric Mann–Kendall test is commonly employed to detect monotonic trends in time series of environmental data, climate, or hydrological data [46,47]. The Mann–Kendall test is a statistical test widely used for trend analysis in climatological and hydrological time series [48]. There are two advantages of using this test: first, it is a nonparametric test and does not require data to be normally distributed. Second, the test has low sensitivity to abrupt breaks due to inhomogeneous time series [49].

The computational procedure for the Mann–Kendall test considers the time series of n data points and Ti and Tj as two subsets of data where i = 1, 2, 3, ... , n − 1 and j = i + 1, i + 2, i + 3, ... , n. The data values are evaluated as an ordered time series. Each data value is compared with all subsequent data values [46,47]. If a data value from a later time period is higher than a data value from an earlier time period, the statistic S is incremented by 1. On the other hand, if the data value from a later time period is lower than a data value sampled earlier, S is decremented by 1. The net result of all such increments and decrements yields the final value of S [50].

The Mann–Kendall's Statistic is computed as follows:

$$S = \sum_{i=1}^{n-1} \sum_{j=i+1}^{n} \mathrm{sign}(Tj - Ti) \tag{5}$$

where Tj and Ti are the annual maximum daily values in years j and i, j > i, respectively.

If n < 10, the value of | S | is compared directly to Mann–Kendall's theoretical distribution of S derived, the two-tailed test is used. At a certain probability level, H0 is rejected in favor of H1 if the absolute value of S equals or exceeds a specified value Sα/2, where Sα/2 is the smallest S, which has the probability less than α/2 to appear in the case of no trend. A positive (negative) value of S indicates an upward (downward) trend. For n $\geq$ 10, the statistic S is approximately normally distributed with the mean and variance as follows: E(S) = 0. The variance (σ2) for the S statistic is defined by the following:

$$\mathrm{sign}(Tj - Ti) = \begin{cases} 1 & \text{if } Tj - Ti > 0 \\ 0 & \text{if } Tj - Ti = 0 \\ -1 & \text{if } Tj - Ti < 0 \end{cases} \tag{6}$$

$$\sigma^2 = \frac{n(n-1)(2n+5) - \sum t_i(i)(ti-1)(2ti+5)}{18} \tag{7}$$

$$Z_s = \begin{cases} \frac{s-1}{\sigma} & \text{for } S > 0 \\ 0 & \text{for } S = 0 \\ \frac{s+1}{\sigma} & \text{for } S < 0 \end{cases} \tag{8}$$

In which ti denotes the number of ties to an extent i. The summation term in the numerator is used only if the data series contains tied values. The standard test statistic Zs is calculated as follows:

Test statistic Z is used as a measure of significance of trend. For example, if −1.96 < Z < 1.96 = No trend, Z > 1.96 = Increase in trend, Z < −1.96 = Decrease in trend [51].

2.3.2. Magnitude of Trend (Sen's Slope)

Sen's slope estimator is a nonparametric, linear slope estimator that works most efficiently on monotonic data. Different linear regression is not significantly affected by gross data errors, outliers, or missing data [47]. Sen's slope method is used to regulate the scale of the trend line. According to Sen's method, this test computes both the slope, i.e., the linear rate of change, and the intercept [51]. First, a set of linear slopes is calculated as follows:

$$dk = Xi - Xi/j - i \tag{9}$$

For $(1 \leq i < j \leq n)$, where d is the slope, X denotes the variable, n is the number of data, and I and j are indices. Sen's slope is then calculated as the median from all slopes:

$$yat = Xt - b \times t \tag{10}$$

b = Median dk. The intercepts are computed for each time step t as given by the following, and the corresponding intercept is as well as the median of all intercepts. This function also computes the Sen's slope's upper and lower confidence limits [47].

2.3.3. Pearson Correlation between Indices and Ecological Parameters

Correlation coefficients were applied for each of NDVI, LST and rainfall, elevation, and latitude for 1998 through 2017. Using bivariate correlation analysis, the strength of the statistical relationships among drought and the individual study variables were computed using SPSS. The correlation matrix allowed us to find the important statistical relationships between NDVI, LST, and the study variables, such as rainfall, elevation, and latitude. A linear relationship between observed and simulated variables was tested by the Pearson correlation coefficient. It has a value range from -1 to $+ 1$ of which the signs indicate the direction of the relationship, where the absolute value indicates the strength, whereas larger absolute values indicate stronger positive or negative associations [52].

2.3.4. Root Mean Square Error (RMSE) and Coefficient of Residual Mass (CRM)

The (RMSE), also called Root Mean Square Deviation (RMSD), is commonly used to quantify the differences between simulated and actual values, which are called residuals. The RMSE estimates the data scattering to be around a 1:1 relationship, which indicates how much the model under or overestimates the measurements. On the other side, the (CRM) value indicates the model's tendency to over or underestimate the measurements, whereas positive values indicate that the model underestimates the measurements, while negative values indicate an overestimation tendency. For an ideal prediction, RMSE and CRM values should equal 0.0 [53–59].

The RMSE of a model prediction with respect to the estimated variable X model is defined as the square root of the mean squared error:

$$RMAS = \sqrt{\frac{\sum_{i=1}^{n} \left(X_{obs,i} - X_{model,i}\right)^2}{n}} \tag{11}$$

where Xobs is the observed value, and Xmodel is the modeled value at time/place i.

$$CMR = \frac{\sum_{i=1}^{N} Pi - \sum_{i=1}^{N} Oi}{\sum_{i=1}^{N} Oi} \tag{12}$$

where Pi is the predicted, Oi is the observed, and (i = 1 to N).

3. Results

To better understand NDVI and LST patterns and their relationships, in this study, the produced thematic images were imported into ArcGIS 10.4.1. The resultant maps presented in the following pages show the spatial pattern of vegetation cover according to NDVI, LST, and the spatial distribution of annual precipitation averages from 1998 to 2017, as shown in Table 1.

3.1. NDVI

The NDVI has been widely used to examine the relationship between spectral vegetation variability and vegetation growth rate changes. This study's results revealed that NDVI values varied from the lowest value of 0.13 in 2008 to the highest value of 0.48 in 2014 (Table 3).

Table 3 shows the variation in vegetation status in the KRI from 1998 to 2017. As noted, significant decreases were observed in the area of vegetation in the KRI from 2000 to 2008 due to the extreme and severe years of drought that hit Iraq, which led to decreased agricultural land area. The total vegetation area based on NDVI in 2000 and 2008 was 7225.1 (14.4%) and 20,609.9 km^2 (41.0%), respectively. The vegetation cover has been shrunk by 39% and 13%, respectively, based on the average vegetation area (54%) over 20 years. This decline can be mainly attributed to the severe drought episodes that hit Iraq, including the KRI in 2000 and 2008, among other factors, in addition to a significant drop in rainfall averages. On the other side, the highest NDVI-based vegetation area was recorded in 2016, 32,315.2 km^2 (64.2%), representing an increase of 10% based on the vegetation cover average. From the viewpoint of NDVI values, the lowest values were recorded in 2000, 2008, and 2012 at 0.196, 0.131, and 0.202, respectively. Table 3 shows the area of the NDVI-based vegetation density classes in KRI from 1998 to 2017. In class 1, the results revealed that the largest class area was recorded in 2000 and 2008 by 6050.0 (83.7%) and 16,453.7 km^2 (79.8%), respectively; in addition, the lowest area from class 2 at values 0.2–0.6 were recorded in 2000 and 2008 at 1175.1 (16.3%) and 4156.1 km^2 (20.2%), respectively.

The NDVI results are presented in Table 3 and Figure 4, which show the spatial variation of the NDVI-based vegetation classes in the study area from 1998 to 2017. The maps show the impact of drought on the vegetation density in the KRI, whereas it severely impacted some parts of the southern KRI, while there was no impact (no drought) or a slight drought in the northeast parts of the study area. The NDVI results showed that the drought intensity in the KRI gradually increases toward the southwest parts. The drought regions belonging to class 1 (values < 0.2) are a large and continuous distribution. Table 3 and Figure 4 display the actual drought status episodes in 2000, 2008, and 2012 in the KRI. Precisely, the NDVI-based low-vegetation class increased in the three drought years to be 6050.0 (83.7%), 16,453.7 (79.8%), and 14,024.1 km^2 (53.1%) in 2000, 2008, and 2012, respectively. The maps in Figure 4 disclose that the years 2000 and 2008 were the drier years in the KRI, particularly in the southern parts.

Table 3. The max, min, mean, and std. dev. of NDVI values and the area of vegetative cover and the NDVI-Based Vegetation Density Classes in KRI from 1997 to 2017.

					Class 1		Class 2		Class 3					
					Values < 0.2		0.2 < Values ≤ 0.6		0.2 < Values < 1					
					Very Low NDVI		Low to Moderately Low NDVI		Moderately High to High NDVI		Total Vegetative Cover			
Years	Max.	Min.	Mean	Std. Dev.	Area (km^2)	Area (%)	Area (km^2)	Area (%)	Area (km^2)	Area (%)	(km^2)	(%)	(±%)	Total Study Area (km^2)
1998	0.99	0.10	0.27	0.13	9890.0	37.5	16,075.4	60.9	417.8	1.6	26,383.2	52.4	−1.6	53,000
1999	0.98	0.10	0.23	0.10	12,881.7	46.1	14,994.2	53.7	70.2	0.3	27,946.0	55.5	1.5	53,000
2000	0.99	0.02	0.20	0.13	6050.0	83.7	1175.1	16.3	0	0.0	7225.1	14.4	−39	53,000
2001	0.73	0.03	0.22	0.13	14,859.6	50.0	14,707.5	49.5	169.3	0.6	29,736.4	59.1	5	53,000
2002	0.73	0.06	0.23	0.12	14,320.6	47.6	15,741.6	52.3	51.3	0.2	30,113.5	59.8	5.8	53,000
2003	0.72	0.05	0.24	0.12	12,635.4	43.6	16,319.1	56.3	49.4	0.2	29,003.9	57.6	3.6	53,000
2004	0.72	0.04	0.21	0.12	15,076.6	49.9	15,109.7	50.0	11.3	0.0	30,197.6	60	6	53,000
2005	0.73	0.06	0.20	0.10	14,704.7	55.7	11,702.8	44.3	10.9	0.0	26,418.4	52.5	−1.5	53,000
2006	0.78	0.02	0.21	0.14	14,744.0	51.7	13,699.3	48.1	67.8	0.2	28,511.1	56.7	2.6	53,000
2007	0.73	0.11	0.29	0.11	7802.9	25.7	22,419.1	73.9	110.2	0.4	30,332.3	60.3	6.2	53,000
2008	0.64	0.02	0.13	0.09	16,453.7	79.8	4156.1	20.2	0.1	0.0	20,609.9	41	−13	53,000
2009	0.85	0.08	0.26	0.11	9091.2	36.3	15,910.7	63.6	35.4	0.1	25,037.3	49.7	−4.3	53,000
2010	0.72	0.13	0.28	0.11	7873.5	27.3	20,994.2	72.7	27	0.1	28,894.7	57.4	3.4	53,000
2011	0.76	0.06	0.22	0.13	15,185.5	56.4	11,670.7	43.4	61.3	0.2	26,917.5	53.5	−0.6	53,000
2012	0.72	0.01	0.20	0.13	14,024.1	53.1	12,340.4	46.7	36.9	0.1	26,401.4	52.5	−1.6	53,000
2013	0.63	0.16	0.29	0.09	4636.1	16.5	23,491.1	83.5	0.3	0.0	28,127.6	55.9	1.8	53,000
2014	1.00	0.29	0.48	0.12	5674.20	18.4	25,152.7	81.6	0.0	0.0	30,826.8	61.3	7.2	53,000
2015	0.64	0.18	0.31	0.08	2076.6	6.5	29,782.4	93.5	2.2	0.0	31,861.2	63.3	9.3	53,000
2016	0.72	0.18	0.30	0.08	2984.6	9.2	29,325.5	90.8	5.1	0.0	32,315.2	64.2	10.2	53,000
2017	0.64	0.18	0.28	0.07	21.4	0.1	26,111.7	96.4	963.6	3.6	27,096.8	53.8	−0.2	53,000

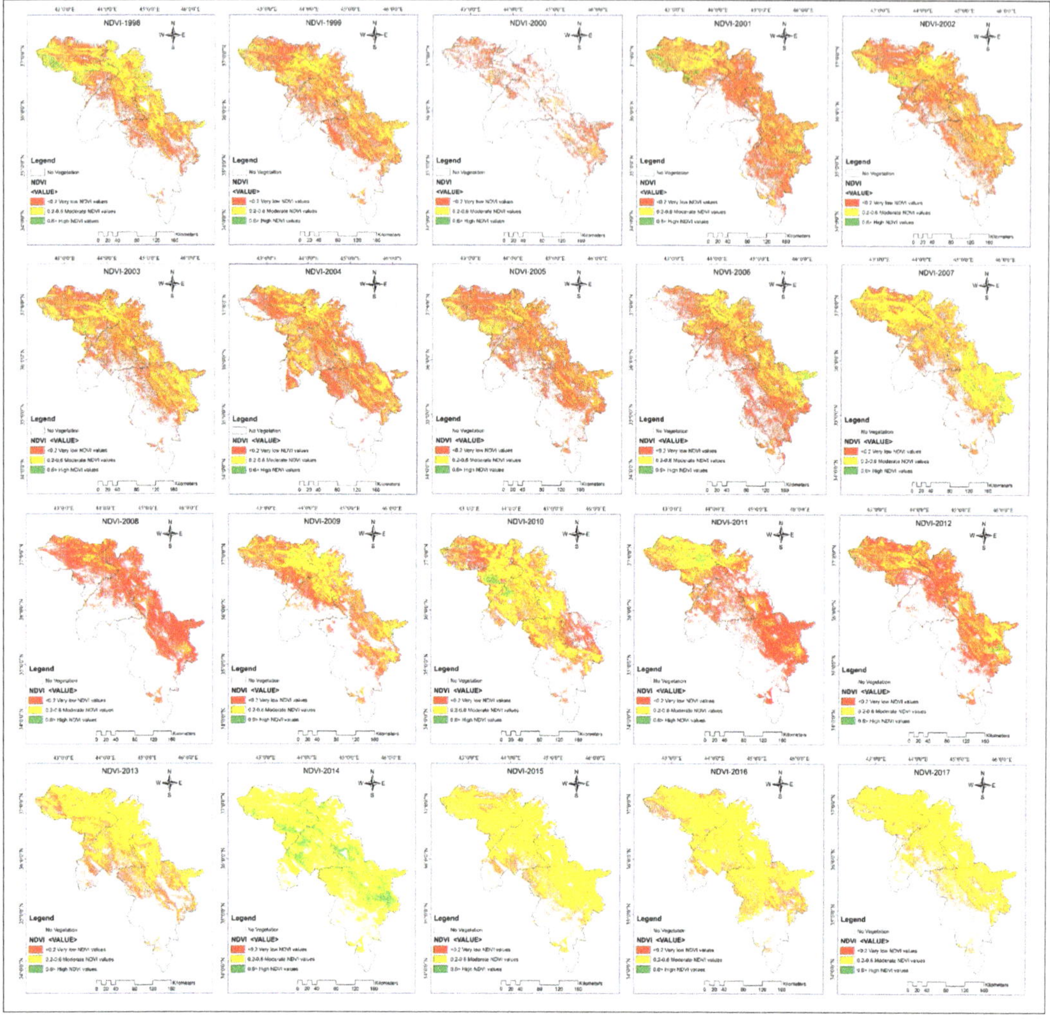

Figure 4. Spatial Variation of the NDVI-Based Vegetation in 1998 to 2017.

The NDVI values were the lowest in the southwest and west parts of the region compared to the northeastern parts, with higher vegetation and a higher NDVI value (Figure 4). There was a significant decline in annual rainfall averages in some sites in KRI compared to rainfall averages of the other studied locations from 1998 to 2017. Figures 5 and 6 shows the minimum values of NDVI-based vegetation cover due to the changes in precipitation rates, whereas precipitation averages were low in some locations (Table 1). On the other side, the precipitation averages were high in some sites, which positively reflected the increase in NDVI values (Figure 4). Low precipitation and high temperature play a negative role in decreasing NDVI values and vegetation cover in the southwest parts of KRI during growing seasons.

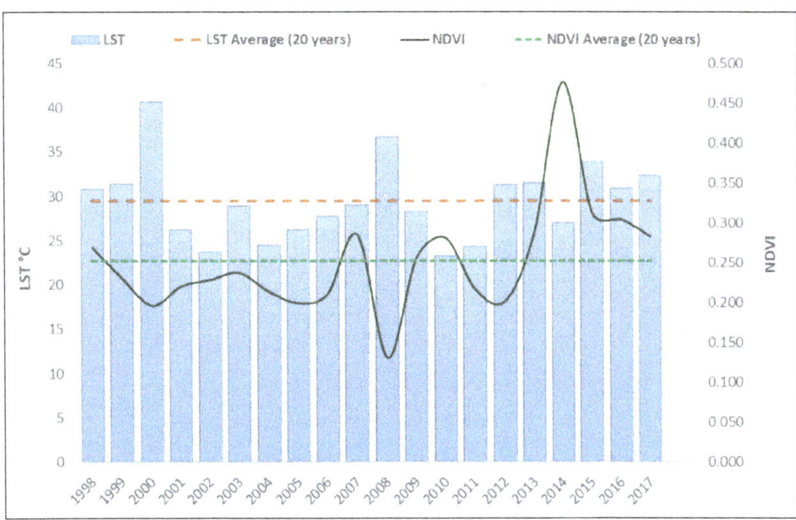

Figure 5. Average values of LST and NDVI in the study area (1998–2017).

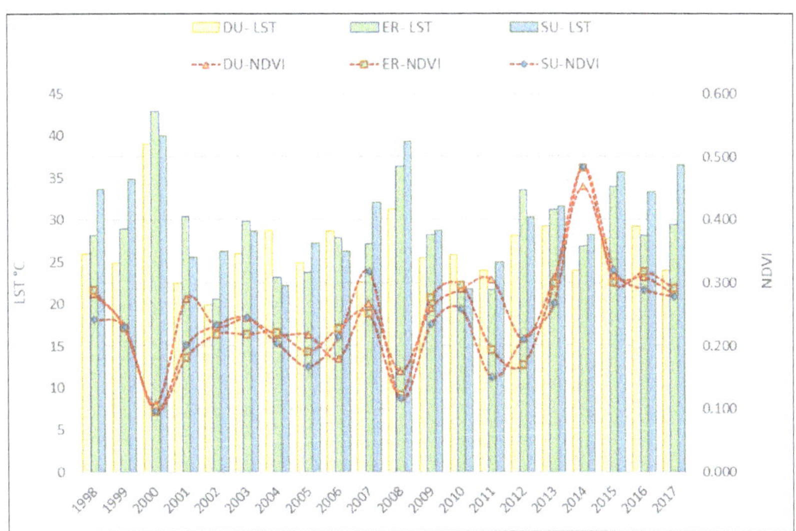

Figure 6. Average values of LST and NDVI in the ER, SU, and DU governorates (1998–2017).

3.2. LST

The LST fraction images were derived using thermal infrared (TIR) of Landsat imagery, which can be utilized to express the land surface temperature and indicate drought status [60]. The LST status of the study area in the period 1998–2017 is given in Figures 5 and 6 and Table 4. Firstly, we calculated the changes in the five classes of droughts for the study area in 20 years (Figure 3). In this study, the changes in the five drought categories were compared. The category in which the greatest change occurred was compared to the other four categories, and it was considered a comparative treatment. Then, we calculated each drought category area to show the temporal changes. Figure 5 shows the LST mean values of each year of the study period compared with the LST average of the 20 years in the KRI from 1998 to 2017. The temperature rate of KRI showed a steady increase, but the degree of temperature in the years 2002 (23 °C) and 2010 (22 °C) experienced a downward trend. On the other hand, the LST degree of 2000 and 2008 in KRI was about 41 and 37 °C, respectively. The LST rate increased sharply throughout the period, exceeding those of the years 2002 and 2010.

Table 4 indicates that in 2000 and 2008, the highest LST value area was in class 5, which was more than 40 °C. The study results revealed that a very severe drought hit 27,660.7 km^2 (55.0%) of the total area in class 5 $\geq$ 40 °C. While in the year 2008, they faced severe heat in 19,261.1 km^2 (38.23%) of class 5 $\geq$ 40 °C. However, the lowest temperatures were recorded in 2002, 2003, 2004, 2010, 2011, and 2014, which was no higher than 0.4% of the total study area.

Table 4. LST Categories Derived from Landsat Thermal Bands for the Years 1998–2000 and Drought Severity Areas (in km^2) and Percentage based on the LST Index.

Year	Class 1 <10 °C		Class 2 10–20 °C		Class 3 20–30 °C		Class 4 30–40 °C		Class 5 > 40 °C	
	Area (km^2)	Area (%)	Area (km^2)	Area (%)	Area (km^2)	Area (%)	Area (km^2)	Area (%)	Area (km^2)	Area (%)
1998	864.3	1.7	2914.5	5.8	19,465.7	38.7	23,901.7	47.5	3181.5	6.3
1999	972.8	1.9	4379.3	8.7	15,646.5	31.1	23,061.8	45.8	6267.1	12.5
2000	424.9	0.8	321.0	0.6	3135.3	6.2	18,785.7	37.3	27,660.7	55.0
2001	589.6	1.2	8546.9	17.0	26,527.3	52.7	14,639.5	29.1	24.2	0.0
2002	1892.9	3.8	14,647.2	29.1	25,793.2	51.3	7,968.2	15.8	26.0	0.1
2003	424.9	0.8	3509.7	7.0	26,701.8	53.1	19,629.0	39.0	62.2	0.1
2004	2106.9	4.2	10,379.9	20.6	31,040.3	61.7	6,588.5	13.1	211.9	0.4
2005	1208.7	2.4	6545.3	13.0	32,586.4	64.7	9,785.2	19.4	202.0	0.4
2006	291.9	0.6	3702.6	7.4	32,097.4	63.8	14,184.1	28.2	51.6	0.1
2007	388.6	0.8	4378.6	8.7	28,483.9	56.6	13,110.3	26.0	3966.2	7.9
2008	881.4	1.8	1547.2	3.1	8150.3	16.2	20,487.5	40.7	19,261.1	38.3
2009	530.7	1.1	6669.1	13.3	25,221.3	50.1	15,938.3	31.7	1968.1	3.9
2010	1471.7	2.9	14,943.1	29.7	29,135.4	57.9	4,642.8	9.2	134.6	0.3
2011	1021.7	2.0	11,501.9	22.9	31,743.5	63.1	6,038.5	12.0	22.0	0.0
2012	223.7	0.4	1737.1	3.5	19,951.8	39.6	22,213.8	44.1	6201.3	12.3
2013	219.6	0.4	1692.6	3.4	17,799.6	35.4	30,154.3	59.9	461.5	0.9
2014	1148.9	2.3	4866.5	9.7	27,964.0	55.6	16,251.3	32.3	96.8	0.2
2015	478.2	1.0	1213.3	2.4	14,941.8	29.7	25,203.0	50.1	8491.2	16.9
2016	810.0	1.6	1483.8	2.9	22,040.9	43.8	21,873.4	43.5	4119.5	8.2
2017	1895.0	3.8	3650.5	7.3	15,964.2	31.7	20,770.0	41.3	8048.0	16.0

Over 20 years, vast areas in the southern part of Erbil and Sulaimaniyah governorates were affected by very severe drought episodes, while most of the other parts of the study area were characterized by slight and moderate droughts based on LST (Figure 7). The southern parts of the study area were warmer compared to the other parts. The mean values of LST in 2000 and 2008 were 40.7 and 36.0 °C, respectively. This significant increase in surface temperature is due to the lack of rainfall, which led to a lack of moisture and lower vegetation cover area that was mostly found in lands with lower elevation. One of the results of the increase in LST surface temperature in the southeast and southwest of the study area is the decrease in vegetation cover represented by NDVI in the study area. This indicates the negative impact of the high LST on the vegetation growth environment that led to the shrinkage in the vegetation area (NDVI) of the region. In the northeast of the study area, only a few sites had increases in precipitation rates and decreases in the LST values, which in turn was reflected in a vegetation increase (NDVI) at those sites (Figures 5 and 6).

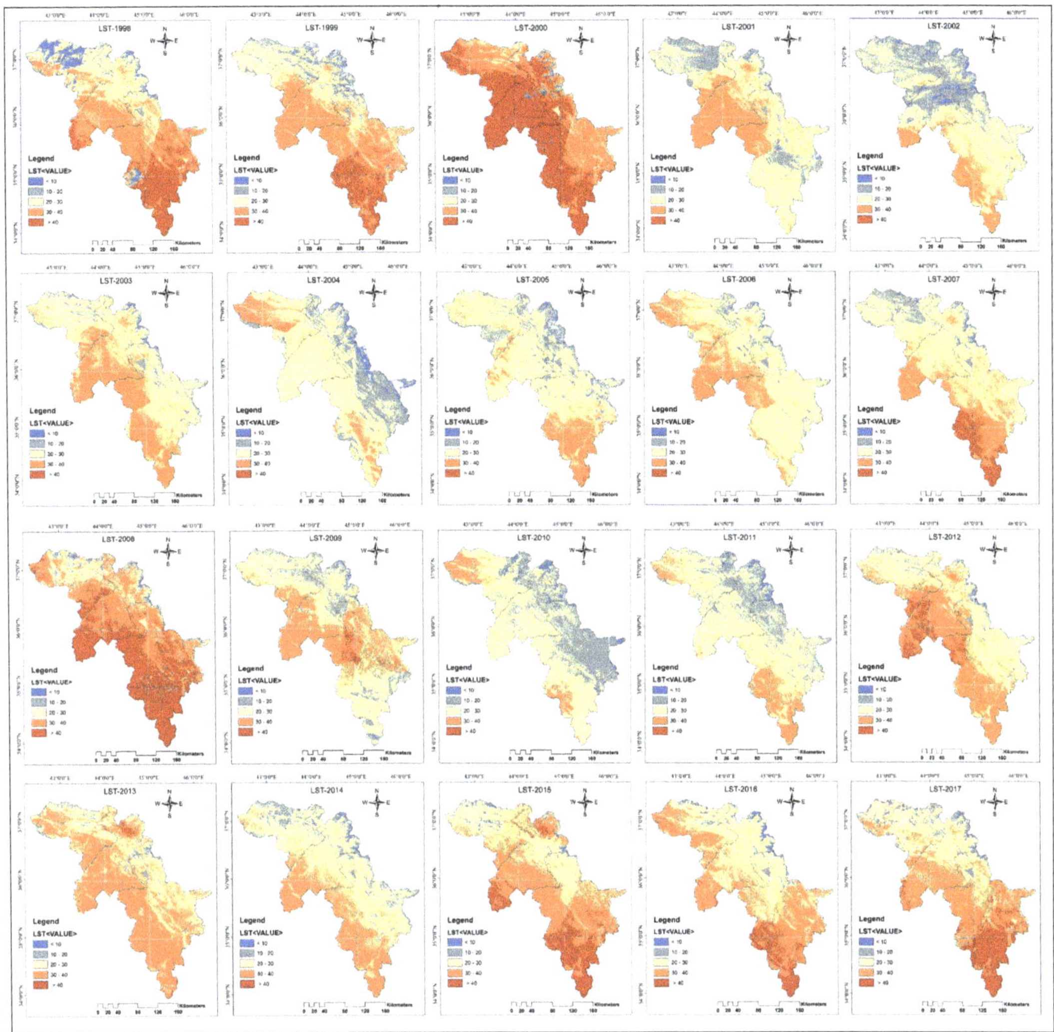

Figure 7. Drought Severity Categories based on LST Index in Years 1998–2017.

3.3. Pearson Correlation Matrix between Indices and Ecological Parameters

Correlation coefficients between latitude, elevation, rainfall, NDVI, and LST during the years from 1998 to 2017 (average of 20 years) are calculated using SPSS and are presented in Table 5. The analysis of variance for the drought indices showed significant differences at $p < 0.01$ and $p < 0.05$ among the analyzed years. The relationship between precipitation, elevation, NDVI, and LST was tested from 1998 to 2017 through Pearson correlation analysis, and the results are presented in Table 5 and Figures 8 and 9. The results showed a significant negative correlation between NDVI and precipitation with LST. On the other hand, there was a positive correlation between NDVI and precipitation (Table 5 and Figures 8 and 9). The correlation between spectral indices based on remote sensing and precipitation was statistically significant. LST and NDVI space's concept refers to the relationship between NDVI with LST, and vegetation abundance was first formulated by Lambing and Ehrlich (1996) with LST plotted as a function of NDVI [61]. In Figures 5 and 6, the lowest values of NDVI were observed in 2000 and 2008 with higher LST during 2000 and 2008 in ER, DU, and SU (Figure 5). The relationship between the mean and area of NDVI and LST is repeatedly negative.

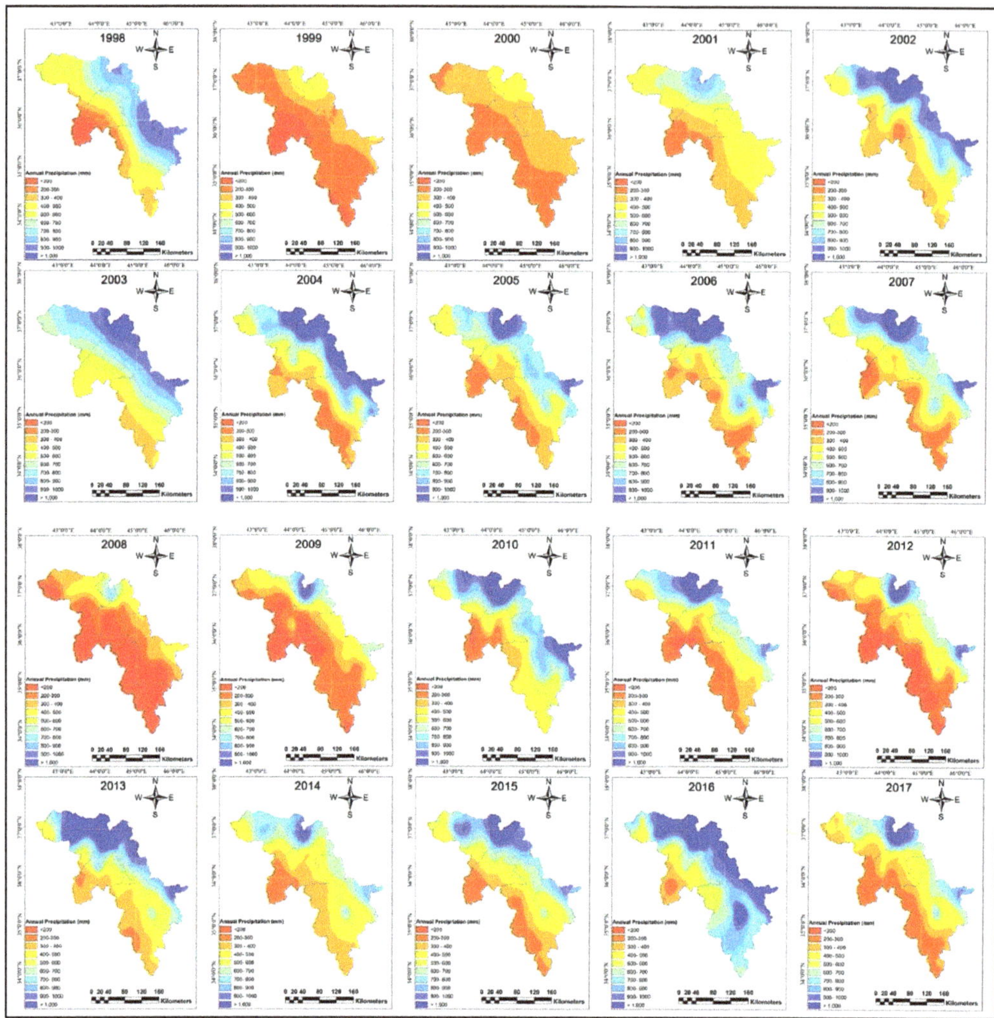

Figure 8. The spatiotemporal distribution of annual precipitation (mm/year) in IKR during the period 2008–2017.

Table 5. Pearson correlation between NDVI, LST, and ecological parameters.

	Longitude	Latitude	Elevation	Rainfall	LST	NDVI
Longitude	1					
Latitude	−0.81 **	1				
Elevation	0.2	0.25	1			
Rainfall	0.14	0.34 **	0.80 **	1		
LST	0.14	−0.59 **	−0.78 **	−0.83 **	1	
NDVI	−0.03	0.53 **	0.76 **	0.83 **	−0.89 **	1

** Correlation is significant at the 0.01 level (2-tailed). * Correlation is significant at the 0.05 level (2-tailed).

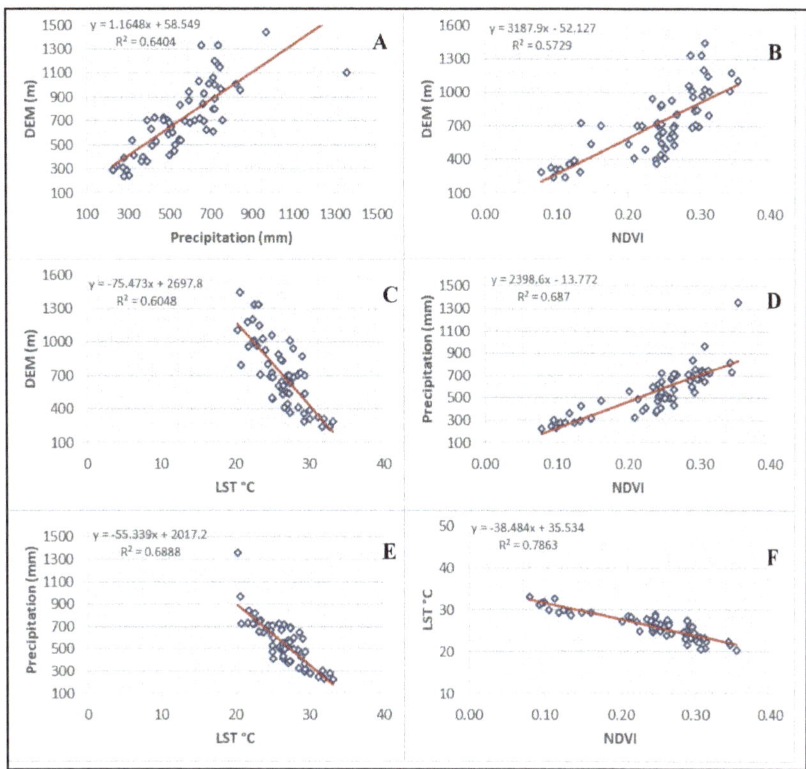

Figure 9. Spatial pattern changes of ecological parameters and drought indices for an average of 20 years in 60 locations. (**A**) Relationships between elevation precipitation. (**B**) Relationships between elevation and NDVI. (**C**) Relationships between elevation and LST. (**D**) Relationships between precipitation and NDVI. (**E**) Relationships between precipitation and LST. (**F**) Relationships between LST and NDVI.

Analysis of variance for the drought indices showed significant differences at $p < 0.01$ and $p < 0.05$ among the analyzed years. The relationship between precipitation, elevation, NDVI, and LST was tested from 1998 to 2017 through Pearson correlation analysis, and the results are presented in Table 5 and Figure 9. The results showed a significant negative correlation between NDVI and precipitation with LST. On the other hand, there was a positive correlation between NDVI and precipitation (Table 5 and Figure 9). The correlation between spectral indices based on remote sensing and precipitation was statistically significant. LST and NDVI space's concept refers to the relationship between NDVI with LST and vegetation abundance was first formulated by Lambing and Ehrlich (1996) with LST plotted as a function of NDVI [61]. In Figures 5 and 6, the lowest values of NDVI were observed in 2000 and 2008 with higher LST during 2000 and 2008 in ER, DU, and SU (Figure 5). The relationship between the mean and area of NDVI and LST is repeatedly negative.

The spatial distributions of the changes in (elevation, precipitation, LST, and NDVI) from 1997 to 2017 are presented in Figure 9A–F. The negative relationships between (NDVI-LST) and (LST-DEM) based on the monthly data from 1998–2017 at 60 different sites were presented in Figure 9C,F. The statistical correlation between (LST-NDVI) was employed to demonstrate the locational variations of temperature effect on vegetation activity. The study's findings also revealed a positive correlation between the LST and NDVI in the northern part of the study area, while the relationship was negative between the mentioned indices at the southern parts.

In general, the three main factors affecting vegetation growth in the study area are the LST, precipitation, and DEM (Figure 8B,D,F). That illustrates the significant shrinking in precipitation averages, the vegetation cover in the southwest, and the considerable increase of vegetation coverage in some of the KRI's northeast parts (Figure 9A,B). Figure 6 shows the significant decreases in LST observed in the country's northeast, caused by increasing NDVI and the precipitation rate. A significant increase was found in the LST values in almost all sites southwest of the KRI, where the vegetation and precipitations are limited. NDVI was sensitive to rainfall and temperature (Figure 9D,E). However, fluctuations were observed in NDVI and LST during the 20 years. The decreases in NDVI were observed during the period LST was increasing in all locations (Figure 9F).

3.4. Trend Analysis of NDVI and LST by Mann–Kendall and Sen's Slope

This study carried out the trend analysis for NDVI and LST in the 60 meteorological stations' locations from 1998 to 2017. MKT and Sen's Slope estimator were used to determine statistical inclining or declining trends. A positive sign indicates an upward slope, while a negative sign represents a downward one. The Sen's slope test results seem to be fairly similar to those obtained from the MKT [52,62].

3.4.1. NDVI

Table 6 indicates the NDVI trends in the KRI through 20 years using the Mann–Kendall test and Sen's slope methods. Out of 60 locations, only 11 recorded significant trends increasing at the 5% level of Sen's estimator of slope following the Mann–Kendall test, which was employed to figure out the change per unit time of trends observed in all NDVI time series. Trends of NDVI have been calculated for each site individually using Sen's magnitude of slope (Q). In the Mann–Kendall test, the Z statistics revealed that the series covers the KRI study area.

The majority of NDVI-based vegetation increases occurred in the northern and northeastern parts. Table 6 reveals the trend analysis results that statistically significant (95% confidence level) positive trends were 2.34, 2.08, 2.21, 2.24, 2.66, 2.24, 3.47, 2.08, 2.17, 2.11, 2.11, 2.5, and 2.17 for Northeast sites, including Khabat, Mergasurer, Barzewa Battle, Zawiya, Mangesh, Kanimasi, Amadea, Bamarni, Bazian, Halabja, Byara, and Mawat.

Table 6. NDVI Trends in the KRI over the 20 Years using the Mann–Kendall Test and Sen's Slope Methods.

	Mann–Kendall Trends						Sen's Slope
Time Series Location Name	First Year	Last Year	N	Test Z	Sen's Slope (Q)	Prop.	Trend (at 95% Level of Significance)
Erbil	1998	2017	20	0.68	0.002	0.7522	no trend
Qushtapa	1998	2017	20	1.52	0.005	0.9364	no trend
Khabat	1998	2017	20	2.34	0.008	0.9903	increasing
Bnaslawa	1998	2017	20	1.91	0.006	0.9722	no trend
harir	1998	2017	20	1.65	0.006	0.9510	no trend
Soran	1998	2017	20	1.52	0.006	0.9364	no trend
Shaqlawa	1998	2017	20	1.72	0.005	0.9572	no trend
Khalifan	1998	2017	20	1.65	0.006	0.9510	no trend
choman	1998	2017	20	1.36	0.003	0.9135	no trend
Sidakan	1998	2017	20	1.40	0.004	0.9185	no trend
Rwanduz	1998	2017	20	1.56	0.005	0.9403	no trend
Mergasur	1998	2017	20	2.08	0.007	0.9811	increasing

Table 6. *Cont.*

| Time Series Location Name | Mann–Kendall Trends | | | | | | Sen's Slope |
	First Year	Last Year	N	Test Z	Sen's Slope (Q)	Prop.	Trend (at 95% Level of Significance)
Dibaga	1998	2017	20	1.20	0.004	0.8850	no trend
Gwer	1998	2017	20	1.04	0.003	0.8504	no trend
barzewa	1998	2017	20	2.21	0.006	0.9863	increasing
Bastora	1998	2017	20	0.97	0.002	0.8348	no trend
Makhmoor	1998	2017	20	1.23	0.004	0.8912	no trend
Koya	1998	2017	20	1.49	0.004	0.9322	no trend
Taqtaq	1998	2017	20	1.91	0.006	0.9722	no trend
Shamamk	1998	2017	20	0.78	0.003	0.7819	no trend
Duhok	1998	2017	20	1.91	0.004	0.9722	no trend
semel	1998	2017	20	1.30	0.005	0.9028	no trend
Zakho	1998	2017	20	1.20	0.003	0.8850	no trend
Batel	1998	2017	20	2.24	0.004	0.9874	increasing
Duhok	1998	2017	20	1.56	0.005	0.9403	no trend
Darkar	1998	2017	20	1.69	0.005	0.9542	no trend
zaxo-farh	1998	2017	20	0.42	0.002	0.6634	no trend
Batifa	1998	2017	20	1.82	0.006	0.9654	no trend
kani masi	1998	2017	20	3.47	0.012	0.9997	no trend
Zaweta	1998	2017	20	2.66	0.007	0.9961	increasing
Mangish	1998	2017	20	2.24	0.008	0.9874	increasing
Deraluke	1998	2017	20	1.98	0.008	0.9761	no trend
Akre	1998	2017	20	1.46	0.004	0.9279	no trend
Amadia	1998	2017	20	2.08	0.005	0.9811	increasing
Sarsink	1998	2017	20	1.20	0.003	0.8850	no trend
Bamarni	1998	2017	20	2.17	0.008	0.9851	increasing
Bardarash	1998	2017	20	0.94	0.003	0.8266	no trend
Qasrok	1998	2017	20	1.78	0.005	0.9628	no trend
SUL	1998	2017	20	1.98	0.005	0.9761	no trend
Bazian	1998	2017	20	2.11	0.005	0.9825	increasing
Halabja	1998	2017	20	2.11	0.005	0.9825	increasing
Penjwen	1998	2017	20	1.91	0.008	0.9722	no trend
Chwarta	1998	2017	20	1.40	0.006	0.9185	no trend
Dukan	1998	2017	20	1.40	0.004	0.9185	no trend
Qaladiza	1998	2017	20	1.27	0.003	0.8971	no trend
Rania	1998	2017	20	1.36	0.003	0.9135	no trend
Said sadiq	1998	2017	20	1.59	0.005	0.9441	no trend
Qaradagh	1998	2017	20	1.33	0.003	0.9083	no trend
Arbat	1998	2017	20	0.91	0.003	0.8182	no trend
mwan	1998	2017	20	1.65	0.004	0.9510	no trend
Byara	1998	2017	20	2.50	0.008	0.9938	increasing
Mawat	1998	2017	20	2.17	0.004	0.9851	increasing
Darbandik	1998	2017	20	1.91	0.005	0.9722	no trend
Chamcha	1998	2017	20	1.20	0.004	0.8850	no trend
Kalar	1998	2017	20	0.97	0.001	0.8348	no trend
Agjalar	1998	2017	20	0.55	0.002	0.7094	no trend
bngrd	1998	2017	20	1.62	0.004	0.9476	no trend
Sangaw	1998	2017	20	1.46	0.005	0.9279	no trend
Bawanor	1998	2017	20	1.49	0.003	0.9322	no trend
Kifri	1998	2017	20	0.71	0.002	0.7623	no trend

Note: $-1.96 < Z < 1.96$ = No trend, $Z > 1.96$ = Increase in trend, $Z < -1.96$ = Decrease in trend.

3.4.2. LST

Table 7 illustrates that a significant trend in LST was 2.04, 2.08, 2.17, 2.01, 1.98, 2.37, 1.98, 2.01, 2.01, and 2.5, for Southwest sites, for Erbil, Qushtapa, Dibaga Gwer, Shamamk, Makhmoor, Mangish, Chamchamal, Kalar, Bawanor and Kifri, respectively. On the other hand, the lower trends were in Northeast sites, including Mangish—2.8, Bamarni—2.11, Penjwen—2.95, Chwarta—2.21, and Byara—2.3.

Table 7. LST trends in the KRI over the 20 years using the Mann–Kendall Test and Sen's slope methods.

	Mann–Kendall Trends						Sen's Slope
Time Series Location Name	First Year	Last Year	N	Test Z	Sen's Slope (Q)	Prop.	Trend (At 95% Level of Significance)
Erbil	1998	2017	20	2.04	0.456	0.9795	increasing
Qushtapa	1998	2017	20	2.08	0.492	0.9811	increasing
Khabat	1998	2017	20	1.10	0.203	0.8650	no trend
Bnaslawa	1998	2017	20	0.71	0.114	0.7623	no trend
harir	1998	2017	20	0.68	0.125	0.7522	no trend
Soran	1998	2017	20	1.07	0.150	0.8578	no trend
Shaqlawa	1998	2017	20	0.58	0.066	0.7204	no trend
Khalifan	1998	2017	20	−0.06	0.000	0.4741	no trend
choman	1998	2017	20	−1.75	−0.242	0.0399	no trend
Sidakan	1998	2017	20	−0.39	−0.051	0.3485	no trend
Rwanduz	1998	2017	20	−0.58	−0.100	0.2796	no trend
Mergasur	1998	2017	20	−1.82	−0.698	0.0346	no trend
Dibaga	1998	2017	20	2.17	0.450	0.9851	increasing
Gwer	1998	2017	20	2.01	0.172	0.9779	increasing
barzewa	1998	2017	20	−1.01	−0.114	0.1573	no trend
Bastora	1998	2017	20	1.85	0.366	0.9678	no trend
Makhmoor	1998	2017	20	2.37	0.264	0.9911	no trend
Koya	1998	2017	20	−1.40	−0.260	0.0815	no trend
Taqtaq	1998	2017	20	0.06	0.010	0.5259	no trend
Shamamk	1998	2017	20	1.98	0.179	0.9761	increasing
Duhok	1998	2017	20	−0.13	−0.025	0.4484	no trend
semel	1998	2017	20	0.13	0.009	0.5516	no trend
Zakho	1998	2017	20	0.10	0.020	0.5388	no trend
Batel	1998	2017	20	−0.03	−0.001	0.4871	no trend
Duhok Dam	1998	2017	20	0.06	0.014	0.5259	no trend
Darkar hajam	1998	2017	20	0.94	0.183	0.8266	no trend
zaxo−farh	1998	2017	20	−1.75	−0.375	0.0399	no trend
Batifa	1998	2017	20	−0.52	−0.087	0.3018	no trend
kani masi	1998	2017	20	−1.52	−0.563	0.0636	no trend
Zaweta	1998	2017	20	−1.33	−0.400	0.0917	no trend
Mangish	1998	2017	20	−2.08	−0.470	0.0189	Decreasing
Deraluke	1998	2017	20	−0.42	−0.065	0.3366	no trend
Akre	1998	2017	20	−1.69	−0.375	0.0458	no trend
Amadia	1998	2017	20	−1.07	−0.240	0.1422	no trend
Sarsink	1998	2017	20	−1.10	−0.285	0.1350	no trend
Bamarni	1998	2017	20	−2.11	−0.717	0.0175	Decreasing
Bardarash	1998	2017	20	0.23	0.031	0.5898	no trend
Qasrok	1998	2017	20	0.29	0.056	0.6149	no trend
Sulaymaniyah	1998	2017	20	0.29	0.045	0.6149	no trend
Bazian	1998	2017	20	−0.78	−0.192	0.2181	no trend
Halabja	1998	2017	20	0.84	0.183	0.8005	no trend
Penjwen	1998	2017	20	−2.95	−0.662	0.0016	Decreasing
Chwarta	1998	2017	20	−2.21	−0.540	0.0137	Decreasing
Dukan	1998	2017	20	0.52	0.065	0.6982	no trend
Qaladiza	1998	2017	20	−1.52	−0.342	0.0636	no trend
Rania	1998	2017	20	−0.32	−0.087	0.3728	no trend
Said sadiq	1998	2017	20	0.42	0.120	0.6634	no trend
Qaradagh	1998	2017	20	−0.23	−0.023	0.4102	no trend
Arbat	1998	2017	20	0.42	0.111	0.6634	no trend
mwan	1998	2017	20	−0.13	−0.034	0.4484	no trend
Byara	1998	2017	20	−2.30	−0.502	0.0106	Decreasing
Mawat	1998	2017	20	−1.85	−0.468	0.0322	no trend
Darbandikhan	1998	2017	20	0.06	0.010	0.5259	no trend
Chamchamal	1998	2017	20	1.98	0.562	0.9761	Increasing
Kalar	1998	2017	20	2.01	0.366	0.9779	Increasing
Agjalar	1998	2017	20	1.43	0.324	0.9233	no trend
bngrd	1998	2017	20	1.27	0.211	0.8971	no trend
Sangaw	1998	2017	20	0.84	0.239	0.8005	no trend
Bawanor	1998	2017	20	2.01	0.454	0.9779	Increasing
Kifri	1998	2017	20	2.50	0.237	0.9938	Increasing

Note: −1.96 < Z < 1.96 = No trend, Z > 1.96 = Increase in trend, Z < −1.96 = Decrease in trend.

3.5. Multiple Regression Statistics, RMSE, and CRM

The Root Mean Square Error (RMSE) value indicates how predicted and observed measurements match, while the Coefficient of Residual Mass (CRM) value measures a model's tendency to over or underestimate the measurements. Positive values for CRM indicate that the model underestimates the measurements, and negative values overestimate [63]. For an ideal fit between the observed and predicted data, RMSE and CRM's values should equal 0.0 [62]. As can be seen from the statistical analysis results, the accuracy of the model in the estimation of NDVI and LST in Tables 8 and 9 for the study periods was tested by calculating the Coefficient of Residual Mass (CRM), Root Mean Square Error (RMSE), and coefficient of determination (R^2), respectively.

The results in Table 8 showed that the NDVI-based vegetation cover was more affected by climatic and topographic factors (precipitation and elevation) in the study area. A high value for multiple regression coefficients indicates strong relationships between the variables, and the low RMSE and CRM values show a reasonable precision and low error of the model. The multiple regression (R), RMSE, and CRM were calculated and presented in Tables 8 and 9. The efficiency and accuracy of the models for predicting drought indices were evaluated using statistical coefficients.

The values of regression parameters were used to predict the drought index (NDVI) in Table 8 from 1998 to 2017. The (R) values ranged from 0.77 in 1998 to 0.87 in 2017, RMSE from 0.039 in 2000 to 0.068 in 2005, and CRM from −0.006 in 2014 to 0.284 in 1998. These results indicate that although the relationship between variables was stronger in 2017, the prediction error was lower in 2008 and 2013. Comparing the observed and simulated measurements, the model gives appropriate predictions of the drought status. Moreover, different time scales were considered in the model. The drought predictions can be more reliable and efficient and ensure that the developed model is suitable and efficient.

The regression analyses in Table 8 showed that the spectral indices were related to total precipitation, geographic elevation, and latitude. The LST values of 1998 to 2017 were (R) ranged from 0.47 in 2001 to 0.85 in 2013, and RMSE were from 2.7 in 1999 to 7.0 in 2000.

Table 8. Parameters of the regression models used for predicting drought index (NDVI) in the KRI.

| | | | $y = \beta 0 + \beta 1\, x1 + \beta 2\, x2 + \beta 3\, x3$ | | | | |
| | | β0 | β1 | β2 | β3 | | |
Year	R	Intercept	x1 Coefficients	x2 Coefficients	x3 Coefficients	RMSE	CRM
1998	0.77	−2.19	6.4×10^{-2}	-9.9×10^{-7}	1.6931×10^{-4}	0.090	0.284
1999	0.80	−1.02	3.0×10^{-2}	9.790×10^{-5}	3.0739×10^{-4}	0.047	0.002
2000	0.80	−1.27	3.5×10^{-2}	5.872×10^{-5}	2.4785×10^{-4}	0.039	0.031
2001	0.72	−0.06	3×10^{-3}	1.0642×10^{-4}	2.1147×10^{-4}	0.062	0.002
2002	0.77	−0.79	2.6×10^{-2}	3.402×10^{-5}	2.0132×10^{-4}	0.056	0.009
2003	0.75	−0.89	2.8×10^{-2}	8.010×10^{-5}	8.339×10^{-5}	0.046	0.000
2004	0.77	−0.94	2.8×10^{-2}	2.258×10^{-5}	2.0384×10^{-4}	0.054	−0.004
2005	0.74	−1.03	2.9×10^{-2}	7.895×10^{-5}	1.7897×10^{-4}	0.068	0.088
2006	0.81	−1.35	4.0×10^{-2}	1.4746×10^{-4}	5.234×10^{-5}	0.050	−0.005
2007	0.76	0.26	-5×10^{-3}	1.1165×10^{-4}	1.6574×10^{-4}	0.056	0.011
2008	0.78	−1.40	3.9×10^{-2}	8.544×10^{-5}	1.2644×10^{-4}	0.043	0.016
2009	0.77	−1.67	4.9×10^{-2}	7.641×10^{-5}	1.7644×10^{-4}	0.057	0.000
2010	0.76	−1.93	5.8×10^{-2}	5.352×10^{-5}	7.833×10^{-5}	0.051	0.006
2011	0.84	−2.12	6.1×10^{-2}	4.172×10^{-5}	1.9596×10^{-4}	0.054	−0.004
2012	0.74	−0.97	2.9×10^{-2}	9.920×10^{-5}	9.931×10^{-5}	0.050	0.004
2013	0.81	−1.84	5.6×10^{-2}	9.731×10^{-5}	4.716×10^{-5}	0.049	0.003
2014	0.73	0.86	-1.9×10^{-2}	1.0719×10^{-4}	2.6338×10^{-4}	0.065	−0.006
2015	0.78	−0.15	8×10^{-3}	9.893×10^{-5}	1.3360×10^{-4}	0.049	0.005
2016	0.84	−0.79	2.7×10^{-2}	1.6592×10^{-4}	4.533×10^{-5}	0.043	0.007
2017	0.87	−0.85	2.6×10^{-2}	1.6069×10^{-4}	1.3260×10^{-4}	0.044	−0.001

Note: When y = NDVI Drought index, x1 = Latitude, x2 = Elevation, x3 = Precipitation.

Figure 10 illustrates the quantile–quantile plots (Q–Q) used to visually examine the degrees of distribution. From the visual point of view, there was little difference when choosing among the various distributions for representing the data used in the study [64]. For instance, the Q–Q plot of observed spectral indices at 60 locations in KRI versus expected values pointed out that NDVI and LST have been fitted to better distributions, as most of the observations fall on and around the straight line and few points are a little bit far away from the fitted line.

Table 9. Parameters of the regression models used for predicting drought index (LST) in KRI.

| | | | $y = \beta0 + \beta1\,x1 + \beta2\,x2 + \beta3\,x3$ | | | | |
| | | $\beta0$ | $\beta1$ | $\beta2$ | $\beta3$ | | |
Years	R	Intercept	x1 Coefficients	x2 Coefficients	x3 Coefficients	RMSE	CRM
1998	0.50	200.326	-4.5090×10^{0}	3.79×10^{-3}	-7.70×10^{-3}	3.366	-0.0019
1999	0.74	162.592	-3.6067×10^{0}	-2.98×10^{-3}	-4.37×10^{-3}	2.736	0.0000
2000	0.63	115.233	-1.8216×10^{0}	-1.62×10^{-3}	-2.774×10^{-2}	7.026	-0.0271
2001	0.47	-92.048	3.2245×10^{0}	3.73×10^{-3}	-2.01×10^{-3}	4.823	-0.0374
2002	0.75	161.735	-3.8018×10^{0}	-3.39×10^{-3}	5.0×10^{-4}	2.709	0.0158
2003	0.59	-8.567	1.0771×10^{0}	-1.60×10^{-3}	-4.29×10^{-3}	3.745	-0.1117
2004	0.68	42.094	-3.318×10^{-1}	-4.21×10^{-3}	-8.13×10^{-3}	3.456	-2.95×10^{-5}
2005	0.67	-51.024	5.825×10^{-1}	1.403×10^{-2}	6.25×10^{-3}	5.988	0.0121
2006	0.58	73.968	-1.1655×10^{0}	-5.41×10^{-3}	-2.93×10^{-3}	3.709	-0.0001
2007	0.66	137.659	-2.9794×10^{0}	1.17×10^{-3}	-7.40×10^{-3}	3.568	0.0001
2008	0.52	-71.482	3.0436×10^{0}	1.90×10^{-3}	-2.42×10^{-3}	3.626	-1.17×10^{-5}
2009	0.55	168.828	-3.8644×10^{0}	4.77×10^{-3}	-9.94×10^{-3}	4.759	0.0003
2010	0.75	61.496	-7.443×10^{-1}	-1.6×10^{-4}	-1.737×10^{-2}	4.132	-0.0002
2011	0.65	110.496	-2.2353×10^{0}	-5.50×10^{-3}	-1.06×10^{-3}	3.189	-0.0001
2012	0.75	88.654	-1.4965×10^{0}	-6.09×10^{-3}	-5.20×10^{-3}	2.774	0.0031
2013	0.85	36.517	1.13×10^{-2}	1.11×10^{-3}	-1.716×10^{-2}	3.840	-0.0109
2014	0.83	130.942	-2.6048×10^{0}	-6.98×10^{-3}	-1.367×10^{-2}	3.522	0.0026
2015	0.80	169.176	-3.6726×10^{0}	-1.752×10^{-2}	3.63×10^{-3}	4.159	-6.24×10^{-5}
2016	0.82	109.215	-1.9879×10^{0}	-5.89×10^{-3}	-1.167×10^{-2}	3.797	-0.0090
2017	0.79	200.224	4.4465×10^{0}	-1.164×10^{-2}	-4.23×10^{-3}	4.413	0.0008

Note: When y = LST Drought Index, x1 = Latitude, x2 = Elevation, x3 = Precipitation.

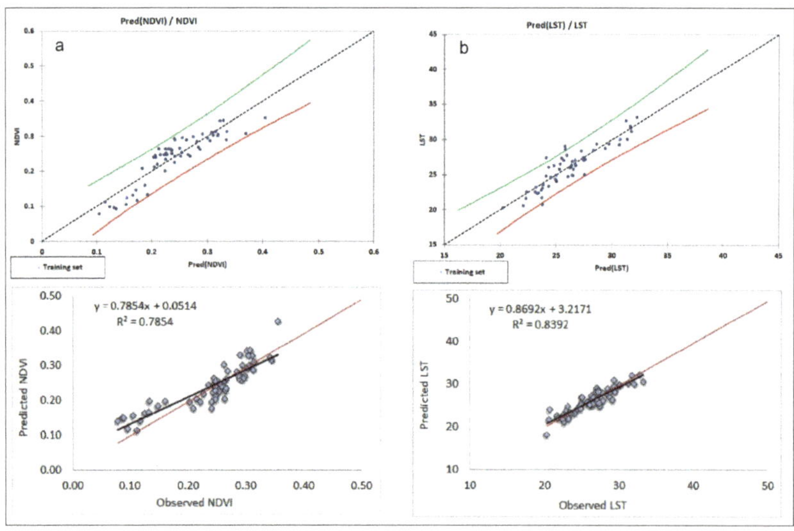

Figure 10. The Quantile–Quantile Plot (Q–Q Plot) of Observed (**a**) NDVI and (**b**) LST at 60 Locations Versus Predicted Values.

4. Discussion

NDVI and LST maps of the consecutive 20 years clearly show the onset and extent of drought. According to the results shown in Table 3 and Figure 4, it can be said that the studied region has faced drought episodes over the study period, especially in the years 2000 and 2008. The Mann–Kendall test assessed changes in drought indices at the 60 sites. The multiple regression for the 20 years was used to understand variable climatic influences on droughts and temporal variations and their relationships with precipitation changes, altitude, and latitude. A considerable change in NDVI values and LST was strongly observed in 2000 and 2008. These results correspond to the study conducted by [65,66], whereas the year 2008 was the driest year during the study period. The occurrence of drought is associated with reduced vegetation area and NDVI mean values. However, LST values increased throughout the 20 years of study, particularly in 2000 and 2008 [12].

The NDVI value indicates very low vegetation, especially in the southern part of the study area (Figure 4 and Table 3). These results correspond to the study conducted by [34], whereas both of them found that year 2008 had the most severe drought year during the study period. Ref. [67] assessed the spatiotemporal changes in drought in Iraq using SPI from 1980 to 2010. They found that drought caused deterioration from normal to extreme levels in Iraq during 2000, 2008, and 2010 (which was the driest year), while NDVI coverage was high in the northern part of the study area in the identified period [66]. The northern part of the study area is characterized by a topography covered with grasses and trees, while the density of NDVI-based vegetation is less in the middle part and lesser in the southern part [25]. The less extensive coverages of precipitation are observed in the southern part of the study area in the KRI (Figures 2C and 8).

The terrain, trees, and shrub area may play an essential role in assessing the vegetation increase due to the progress of laws on the protection of the environment and the prevention of logging in the last ten years, as well as the improvement of the living costs and raising awareness of the people living in these areas and continuous artificial afforestation in that area [68]. After investigating drought vicissitudes in KRI changes with NDVI and LST variation, we observed droughts' vicissitudes in terms of frequency, duration, and intensity. Specifically, we first calculated the changes in classes (classes 1, 2, and 3) for NDVI and classes 1, 2, 3, 4, and 5 for LST. Then, we compared the changes among the five drought categories and selected the one that shows the largest change compared to the other four categories as the dominant one. The percentage of lands dominated by each drought category was counted for each period to show the temporal evolution.

Based on the drought index scale, areas affected by drought have low NDVI [41]. NDVI values indicated that the southwestern and western parts of KRI experienced drought (Figures 4 and 5). However, its magnitude and spatial extension varied. Low values indicated the dry season, while high values indicated the wet season [69]. The values varied spatially and temporally across the region from 1998 to 2017 (Figures 3 and 4). This was mainly because of the precipitation amount, frequency, and intensity [70]. The variation in NDVI is controlled by meteorological variables, such as precipitation, temperature, and relative humidity [41]. The NDVI values were spatially varied due to climate, soil, and topographic variability.

The southern region shows a net decrease of its vegetative cover during the considered time range. It seems to be affected by land-degradation processes caused by droughts, which are an issue in this area [13]. It cannot be expected to perceive differences in the vegetation or physiological (decrease or increase of certain strata) density with slight accuracy. However, it seems to have been shown that the main decreased features of vegetation are in the southern part of KRI. All correlations between the monthly index values and meteorological parameters from the different locations are statistically significant at 95%. The spatial patterns of droughts for the growing seasons in April showed rainfall decreases, and LST values increase in the growing seasons led to an NDVI value reduction in the southwest parts of the KRI. Thus, the increase in rainfall and the slight increase in LST caused NDVI to rise at a few locations in the northeast. The direction or absence of vegetation trends often matches the precipitation trend. This indicates that the combined effect of precipitation and temperature played an important role in decreasing NDVI values and the area of vegetation in the southwest of KRI during the growing season [12].

The increases in LST in the south and southeast parts were due to the lack of vegetation cover and the soil's relative humidity due to the rainfall deficiency in these areas. These areas have led to continuous land and crop degradation and yield losses. The increased temperatures can cause severe natural effects on the environment, including hydrological drought (IPCC, 2012) [71]. The increase of LST is assumed to negatively affect vegetation strength and cause plant stress [69]. The mean LST increases in almost all parts of the KRI, and the minimum LST increases in the mountains. In the eastern parts, maximum LST increases in the southwestern parts of KRI. This trend pattern corresponds with the variation in NDVI (see Figures 5–7). While the lack of precipitation is often the primary cause of drought, increased potential evapotranspiration is linked to temperature and relative humidity [72]. Actual evapotranspiration is additionally controlled by soil moisture, which constitutes a limiting factor for further drying under drought conditions and other processes impacting vegetation development and phenology; for instance, the temperature is also relevant [70,73,74]. Another LST study by Robaa and AL-Barazanji [75] showed that after 1995, the rising trend of the annual mean temperature over Iraq was about 0.5 °C/decade.

In Table 5, the correlation matrix investigates the generality of the represented LST and NDVI relationship with respect to drought monitoring and assessment; LST and NDVI relations show negative correlations in the study area. Usually, the relationship between NDVI and LST is negative, as the value of NDVI increases with decreases in LST. The LST–NDVI correlations are generally negative [39,76]. In the northeast and south parts, where precipitation increased at some locations, there were significant changes in NDVI, except for an insignificant increase at some sites. Increases were observed in the NDVI for the study period except in 2000, 2008, and 2012. A sharp increase between 2000 and 2008 was observed for LST.

Precipitation was gradually increased with the increase in NDVI in the northern parts of KRI. This increase has resulted in an increasing trend in the northeast's NDVI values and vegetation area. On the other hand, continuous increases in LST and decreasing precipitation resulted in continuous decreases in NDVI in the region's southern locations. This indicates that LST was crucial to decreasing NDVI in these areas during the growing season. Figures 4–7 show a constant variability in terrestrial ecosystems at different spatial and temporal scales because of natural and/or anthropogenic causes. The droughts in semi-arid areas significantly contribute to environmental degradation, as they limit the development of vegetation cover and expose the soil to erosion [77].

The spatiotemporal variability of the LST–NDVI relationship on continental or global scales has been investigated in several studies [78,79] and was based on the assumption that complementary information in these studies may provide a more robust characterization for different phenomena at the land's surface. Studies have revealed a strong negative correlation between NDVI and LST resulting from canopy transpiration's cooling effects. This study's NDVI varied due to temporal and spatial variability in rainfall [80,81]. Therefore, NDVI is a relatively good indicator of drought in KRI, and warmer temperatures are more favorable for vegetation growth [82]. Therefore, the application of empirical NDVI–LST-based indices must be limited to areas and periods where negative correlations are observed and not on a global scale. The mean NDVI indicating the vegetation's greenness was strongly related to seasonal rainfall, which indicates the possibility of using NDVI to predict drought [83]. In general, prior studies suggest that the LST–NDVI slope sign may be governed by whether vegetation growth is water-limited (negative slope) or energy-temperature limited (positive slope). The latter is prevalent at high latitudes or in the evergreen tropical forests, whereas the latter may occur at lower latitudes, especially in dry lands [84,85]. A statistical trend test provides more reliable ways to describe trends in long time series than linear regression. Moreover, the *p*-values for the Mann–Kendall test are calculated. The Mann–Kendall statistics are shown in Table 5. Only 11 locations (Khabat, Mergasur, Barzewa, Batel, Mangish, Deraluke, Amadia, Bazian, Halabja, Mawat, and Darbandikhan) out of 60 showed an upward trend in NDVI.

The Mann–Kendall statistics are shown in Table 7. Nine locations show statistically significant upward trends in LST (Erbil, Qushtapa, Dibaga, Gwer, Shamamk, Chamchamal, Kalar, Bawanor, and Kifri). These results are in accordance with the findings of Razvanchy [86]. Generally, the southern parts of the study area are warmer. In particular, the temperature rise in the southern zone of the region is the lowest of precipitation and low-vegetation cover and elevation and correlated to a statistically downward trend in annual precipitation Figure 9 and Table 7. Multiple regression analysis revealed that the correlation between rainfall, elevation, and latitude with spectral indices is significant during the beginning of the growing season, whereas other biophysical variables play a lesser role. One of the study objectives was to examine the feasibility of regression analysis to make NDVI and LST forecasts.

A time series of drought indices provides a framework for evaluating drought parameters of interest. In order to quantify the prediction accuracy and precision of the model, the (R), RMSE, and CRM were calculated (see Tables 7 and 8). Some statistical coefficients evaluated the efficiency and accuracy of models used for predicting drought indices. Weather and climate phenomena reflect the interaction of dynamic and thermodynamic processes over a wide range of spatial and temporal scales. This complexity results in highly variable atmospheric conditions, including temperatures, motions, and precipitation; events include the persistence of drought conditions over decades of timescales. Thus, rainfall is associated with both altitude and position [25]. However, this explanation for the spatial distribution of precipitation was supported by dense vegetation in mountain areas, where oak forests were more intense than the other areas. In contrast, the northeastern region dominated the study area at a certain height. However, it became less after the first or second hill, and in the plains of Erbil and Sulaimaniyah, where winter crops are cultivated, which are sensitive to high temperatures and low precipitation [87,88].

5. Conclusions

This study assessed droughts status changes during 20 years of growing seasons in the KRI. This study contributes to drought severity assessment by quantifying NDVI decrease and LST increase during long-term climate. The resultant maps show the change pattern in the relationship between remote sensing-based drought indices and climate factors. From this study, the following points can be concluded:

Severe drought circumstances prevailed during 2000 and 2008 over a large KRI area. The onset and extent of drought can be clearly observed through NDVI, LST, and precipitation maps for the studied 20 years. The land-cover classification shows that the vegetation coverage area was more seriously affected by climatic factors (precipitation and temperature), especially in 2000 and 2008. Considering the significant recurrence of drought, it is crucial to satisfy the water needs of the study area by using other available water resources, such as groundwater, for supplementary irrigation in the rainfed areas of the southern part of KRI. The correlation between LST and NDVI in the same measured year was significant, likely due to the delayed effect of scarce precipitation on vegetation. More detailed investigations are needed to understand the frequency of drought and its relationship to factors affecting it.

Landsat-based spectral drought indices were significantly correlated with precipitation, geographical elevation, and latitude. High values of multiple correlation and regression indicate strong relationships between the variables, and the low RMSE and CRM values show a reasonable precision and a low error of the resultant model. Comparing the results obtained for the modeling indicates that the presented model gives appropriate predictions of the drought situation. Moreover, different time scales were considered in the model so that the drought predictions can be more reliable and efficient, and the developed model is ensured to be suitable and efficient.

Acute water stress was evident all over the study area in 2000, 2008, and 2012. Despite the prevalence of drought conditions over a large area of KRI during the mentioned years, some areas in the eastern part of the region remained unaffected by the lack of precipitation and water stress. Those areas are characterized by humid and sub-humid climate types, which helps keep the area green even during the drought year.

Spatial variation in the NDVI and LST resulted from the uneven distribution of rainfall and geographical elevation effects in the study area. Since the region receives much higher monsoonal rainfall than the western part, even in the drought year, it remains suitable for tree and shrub growth. Unlike the meteorological data available from sparsely distributed meteorological stations, remote sensing meteorological data and remote sensing-based indices can be successfully used to delineate the spatiotemporal extent of drought. In the future, studies may incorporate agricultural production and surface evaporation data to evaluate further the mechanisms by which these factors interact during periods of drought. Due to the large local spatial variation in rainfall, NDVI values also show a high variation, ranging from a low area of 14.4% (7225.1 km^2) in 2000 to 64.2% (32,315.2 km^2) in 2016 (Table 3). On the other hand, LST indicates an upward slope in 2000, 2008, and 2012. The regression model parameters for predicting drought indices from this dataset were disabled to determine the annual precipitation or elevation playing a significant role in the yearly trends in NDVI and LST.

Supplementary Materials: The following are available online at https://www.mdpi.com/article/10.3390/w14060927/s1, Table S1. Landsat data chosen for analysis were a mixture of Landsat TM5, TM7 and OLI8.

Author Contributions: Conceptualization, A.M.F.A.-Q., H.A.A.G. and K.M.; Data curation, H.A.A.G. and A.M.F.A.-Q.; Formal analysis, H.A.A.G. and A.M.F.A.-Q.; Investigation, A.M.F.A.-Q., C.R. and H.A.A.G.; Methodology, H.A.A.G. and A.M.F.A.-Q.; Resources, A.M.F.A.-Q., C.R., K.M. and H.A.A.G.; Supervision, A.M.F.A.-Q.; Validation, H.A.A.G. and A.M.F.A.-Q.; Visualization, A.M.F.A.-Q., C.R. and H.A.A.G.; Writing—original draft, A.M.F.A.-Q. and H.A.A.G.; Review, editing, improving, A.M.F.A.-Q.,C.R and K.M. All authors have read and agreed to the published version of the manuscript.

Funding: This study has been funded by Nuffic, the Orange Knowledge Programme, through the OKP-IRA-104278 project titled "Efficient water management in Iraq switching to climate smart agriculture: capacity building and knowledge development" coordinated by Wageningen University & Research, The Netherlands and Salahaddin University, Erbil, Kurdistan Region, Iraq.

Institutional Review Board Statement: Not applicable.

Informed Consent Statement: Not applicable.

Data Availability Statement: Some data in this manuscript were obtained from the Ministry of Agriculture and Water Resources, Kurdistan Region, Iraq, and the other data from the United States Geological Service (USGS) for providing the Landsat images freely on their web-site, and Statistical analysis of the parameters.

Acknowledgments: The authors would like to thank the United States Geological Service (USGS) for providing the Landsat images freely on their website. The authors would like to thank Tariq Hamakareem Kakahama and Fuad Mohamed Ahmad from the College of Agricultural Engineering Sciences, Salahaddin University, for their sincere assistance. We are also thankful to Nuffic, the Orange Knowledge Programme, through the OKP-IRA-104278, Wageningen University & Research, The Netherlands, the Ministry of Agriculture and Water Resources, Salahaddin University, and the Tishk International University, Erbil, Kurdistan Region, Iraq, for their valuable support.

Conflicts of Interest: The authors declare no conflict of interest.

References

1. Aadhar, S.; Mishra, V. Data Descriptor: High-resolution near real-time drought monitoring in South Asia. *Sci. Data* **2017**, *4*, 170145. [CrossRef] [PubMed]
2. Dubovyk, O. The role of Remote Sensing in land degradation assessments: Opportunities and challenges. *Eur. J. Remote Sens.* **2017**, *50*, 601–613. [CrossRef]

3. Rajsekhar, D.; Singh, V.P.; Mishra, A.K. Multivariate drought index: An information theory based approach for integrated drought assessment. *J. Hydrol.* **2015**, *526*, 164–182. [CrossRef]
4. Lloyd-Hughes, B. The impracticality of a universal drought definition. *Theor. Appl. Climatol.* **2014**, *117*, 607–611. [CrossRef]
5. Yan, G.; Liu, Y.; Chen, X. Evaluating satellite-based precipitation products in monitoring drought events in southwest China. *Int. J. Remote Sens.* **2018**, *39*, 3186–3214. [CrossRef]
6. Al-Quraishi, A.M.F.; Negm, A.M. *Environmental Remote Sensing and GIS in Iraq*; Springer Water; Springer: Cham, Switzerland, 2019; ISBN 9783030213435.
7. Schucknecht, A.; Erasmi, S.; Niemeyer, I.; Matschullat, J. Assessing vegetation variability and trends in north-eastern Brazil using AVHRR and MODIS NDVI time series. *Eur. J. Remote Sens.* **2013**, *46*, 40–59. [CrossRef]
8. Van Loon, A.F.; Gleeson, T.; Clark, J.; Van Dijk, A.I.J.M.; Stahl, K.; Hannaford, J.; Di Baldassarre, G.; Teuling, A.J.; Tallaksen, L.M.; Uijlenhoet, R.; et al. Drought in the Anthropocene. *Nat. Geosci.* **2016**, *9*, 89. [CrossRef]
9. Keyantash, J.A.; Dracup, J.A. An aggregate drought index: Assessing drought severity based on fluctuations in the hydrologic cycle and surface water storage. *Water Resour. Res.* **2004**, *40*, 1–14. [CrossRef]
10. Wilhite, D.A.; Pulwarty, R.S. *Drought and Water Crises*; Routledge: London, UK, 2017; ISBN 9781138035645.
11. UNESCO (United Nations Educational, Scientific and Cultural Organization). *Integrated Drought Risk Management*; UNESCO: Paris, France, 2014.
12. Fadhil, A.M. Drought mapping using Geoinformation technology for some sites in the Iraqi Kurdistan region. *Int. J. Digit. Earth* **2011**, *4*, 239–257. [CrossRef]
13. Andrieu, J. Phenological analysis of the savanna–forest transition from 1981 to 2006 from Côte d'Ivoire to Benin with NDVI NOAA time series. *Eur. J. Remote Sens.* **2017**, *50*, 588–600. [CrossRef]
14. IdRC. *Capacity Assessment for Drought Risk Management in Iraq*; Final Report; Submitted to: United Nations Development Programme (UNDP); Interdisciplinary Research Consultants (IdRC): Amman, Jordan, 2012.
15. Al-Quraishi, A.M.F.; Gaznayee, H.A.A.; Messina, J. Drought severity trend analysis based on the Landsat time-series dataset of 1998–2017 in the Iraqi Kurdistan Region. *IOP Conf. Ser. Earth Environ. Sci.* **2021**, *779*, 012083. [CrossRef]
16. Wu, W.; Muhaimeed, A.S.; Al-Shafie, W.M.; Al-Quraishi, A.M.F. Using L-band radar data for soil salinity mapping—A case study in Central Iraq. *Environ. Res. Commun.* **2019**. [CrossRef]
17. Haktanir, K. *Environmental Challenges in the Mediterranean 2000–2050*; Springer: Berlin/Heidelberg, Germany, 2004. [CrossRef]
18. Harun, R.; Muresan, I.C.; Arion, F.H.; Dumitras, D.E.; Lile, R. Analysis of factors that influence the willingness to pay for irrigation water in the Kurdistan Regional Government, Iraq. *Sustainablity* **2015**, *7*, 9574–9586. [CrossRef]
19. Fadhil, A.M. Sand dunes monitoring using remote sensing and GIS techniques for some sites in Iraq. In Proceedings of the 3rd International Conference on Photonics and Image in Agriculture Engineering (PIAGENG 2013), Sanya, China, 27–28 January 2013. [CrossRef]
20. Awchi, T.A.; Jasim, A.I. Rainfall Data Analysis and Study of Meteorological Draught in Iraq for the Period 1970–2010. *TKRIit J. Eng. Sci.* **2017**, *24*, 110–121. [CrossRef]
21. Hameed, H. Water Harvesting in Erbil Governorate, Kurdistan Region, Iraq Detection of Suitable Sites Using Geographic Information System and Remote Sensing. Master's Thesis, Lund University, Lund, Sweden, 2013.
22. Mohammed, R.; Scholz, M. Adaptation Strategy to Mitigate the Impact of Climate Change on Water Resources in Arid and Semi-Arid Regions: A Case Study. *Water Resour. Manag.* **2017**, *31*, 3557–3573. [CrossRef]
23. Yao, Y.; Qin, Q.; Fadhil, A.M.; Li, Y.; Zhao, S.; Liu, S.; Sui, X.; Dong, H. Evaluation of EDI derived from the exponential evapotranspiration model for monitoring China's surface drought. *Environ. Earth Sci.* **2011**, *63*, 425–436. [CrossRef]
24. Zewdie, W.; Csaplovics, E. Remote sensing-based multi-temporal land cover classification and change detection in northwestern Ethiopia. *Eur. J. Remote Sens.* **2015**, *48*, 121–139. [CrossRef]
25. Al-Quraishi, A.M.F.; Gaznayee, H.A.A.; Crespi, M. Drought Trend Analysis in Sulaimaniyah (Iraqi Kurdistan Region) for the Period 1998–2017 using Remote Sensing and GIS. *J. Arid. Land* **2020**, *3*, 413–430.
26. Dabrowska-Zielinska, K.; Kogan, F.; Ciolkosz, A.; Gruszczynska, M.; Kowalik, W. Modelling of crop growth conditions and crop yield in Poland using AVHRR-based indices. *Int. J. Remote Sens.* **2002**, *23*, 1109–1123. [CrossRef]
27. Abdel-Hamid, A.; Dubovyk, O.; Graw, V.; Greve, K. Assessing the impact of drought stress on grasslands using multi-temporal SAR data of Sentinel-1: A case study in Eastern Cape, South Africa. *Eur. J. Remote Sens.* **2020**, *53*, 1–14. [CrossRef]
28. Alqasemi, A.S.; Hereher, M.E.; Al-Quraishi, A.M.F.; Saibi, H.; Aldahan, A.; Abuelgasim, A. Retrieval of monthly maximum and minimum air temperature using MODIS aqua land surface temperature data over the United Arab Emirates. *Geocarto Int.* **2021**. *ahead of print*. [CrossRef]
29. Vicente-Serrano, S.M.; Beguería, S.; López-Moreno, J.I. A multiscalar drought index sensitive to global warming: The standardized precipitation evapotranspiration index. *J. Clim.* **2010**, *23*, 1696–1718. [CrossRef]
30. Owrangi, M.A.; Adamowski, J.; Rahnemaei, M.; Mohammadzadeh, A.; Sharifan, R.A. Drought Monitoring Methodology Based on AVHRR Images and SPOT Vegetation Maps. *J. Water Resour. Prot.* **2011**, *3*, 325–334. [CrossRef]
31. Wan, Z.; Wang, P. International Journal of Remote Using MODIS Land Surface Temperature and Normalized Difference Vegetation Index products for monitoring drought in the southern Great Plains, USA. *Int. J. Remote Sens.* **2004**, *25*, 37–41. [CrossRef]
32. Heydari, H.; Zoej, M.J.V.; Maghsoudi, Y.; Dehnavi, S. An investigation of drought prediction using various remote-sensing vegetation indices for different time spans. *Int. J. Remote Sens.* **2018**, *39*, 1871–1889. [CrossRef]

33. Karim, T.H.; Keya, D.R.; Amin, Z.A. Temporal and spatial variations in annual rainfall distribution in erbil province. *Outlook Agric.* **2018**, *47*, 59–67. [CrossRef]

34. Eklund, L.; Persson, A.; Pilesjö, P. Cropland changes in times of conflict, reconstruction, and economic development in Iraqi Kurdistan. *Ambio* **2016**, *45*, 78–88. [CrossRef] [PubMed]

35. Eklund, L.; Abdi, A.; Islar, M. From Producers to Consumers: The Challenges and Opportunities of Agricultural Development in Iraqi Kurdistan. *Land* **2017**, *6*, 44. [CrossRef]

36. Saeed, M.A. Analysis of Climate and Drought Conditions in the Fedral. *J. Int. Sci. Environ.* **2012**, *2*, 953.

37. Ministry of Planning-KRG. *Regional Development Strategy for Kurdistan Region. 2012–2016 Final report 2017*; Ministry of Planning-KRG: Erbil, Iraq, 2017.

38. *Module ENVI Atmospheric Correction Module: QUAC and FLAASH User's Guide*; Module Version; ITT Visual Information Solutions: Boulder, CO, USA, 2009; 44p.

39. Sun, Q.; Tan, J.; Xu, Y. An ERDAS image processing method for retrieving LST and describing urban heat evolution: A case study in the Pearl River Delta Region in South China. *Environ. Earth Sci.* **2009**, *59*, 1047–1055. [CrossRef]

40. Rouse, W.; Haas, H.; Deering, W. Monitoring Vegetation Systems in the Great Plains With Erts. In Proceedings of the 3rd Earth Resource Technology Satellite (ERTS) Symposium, Washington, DC, USA, 10–14 December 1973; Volume 1, pp. 48–62.

41. Aquino, D.d.N.; Neto, O.C.d.R.; Moreira, M.A.; dos S. Teixeira, A.; de Andrade, E.M. Use of remote sensing to identify areas at risk of degradation in the semi-arid region. *Rev. Cienc. Agron.* **2018**, *49*, 420–429. [CrossRef]

42. Nath, B.; Acharjee, S. Forest Cover Change Detection using Normalized Difference Vegetation Index (NDVI): A Study of Reingkhyongkine Lake's, Adjoining Areas. *Indian Cartogr.* **2013**, *XXXIII*, 348–403.

43. Di Gregorio, A. *Land Cover Classification System*; FAO: Rome, Italy, 2005; ISBN 9251053278.

44. Dash, P.; Göttsche, F.M.; Olesen, F.S.; Fischer, H. Land surface temperature and emissivity estimation from passive sensor data: Theory and practice-current trends. *Int. J. Remote Sens.* **2002**, *23*, 2563–2594. [CrossRef]

45. Sun, D.; Kafatos, M. Note on the NDVI-LST relationship and the use of temperature-related drought indices over North America. *Geophys. Res. Lett.* **2007**, *34*, 1–4. [CrossRef]

46. Qutbudin, I.; Shiru, M.S.; Sharafati, A.; Ahmed, K.; Al-Ansari, N.; Yaseen, Z.M.; Shahid, S.; Wang, X. Seasonal Drought Pattern Changes Due to Climate Variability: Case Study in Afghanistan. *Water* **2019**, *11*, 1096. [CrossRef]

47. Pohlert, T. Package 'trend': Non-Parametric Trend Tests and Change-Point Detection. *R Packag.* **2016**, *26*. [CrossRef]

48. Lu, Y.; Jiang, S.; Ren, L.; Zhang, L.; Wang, M.; Liu, R.; Wei, L. Spatial and Temporal Variability in Precipitation Concentration over Mainland China, 1961–2017. *Water* **2019**, *11*, 881. [CrossRef]

49. U.S. Army Corps of Engineers. *Mann-Kendall Analysis Mann-Kendall Analysis for the Fort Ord Site*; HydroGeoLogic, Inc.—OU-1 Annual Groundwater Monitoring Report—Former Fort Ord, California; HydroGeoLogic, Inc.: Herndon, VA, USA, 2005.

50. Gilbert, R.O. Statistical Methods for Environmental Pollution Monitoring. *Stat. Methods Environ. Pollut. Monit.* **1987**, 204–224.

51. Partal, T.; Kahya, E. Trend analysis in Turkish precipitation data. *Hydrol. Process.* **2006**, *20*, 2011–2026. [CrossRef]

52. Batool, N.; Shah, S.A.; Dar, S.N.; Skinder, S. Rainfall variability and dynamics of cropping pattern in Kashmir Himalayas: A case study of climate change and agriculture. *SN Appl. Sci.* **2019**, *1*, 606. [CrossRef]

53. Jiao, W.; Tian, C.; Chang, Q.; Novick, K.A.; Wang, L. A new multi-sensor integrated index for drought monitoring. *Agric. For. Meteorol.* **2019**, *268*, 74–85. [CrossRef]

54. Yao, X.; Yao, X.; Jia, W.; Tian, Y.; Ni, J.; Cao, W.; Zhu, Y. Comparison and intercalibration of vegetation indices from different sensors for monitoring above-ground plant nitrogen uptake in winter wheat. *Sensors* **2013**, *13*, 3109–3130. [CrossRef] [PubMed]

55. Dutta, D.; Kundu, A.; Patel, N.R. Predicting agricultural drought in eastern Rajasthan of India using NDVI and standardized precipitation index. *Geocarto Int.* **2013**, *28*, 192–209. [CrossRef]

56. Gessner, U.; Machwitz, M.; Conrad, C.; Dech, S. Estimating the fractional cover of growth forms and bare surface in savannas. A multi-resolution approach based on regression tree ensembles. *Remote Sens. Environ.* **2013**, *129*, 90–102. [CrossRef]

57. Lehnert, L.W.; Meyer, H.; Wang, Y.; Miehe, G.; Thies, B.; Reudenbach, C.; Bendix, J. Retrieval of grassland plant coverage on the Tibetan Plateau based on a multi-scale, multi-sensor and multi-method approach. *Remote Sens. Environ.* **2015**, *164*, 197–207. [CrossRef]

58. Li, X.; Zhang, Y.; Bao, Y.; Luo, J.; Jin, X.; Xu, X.; Song, X.; Yang, G. Exploring the best hyperspectral features for LAI estimation using partial least squares regression. *Remote Sens.* **2014**, *6*, 6221–6241. [CrossRef]

59. Peres, L.F.; DaCamara, C.C. Land surface temperature and emissivity estimation based on the two-temperature method: Sensitivity analysis using simulated MSG/SEVIRI data. *Remote Sens. Environ.* **2004**, *91*, 377–389. [CrossRef]

60. Gutman, G.G. Towards Monitoring Droughts from Space. *J. Clim.* **2002**, *3*, 282–295. [CrossRef]

61. Lambin, E.F.; Ehrlich, D. The surface temperature-vegetation index space for land cover and land-cover change analysis. *Int. J. Remote Sens.* **1996**, *17*, 463–487. [CrossRef]

62. Aliyu, S.; Bello, S. Performance Assessment of Hargreaves Model in Estimating Global Solar Radiation in Sokoto, Nigeria. *Int. J. Adv. Sci. Res. Eng.* **2018**, *3*, 6–15. [CrossRef]

63. Park, H.; Kim, K.; Lee, D.K. Prediction of severe drought area based on random forest: Using satellite image and topography data. *Water* **2019**, *11*, 705. [CrossRef]

64. Population, N. Quantile-Quantile Plot (QQ-plot) and the Normal Probability Plot Section 6-6. *Norm. Probab. Plot.* **2012**, *2377*, 1–8.

65. Almamalachy, Y. Utilization of Remote Sensing in Drought Monitoring Over Iraq. Master's Thesis, Portland State University, Portland, OR, USA, 2017. Paper 3996. [CrossRef]
66. Mustafa, Y.T. Spatiotemporal Analysis of Vegetation Cover in Kurdistan Region-Iraq using MODIS Image Data. *J. Appl. Sci. Technol. Trends* **2020**, *1*, 1–6. [CrossRef]
67. Al-timimi, Y.K.; George, L.E.; Al-jiboori, M.H. Drought Risk Assessment in Iraq using Remote Sensing and GIS Techniques. *Iraqi J. Sci.* **2012**, *53*, 1078–1082.
68. Mustafa, Y.T. *Mapping and Estimating Vegetation Coverage in Iraqi Kurdistan Region using Remote Sensing and GIS*; Iraq-Final Report; Prepared and Implemented by Applied Remote Sensing & Gis (Ars&Gis) Center, University Of Zakho, Salahaddin University, University of Garmian, and Directorate of Forests and Range in Duhok; General Directorate of Horticulture, Forestry and Rangeland, Ministry of Agriculture and Water Resources: Erbil, Kurdistan Region, Iraq, 2015; pp. 1–136.
69. Anyamba, A.; Tucker, C.J. Historical perspectives on AVHRR NDVI and vegetation drought monitoring. *Remote Sens. Drought Innov. Monit. Approaches.* **2012**, 23–49. [CrossRef]
70. Wang, J.; Wang, J.; Rich, P.M.; Price, K.P. Temporal responses of NDVI to precipitation and temperature in the central Great Plains, USA. *Int. J. Remote Sens.* **2003**, *24*, 2345–2364. [CrossRef]
71. AlJawa, S.B.; Al-Ansari, N. Open Access Online Journal of the International Association for Environmental. *Hydrol. Assess. Groundw. Qual. Using* **2017**, *26*, 1–14.
72. Oku, Y.; Ishikawa, H.; Haginoya, S.; Ma, Y. Recent trends in land surface temperature on the Tibetan Plateau. *J. Clim.* **2006**, *19*, 2995–3003. [CrossRef]
73. Koundouri, P.; Karousakis, K.; Assimacopoulos, D.; Jeffrey, P.; Lange, M. Water Management in Arid and Semi-Arid Regions. In *Water Management in Arid and Semi-Arid Regions*; Northampton, Mass Edward Elgar: Cheltenham, UK, 2006; ISBN 9781845429973.
74. Zhang, Y.; Wang, X.; Li, C.; Cai, Y.; Yang, Z.; Yi, Y. NDVI dynamics under changing meteorological factors in a shallow lake in future metropolitan, semi-arid area in North China. *Sci. Rep.* **2018**, *8*, 1–13. [CrossRef]
75. Robaa, S.M.; Zhian, J. AL-Barazanji Trends of Annual Mean Surface Air Temperature over Iraq. *Nat. Sci.* **2013**, *11*, 138–145.
76. Yang, J.; Wang, Y.Q.; Sun, Z.; Wang, Q.Q.; Batkhishig, O.; Ouyang, Z.; Cammalleri, C.; Sepulcre-Cantó, G.; Vogt, J.; Mavi, H.S. Mapping Drought Hazard Using SPI index And GIS (A Case study: Fars province, Iran). *Remote Sens. Environ.* **2017**, *7*, 1–4. [CrossRef]
77. Kogan, F.N. Global drought watch from space. *Bull. Am. Meteorol. Soc.* **1997**, *78*, 621. [CrossRef]
78. Lambin, E.F. Change detection at multiple temporal scales: Seasonal and annual variations in landscape variables. *Photogramm. Eng. Remote Sensing* **1996**, *62*, 931–938.
79. Julien, Y.; Sobrino, J.A. Comparison of cloud-reconstruction methods for time series of composite NDVI data. *Remote Sens. Environ.* **2010**, *114*, 618–625. [CrossRef]
80. Friedl, M.A.; Davis, F.W. Sources of variation in radiometric surface temperature over a tallgrass prairie. *Remote Sens. Environ.* **1994**, *48*, 1–17. [CrossRef]
81. Goward, S.N.; Xue, Y.; Czajkowski, K.P. Evaluating land surface moisture conditions from the remotely sensed temperature/vegetation index measurements: An exploration with the simplified simple biosphere model. *Remote Sens. Environ.* **2002**, *79*, 225–242. [CrossRef]
82. Gupta, R.K.; Prasad, T.S.; Vijayan, D. Estimation of roughness length and sensible heat flux from wiFS and NOAA AVHRR data. *Adv. Sp. Res.* **2002**, *29*, 33–38. [CrossRef]
83. Gaikwad, S.V.; Kale, K.V.; Kulkarni, S.B.; Varpe, A.B.; Pathare, G.N. Agricultural Drought Severity Assessment using Remotely Sensed Data: A Review. *Int. J. Adv. Remote Sens. GIS.* **2015**, *4*, 1195–1203. [CrossRef]
84. Nemani, R.R.; Running, S.W. Estimation of Regional Surface Resistance to Evapotranspiration from NDVI and Thermal-IR AVHRR Data. *J. Appl. Meteorol.* **1989**, *28*, 276–284. [CrossRef]
85. Karnieli, A.; Bayasgalan, M.; Bayarjargal, Y.; Agam, N.; Khudulmur, S.; Tucker, C.J. Comments on the use of the Vegetation Health Index over Mongolia. *Int. J. Remote Sens.* **2006**, *27*, 2017–2024. [CrossRef]
86. Al-Quraishi, A.M.F.; Sadiq, H.A.; Messina, J.P. Characterization and Modeling Surface Soil Physicochemical Properties Using Landsat Images: A Case Study in the Iraqi Kurdistan Region. *Int. Arch. Photogramm. Remote Sens. Spatial Inf. Sci.* **2019**, *XLII-2/W16*, 21–28. [CrossRef]
87. Gaznayee, H.A.A.; Al-Quraishi, A.M.F. Analysis of agricultural drought, rainfall, and crop yield relationships in Erbil province, the Kurdistan region of Iraq based on Landsat time-series MSAVI2. *J. Adv. Res. Dyn. Control Syst.* **2019**, *11*, 536–545. [CrossRef]
88. Gaznayee, H.A.A.; Al-Quraishi, A.M.F. Analysis of Agricultural Drought's Severity and Impacts in Erbil Province, the Iraqi Kurdistan Region based on Time Series NDVI and TCI Indices for 1998 through 2017. *J. Adv. Res. Dyn. Control Syst.* **2019**, *11*, 287–297. [CrossRef]

Article

Integrating Remote Sensing Techniques and Meteorological Data to Assess the Ideal Irrigation System Performance Scenarios for Improving Crop Productivity

Heman Abdulkhaleq A. Gaznayee [1,*], **Sara H. Zaki** [1], **Ayad M. Fadhil Al-Quraishi** [2,*], **Payman Hussein Aliehsan** [1], **Kawa K. Hakzi** [3,*], **Hawar Abdulrzaq S. Razvanchy** [3], **Michel Riksen** [4] **and Karrar Mahdi** [4]

[1] Department of Forestry, College of Agricultural Engineering Science, Salahaddin University-Erbil, Erbil 44003, Iraq; sarah.haiman90@gmail.com (S.H.Z.); payman.aliehsan@su.edu.krd (P.H.A.)

[2] Petroleum and Mining Engineering Department, Faculty of Engineering, Tishk International University, Erbil 44001, Iraq

[3] Department of Soil and Water, College of Agricultural Engineering Science, Salahaddin University-Erbil, Erbil 44003, Iraq; hawar.sadiq@su.edu.krd

[4] Soil Physics and Land Management Group, Wageningen University & Research, 6700 AA Wageningen, The Netherlands; michel.riksen@wur.nl (M.R.)

* Correspondence: heman.ahmed@su.edu.krd (H.A.A.G.); ayad.alquraishi@tiu.edu.iq (A.M.F.A.-Q.); kawahakzy@gmail.com (K.K.H.)

Citation: Gaznayee, H.A.A.; Zaki, S.H.; Al-Quraishi, A.M.F.; Aliehsan, P.H.; Hakzi, K.K.; Razvanchy, H.A.S.; Riksen, M.; Mahdi, K. Integrating Remote Sensing Techniques and Meteorological Data to Assess the Ideal Irrigation System Performance Scenarios for Improving Crop Productivity. *Water* **2023**, *15*, 1605. https://doi.org/10.3390/w15081605

Academic Editor: Guido D'Urso

Received: 16 March 2023
Revised: 13 April 2023
Accepted: 17 April 2023
Published: 20 April 2023

Abstract: To increase agricultural productivity and ensure food security, it is important to understand the reasons for variations in irrigation over time. However, researchers often avoid investigating water productivity due to data availability challenges. This study aimed to assess the performance of the irrigation system for winter wheat crops using a high-resolution satellite, Sentinel 2 A/B, combined with meteorological data and Google Earth Engine (GEE)-based remote sensing techniques. The study area is located north of Erbil city in the Kurdistan region of Iraq (KRI) and consists of 143 farmer-owned center pivots. This study also aimed to analyze the spatiotemporal variation of key variables (Normalized Difference Moisture Index (NDMI), Normalized Difference Vegetation Index (NDVI), Precipitation (mm), Evapotranspiration (ETo), Crop evapotranspiration (ETc), and Irrigation (Hours), during the wheat-growing winter season in the drought year 2021 to understand the reasons for the variance in field performance. The finding revealed that water usage fluctuated significantly across the seasons, while yield gradually increased from the 2021 winter season. In addition, the study revealed a notable correlation between soil moisture based on the (NDMI) and vegetation cover based on the (NDVI), and the increase in yield productivity and reduction in the yield gap, specifically during the middle of the growing season (March and April). Integrating remote sensing with meteorological data in supplementary irrigation systems can improve agriculture and water resource management by boosting yields, improving crop quality, decreasing water consumption, and minimizing environmental impacts. This innovative technique can potentially enhance food security and promote environmental sustainability.

Keywords: irrigation system; NDVI; NDMI; meteorological data; center pivot

1. Introduction

Water stress is a pressing issue that affects many regions of the world, particularly in arid and semi-arid areas where water resources are scarce. In recent years, governments and humanitarian organizations have attempted to address this issue by improving access to water for those living in water-stressed areas. However, as climate change and population growth continue to pressure water resources, the problem is expected to worsen in the coming years. In addition, water stress can impact agricultural production, which can have an effect on the economy and food security [1,2]. Water stress can result in conflicts and mass migrations in certain cases. Moreover, the migration of people due to water stress can pose new challenges for governments and humanitarian organizations. Mitigating the challenges associated with water stress necessitates a collective endeavor from all stakeholders. Global collaboration in water management, in conjunction with efficacious policies and practices at the regional level, can facilitate the provision of secure and hygienic water access for all, and avert the societal, economic, and ecological repercussions stemming from water stress [3]. Countries are being encouraged to implement innovative and sustainable measures such as rainwater harvesting and enhanced irrigation techniques [2]. The Middle East has faced a severe water scarcity problem in recent decades due to climate change and inadequate management of water resources [4,5]. Water scarcity considerably limits wheat production in Northern Iraq, especially in regions with low and erratic rainfall during the crop growth period. In such circumstances, supplemental irrigation can boost and sustain yields over time, particularly if applied during critical growth stages [6].

Supplementary irrigation has been identified as a plausible measure to address this quandary. This entails the provision of supplementary water by farmers to their crops to offset the shortfall in precipitation, especially during crucial growth phases, since wheat necessitates varying quantities of water at distinct stages of its growth. The application of supplementary irrigation at vital growth stages warrants the provision of adequate water supply to the crop during its peak water requirements, and thereby augments and preserves yields over a protracted time period, even in regions confronted with a dearth of or erratic precipitation [1,7].

The inaccessibility of sufficient water resources can pose a formidable obstacle to farmers, particularly during periods of drought [8]. Shallow and deep tube wells are both employed as means to provide irrigation water. In 2007, the Ministry of Agriculture and Water Resources of the KRI initiated a program to assist farmers with contemporary irrigation systems, such as the center pivot, to enhance crop production after experiencing droughts on several occasions in previous years. Nevertheless, inadequate awareness regarding water scarcity and efficient water utilization has led some farmers to utilize the center pivot for prolonged durations and over-irrigate by extending the center pivot revolution time to amplify the water application depth. The assimilation of crop models and remotely sensed data via optimization algorithms has emerged as an efficacious and prospective technique for monitoring crop growth status and appraising crop yields, as it ameliorates specific deficiencies and amalgamates the benefits of individual methods [9,10].

In the KRI region, groundwater is abundant and generally of reasonable quality, but it has been heavily extracted for domestic, agricultural, and industrial purposes, resulting in a depletion of the groundwater table level. Mitigating the effects of climate change and rehabilitating nature requires more water management projects to ensure the efficient use of available water. Groundwater is the primary source of inaccessible surface water, and rainfall, groundwater, and rivers account for 51%, 48%, and 1% of crop production water, respectively [11]. Despite an increasing number of wells and groundwater abstraction in the Erbil plain, the aquifer is being harmed, with an average drop in groundwater levels of 50 m [2].

Agriculture is the largest water-consuming sector globally, accounting for 78–90% of water use, followed by domestic and industrial use [12]. In Iraq, the agricultural sector is the primary consumer of water resources. Irrigation in Iraq relies on three main sources of water: surface water, rainwater, and groundwater, and farmers use drip, sprinkler, and central pivot irrigation techniques [13]. Wheat and barley are the dominant crops in Iraq, covering 73% of the total cultivated land [14]. Wheat productivity under irrigation exceeds rainfed productivity by a factor of two to three, demonstrating the high productivity potential of wheat varieties under irrigation [2]. Given the increasing scarcity of freshwater, it is critical to optimize water use, particularly in irrigated agriculture [15]. Rainfall is the primary source of water for agriculture in Iraq, accounting for 51% of water supply during winter, while groundwater and rivers contribute 48% and 1%, respectively [16], as per the joint report of UNDP, USAID, and FAO (2019). Farmers in arid regions are confronted with a multitude of challenges, such as low agricultural productivity, frequent droughts, climate variability, high soil erosion rates, and deforestation [17]. Nevertheless, a comprehensive study of drought must be conducted for each area to develop a research framework [18,19]. Understanding the spatial and temporal variations in crop growth is crucial for effective crop management and achieving food security. The combination of remote sensing data and crop growth models has proven to be a valuable approach for monitoring crop growth and estimating crop yields.

The impact of climate change, combined with reduced water discharge in the Tigris and Euphrates rivers, has led to more frequent droughts in Iraq, which have particularly affected agriculture, resulting in significant reductions in crop yields and vegetation cover [20]. The Erbil province, KRI, is particularly vulnerable to severe drought, with reports indicating a reduction in precipitation and vegetation cover in 2000, 2008, 2012, and 2021 [21,22]. Southern and western parts of the KRI have been the hardest hit by rainfall shortages, whereas the north and east have seen increased moisture levels. The FAO has predicted that wheat crop production in northern Iraq will be approximately 50% lower in 2021 than the previous year, according to KRI authorities [8,23]. The use of secondary agricultural data in combination with remote sensing can provide more accurate estimates of irrigation performance [24]. The NDVI is a useful parameter for crop monitoring due to its unambiguous response to irrigation interventions. Modern irrigation management strategies aim to enhance agricultural productivity while minimizing water consumption [25], which is crucial given the limited water supply in changing environmental conditions. The NDMI can be used to describe a crop's water stress level and is calculated as the ratio between the difference and the sum of the refracted radiation in the near-infrared and SWIR spectrums [26]. Effective water resource management and enhancement of water productivity require a sound understanding and management of evapotranspiration [27]. Monitoring significantly varying Normalized Difference Vegetation Index (NDVI) values in a region can provide valuable insight into potential pest or disease concerns and the presence of weeds in areas of a field [28]. NDVI data can be used to display changes in vegetation cover and analyze patterns of drought occurrences. However, the performance of NDVI may be influenced by mistakes during the growing season and saturation effects on dense vegetation [29]. Therefore, additional factors should be considered to improve the precision of the results [30]. Various alternative vegetation and moisture indices are available to monitor crop growth, including the Enhanced Vegetation Index (EVI), Normalized Difference Water Index (NDWI), Leaf Area Index (LAI), Soil Adjusted Vegetation Index (SAVI), Thermal Vegetation Index (TVI), Crop Water Stress Index (CWSI), and Temperature Vegetation Dryness Index (TVDI). The most appropriate index selection depends on various factors, such as the type of crop, growth stage, and environmental conditions. NDVI have been widely used in semi-arid areas and have shown good performance in detecting vegetation changes [31–33]. The Normalized Difference Moisture Index (NDMI) is a reliable indicator of vegetation water content and can be used to monitor water stress in crops or other vegetation, particularly in areas where water stress is a concern [34,35]. NDMI is derived from the difference between the near-infrared and mid-infrared bands of satellite imagery. By monitoring NDMI values over time, changes in vegetation water content can be detected and the level of water stress experienced by crops or other vegetation can be assessed [15].

Inefficient water application occurs when the irrigation system is not appropriately managed, resulting in uneven water distribution across the field, reduced crop yields, and soil erosion. Such inefficiency happens when water evaporates or is lost through other means. Minimizing water loss requires proper irrigation scheduling, use of efficient irrigation systems, and managing the system appropriately based on remote sensing. Monitoring through the Sentinel satellite every five days and ETo, ETc data helps ensure proper irrigation scheduling based on remote sensing and GIS techniques. The capability to develop agriculture effectively with limited water resources is a crucial strategic objective for addressing future climate change and fulfilling Sustainable Development Goal 2 of the United Nations (SDG2). The primary objective of this research is to comprehend the impact of drought and compare wheat production between rainfed and center pivot systems during the study period. Additionally, it seeks to analyze the key factors that affect the spatial and temporal variation of NDVI and NDWI in wheat cultivated on center pivots. The study aims to investigate irrigation performance and water productivity at various scales to develop suitable water management strategies, especially considering decreasing water availability, rising threats from climate change, and growing population and food demand. This study provides a new strategy for agricultural resource management by providing consistent estimations of winter wheat water requirements and yield. This information can be used to optimize irrigation practices and improve crop management, particularly in arid and semi-arid regions where water availability is limited.

2. Materials and Methods

2.1. Study Area

The study area, comprising the Ain Kawa sub-district in the Erbil Governorate of northern Iraq, is illustrated in Figure 1 and spans an estimated area of 25,525 hectares. North Erbil was chosen as the case study region for a combination of reasons. Firstly, the wheat, which is a crucial crop that contributes significantly to the Kurdistan Region of Iraq's economy, is extensively grown in the region. Secondly, Erbil is a vital agricultural center, known as Iraq's breadbasket, and is responsible for a substantial portion of the country's food supply. Finally, the need to enhance and expand the existing water management sector in Iraq is pressing, and there are related challenges that require immediate attention.

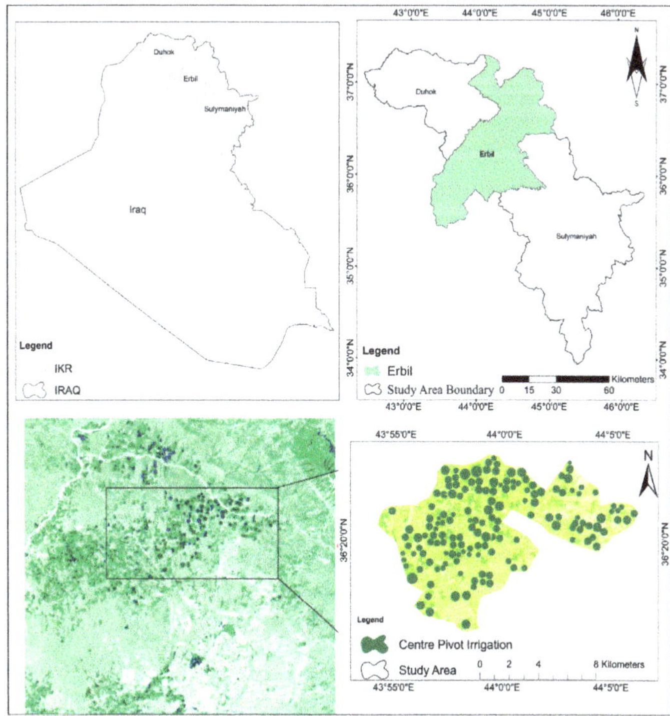

Figure 1. Site map of the study area in Erbil, KRI.

2.2. Climate

The climatic conditions prevailing in Erbil, KRI, are predominantly continental, subtropical, and semi-arid, while a Mediterranean climate characterizes the mountainous parts. Rainfall in the mountainous areas usually occurs between December to February or November to April. The average daytime temperature during winter is approximately 16 degrees Celsius, which drops to around 2 °C at night, with a possibility of frost. Conversely, summers are characterized by high temperatures, with an average temperature exceeding 45 °C in July and August and falling to around 25 °C at night [36,37]. The period of precipitation in the region is typically between October to May, with the annual rainfall ranging between 100 and 200 mm. The Ministry of Agriculture and the Department of Meteorology in the KRI have validated the accuracy of the average precipitation recorded during this period. It is worth noting that farmers in the study region rely on rainfall to irrigate their crops, although they often supplement it with irrigation to enhance crop productivity.

Meteorological Data

Figure 2 presents the data on the mean daily maximum and minimum temperature, relative humidity, wind speed, sunshine, and hours, collected from the Ministry of Agriculture and Water Resources of Erbil for two stations during the year 2021.

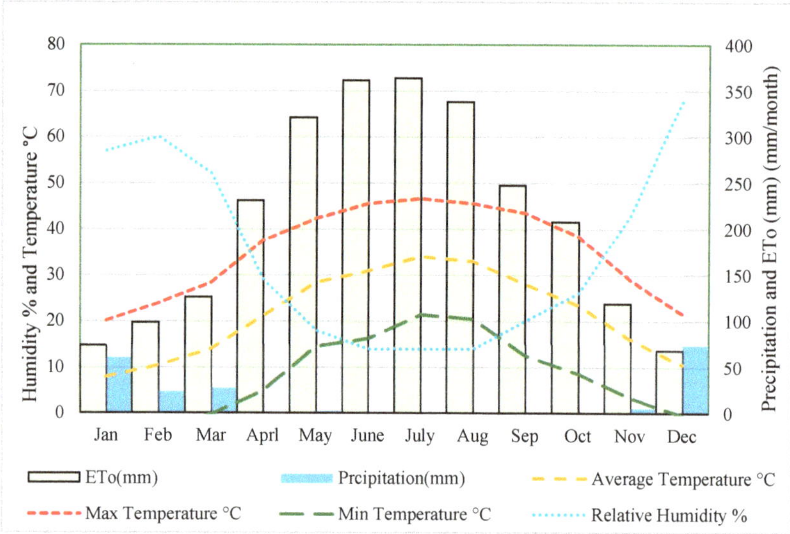

Figure 2. Monthly precipitation, relative humidity, actual evaporation, maximum, minimum, and mean temperature of North Erbil and surrounding areas recorded for 2021 years.

2.3. Sentinel 2 Satellite Imagery Acquisition

This study used open-access satellite data from the Sentinel 2 missions to calculate the vegetation and water indices as a proxy for factors. Sentinel-2 is a 10 m high-resolution, wide-swath, multispectral imaging mission that supports Copernicus Land Monitoring investigations, such as plant, soil, and water cover monitoring, and observation of both inland waterways and coastal regions [38]. Thirteen spatial resolution spectral bands captured by Sentinel-2 represent TOA reflectance, including four bands at 10 m, six at 20 m, and three at 60 m spatial resolution. In addition to data from the Sentinel 2 missions, detailed meteorological and secondary data from the literature were required, including corn winter wheat crops for monthly, seasonal periods in 2021, with season intervals ranging from January to May. The Copernicus Open Access Hub was used for this investigation to obtain Level-1C products of Sentinel-2 A/B satellite data. These products contain top-of-atmosphere (TOA) reflectance values for 13 spectral bands with spatial resolutions ranging from 10 to 20 m. The Sentinelsat Python API was used to obtain the data for the study area and time period of interest, which were subsequently preprocessed in Google Earth Engine (GEE) to acquire surface reflectance values. The preprocessing procedure involved resampling the data to a standardized spatial resolution of 10 m, masking out clouds and cloud shadows utilizing the SCL band, and applying a bi-directional reflectance distribution function (BRDF) correction to consider the impact of directional effects induced by surface roughness and slope.

In order to enable more in-depth analysis and visualization, the classified Sentinel 2 data underwent a process of division into smaller tiles. This was accomplished utilizing the ee.data.getTileUrl function in the GEE cloud-based platform, which allowed the extraction of individual tiles based on their geographical coordinates and zoom level. The resulting tiles were then saved in the GeoTIFF format, which can be easily imported into QGIS and other GIS software for further processing and analysis. The classified and divided Sentinel 2 data were subsequently subjected to a range of statistical analyses, including frequency distributions, cross-tabulations, and spatial autocorrelation analysis, which were performed using both GEE and QGIS software. These analyses were employed to quantify the spatial patterns and relationships between different meteorological variables in the study area. Additionally, it is imperative to comprehend the current status of the center pivot's performance for the purpose of determining its potential for improvement. As such, a manual selection process was undertaken, which involved identifying 143 center pivot farms in the vicinity for further investigation.

The biophysical factors were computed using the formula derived from Sentinel 2 Indices, which comprises two satellite sensors (S2A and S2B) that have generated products with a 5-day temporal resolution (at the equator) and a 10 m spatial grid-cell resolution since January 2021. The satellite products from the Sentinel 2 mission were obtained and analyzed using GEE. The QGIS (v 24.3) geographic information system program utilized the Semi-Automatic Classification Tool plugin, with the following options defined in the plugin's menu: (a) Band 4-RED and Band 8-NIR from Sentinel 2; (b) the study area coordinates; (c) the time search windows from 1 January to 21 May in 2021, as well as (Figure 3); and (d) an acceptable imagery cloud cover set to 100%. Figure 1 illustrates a Google Earth satellite image of a farm utilizing a center-pivot irrigation system in the study area.

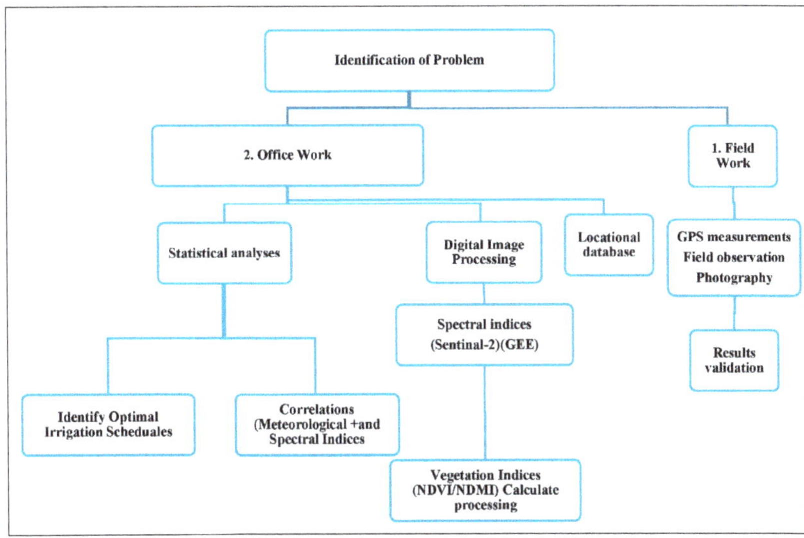

Figure 3. Flowchart of the methodology adopted in this study.

2.4. Validation Fields

The irrigation hours and actual evapotranspiration estimates were validated using available measurements from 14 wheat fields (C1, C2, C3, . . . , C14) from January to May (Figure 4). Comparison has been made the weekly evolution of irrigation, observed actual ET, and the amount of irrigation and rainfall for the 14 wheat fields. A general coherence was observed, suggesting the good performance of irrigation hours in mitigating crop water stress through supplementary irrigation. The center pivot of the study area, characterized by higher production and a regular irrigation system, had higher NDVI-based vegetation density in fields C1 to C14, while fields in other center pivots had lower density.

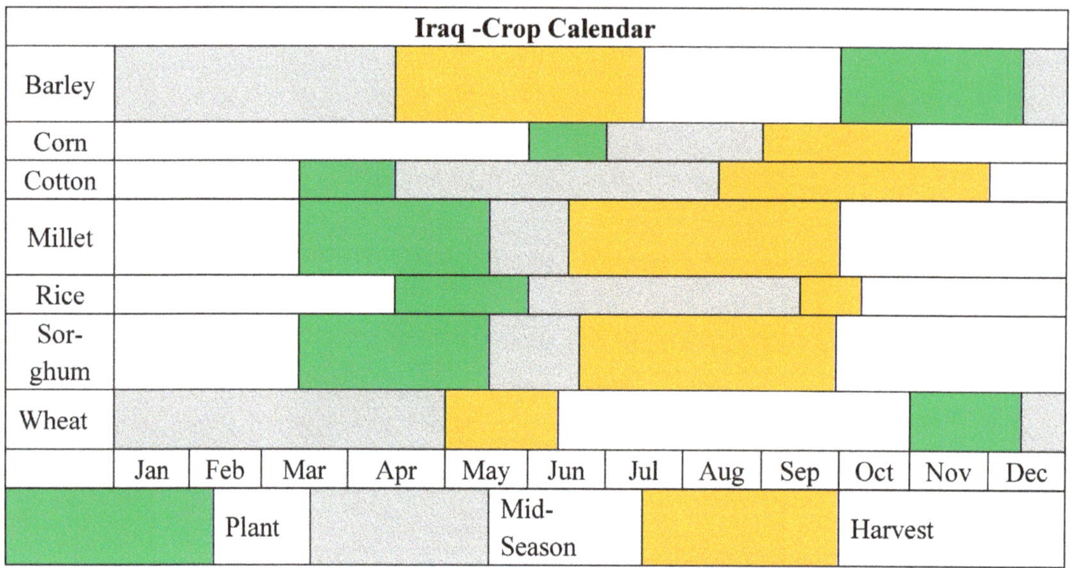

Figure 4. Crop Calendar in Iraq [7].

2.5. Spectral Indices

2.5.1. NDVI

The Near-Infrared (NIR) and Red bands were also applied to calculate the NDVI images [39]. The NDVI index is calculated with the aid of the red (Red) and the Near-Infrared (NIR) bands of Sentinel-2 images, using Equation (1), as follows [39]:

$$\text{NDVI} = (\text{B8} - \text{B4})/(\text{B8} + \text{B4}) \tag{1}$$

where B4 is RED, 664.5 nm, and B8 is NIR, 835.1 nm.

Theoretically, NDVI values ranged between −1.0 and +1.0. However, the typical range of NDVI gauged from vegetation and other earth surface materials is between approximately −0.1 (NIR less than Red) for non-vegetated surfaces and as high as 0.9 for dense vegetative cover. To better understand NDVI values, they are often classified into different ranges. Typically, values between 0.1 and 0.4 indicate low vegetation density or sparse vegetation, while values between 0.4 and 0.6 indicate moderate density or healthy vegetation. Values above 0.6 indicate high density or very healthy vegetation. These NDVI density classes can be beneficial when monitoring vegetation growth over time or comparing different areas. For instance, if an area consistently has NDVI values within the low-density range, it may indicate the need for irrigation or other interventions to promote plant growth. Conversely, if an area consistently exhibits high NDVI values, it may suggest a healthy and productive ecosystem [40].

2.5.2. NDMI

The NDMI normalizes the different moisture response bands between near-infrared (NIR) and shortwave infrared (SWIR) (Equation (2)). The linear correlation between the NIR/SWIR ratio and leaf relative water content was discovered by Hunt Jr and Rock [41]. He calculated NDMI using the following equation:

$$\text{NDMI} = \text{NIR} - \text{SWIR}/\text{NIR} + \text{SWIR} \tag{2}$$

where NIR is the Near-Infrared band and SWIR is the Shortwave Infrared band. The values of these bands can be obtained from remote sensing data, such as satellite imagery.

The NDMI ranges from -1 to 1, where values closer to 1 indicate high moisture content in vegetation, and values closer to -1 indicate low moisture content. In general, vegetation with high moisture content reflects more NIR and absorbs more SWIR, resulting in a higher NDMI value. NDMI is often used in agriculture and environmental monitoring to assess vegetation health and drought conditions. It can be used to detect areas of vegetation stress, monitor changes in soil moisture, and predict crop yields [41].

2.6. Meteorological Indices

Reference Evapotranspiration (ETo), and Crop Evapotranspiration (ETc)

The FAO Penman–Monteith equation is a close, simple representation of the physical and physiological factors governing the evapotranspiration process. Using the FAO Penman–Monteith definition for (ETo), the formula for ETo using the Penman–Monteith equation is:

$$\text{ETo} = (0.408 \times \text{delta} \times \text{Rn} + \text{gamma} \times (900/(T + 273)) \times U2 \times (es - ea))/(\text{delta} + \text{gamma} \times (1 + 0.34 \times U2)) \tag{3}$$

where:

ETo = potential evapotranspiration (mm/day).
delta = slope of the saturation vapor pressure-temperature curve (kPa/°C).
Rn = net radiation at the crop surface (MJ/m2/day).
T = mean daily air temperature at 2 m height (°C).
U2 = wind speed at 2 m height (m/s).
es = saturation vapor pressure (kPa).
ea = actual vapor pressure (kPa).
gamma = psychrometric constant (kPa/°C) [42].

Crop evapotranspiration (ETc) is the amount of water lost from a crop due to evaporation from the soil surface and transpiration from the crop itself. It is a measure of the amount of water required for a specific crop to achieve optimum growth and yield. Various factors, including the climate, soil characteristics, crop type, and stage of growth, influence ETc. The most common method used to estimate ETc is using reference evapotranspiration (ETo), which is the amount of water lost from a reference crop under standardized conditions. Once the ETo is calculated, crop coefficients are applied to adjust the ETo for the specific crop being grown [42].

The resulting value is the crop evapotranspiration (ETc) for that crop at that location and time. ETc is an important parameter in irrigation management, as it determines the amount of water that must be applied to a crop to maintain optimal growth and yield. Over-irrigation can lead to waterlogging and the leaching of nutrients, while under-irrigation can reduce crop yield and quality. Therefore, an accurate estimation of ETc is crucial for the efficient and sustainable use of water resources in agriculture [42]. The ETc equation is expressed as:

$$\text{ETc} = \text{Kc} \times \text{ETo} \tag{4}$$

where:

ETc = crop evapotranspiration (mm/day).
Kc = crop coefficient (dimensionless).

$$ETo = potential\ evapotranspiration\ (mm/day)\ [42].$$

3. Results and Discussion

3.1. NDVI and NDMI

Based on remote sensing data, the NDMI values show more pronounced changes between February and March, potentially due to local farmers perceiving irrigation as unnecessary during January and February's cold weather. The higher variability in NDMI values indicates that it responds to changes in soil moisture content. Similarly, the greater vertical variability in biomass suggests that wheat biomass reacts more significantly to moisture stress during the initial growth stage (February). This implies that crops are more vulnerable to stress in March and April than in other months. A comparison of NDVI- and NDMI-designated fields indicates that irrigation's impact on crop yield varies geographically, with a substantial yield increase observed in areas with abundant light. To maximize yield, it is crucial to avoid water stress during the wheat's most susceptible seasons. However, there was a moderately negative correlation between NDVI and NDMI in 2021. Moreover, the majority of hotspot fields exhibited higher NDMI values, while bright spots showed early growth and greater biomass accumulation throughout the growing stages. This difference may be related to irrigation methods and soil moisture stress, particularly in March and April when the center pivot is essential for reducing temperature and relieving soil moisture stress. However, some farmers do not irrigate during January and February, which may lead to water stress. Remote sensing can detect the effects of water stress on plants with a lag time.

The NDVI has been widely used to investigate the correlation between spectral vegetation variability and growth rate changes. The findings of this investigation have demonstrated that the NDVI values fluctuated between 174.5 area/ha in January, the lowest area of dense NDVI, and 6028.3 ha in April, the highest area. The NDVI Value is a numerical index employed to assess vegetation health and growth. It indicates denser and healthier vegetation, with higher values ranging from -1 to $+1$. The NDVI value categorizes vegetation density into different classes such as Dense, Moderate, Sparse, and Open vegetation. The area of each category is measured in hectares (ha), a common unit in agriculture. Table 1 shows the NDVI Area/ha for Dense Vegetation Classes for different months, which include 1-Jan, 6-Jan, 11-Jan, 26-Jan, 15-Feb, 25-Feb, 7-Mar, 1-Apr, 11-Apr, 16-Apr, 21-Apr, 1-May, 6-May, 11-May, 16-May, and 21-May. The values of the NDVI area/ha for the Dense category during those months are 1174.49, 386.00, 372.47, 549.33, 3191.68, 4536, 5334.15, 6028.25, 3850.27, 3327.33, 2362.02, 1789.78, 1702.87, 801.12, 171.77, and 58.61, respectively. The Dense category signifies areas with high vegetation density, indicating the presence of very healthy vegetation.

According to the data presented in Table 1, there was a noticeable change in the vegetation status of the Center pivot area from January to May. Specifically, the NDVI value of the Center pivot area decreased significantly in May due to the maturity stage of the crops and the increase in ETo, which resulted in a reduction in the agricultural land area. Moreover, the rise in temperature was detectable earlier, particularly during the drought season. Thus, NDVI with NDMI can help improve our understanding of how irrigation and climate change affect the yield and can be used as an early warning for Moisture stress. Moreover, this information can assist farmers and policymakers in making more informed management decisions. In order to achieve the greatest range of NDVI and NDMI values for wheat crops in 143 central pivot areas in north Erbil, we utilized Sentinel 2 images on coincident days. The dataset used for NDVI and NDMI was then split into two subsets: a training dataset and a validation dataset. These datasets were collected from 1 January 2021 to 21 May 2021, covering fourteen Center pivot areas throughout the phonological cycle to monitor various crops. The analysis of Tables 1 and 2 and Figures 5 and 6 revealed that the study region experienced drought episodes over the study period, with March and April 2021 being particularly affected.

Table 1. Statistical indices of measured NDVI value, classes density, and area of each class.

Vegetation Classes	1-Jan-21	6-Jan-21	11-Jan-21	26-Jan-21	15-Feb-21	25-Feb-21	7-Mar-21	1-Apr-21
DENSE	0.60–1.00	0.60–1.00	0.60–1.00	0.60–1.00	0.60–1.00	0.60–1.00	0.60–1.00	0.60–1.00
Area/ha	174.49	386.00	372.47	549.33	3191.68	4536	5334.15	6028.25
MODERATE	0.40–0.60	0.40–0.60	0.40–0.60	0.40–0.60	0.40–0.60	0.40–0.60	0.40–0.60	0.40–0.60
Area/ha	411.69	1294.53	1171.53	2028.78	3269.82	3154.02	3043.28	2296.82
SPARCE	0.20–0.40	0.20–0.40	0.20–0.40	0.20–0.40	0.20–0.40	0.20–0.40	0.20–0.40	0.20–0.40
Area/ha	3601.15	5261.86	4639.91	4554.1	2880.12	1410.72	1161.43	762.6
OPEN SOIL	−1.00–0.20	−1.00–0.20	−1.00–0.20	−1.00–0.20	−1.00–0.20	−1.00–0.20	−1.00–0.20	−1.00–0.20
Area/ha	5055.9	3031.05	3182.37	1495.82	664.35	542.26	443.31	481.79

Vegetation Classes	11-Apr-21	16-Apr-21	21-Apr-21	1-May-21	6-May-21	11-May-21	16-May-21	21-May-21
DENSE	0.60–1.00	0.60–1.00	0.60–1.00	0.60–1.00	0.60–1.00	0.60–1.00	0.60–1.00	0.60–1.00
Area/ha	3850.27	3327.33	2362.02	1789.78	1702.87	801.12	171.77	58.61
MODERATE	0.40–0.60	0.40–0.60	0.40–0.60	0.40–0.60	0.40–0.60	0.40–0.60	0.40–0.60	0.40–0.60
Area/ha	2825.15	1535.55	1229.22	295.54	708.61	1184.93	409.53	126.65
SPASE	0.20–0.40	0.20–0.40	0.20–0.40	0.20–0.40	0.20–0.40	0.20–0.40	0.20–0.40	0.20–0.40
Area/ha	2787.42	3660.23	4406.41	919.25	1600.48	2179.02	2961.71	2577.09
OPEN SOIL	−1.00–0.20	−1.00–0.20	−1.00–0.20	−1.00–0.20	−1.00–0.20	−1.00–0.20	−1.00–0.20	−1.00–0.20
Area/ha	442.22	626	763.32	1662.93	4208.1	5691.78	6222.15	7201.58

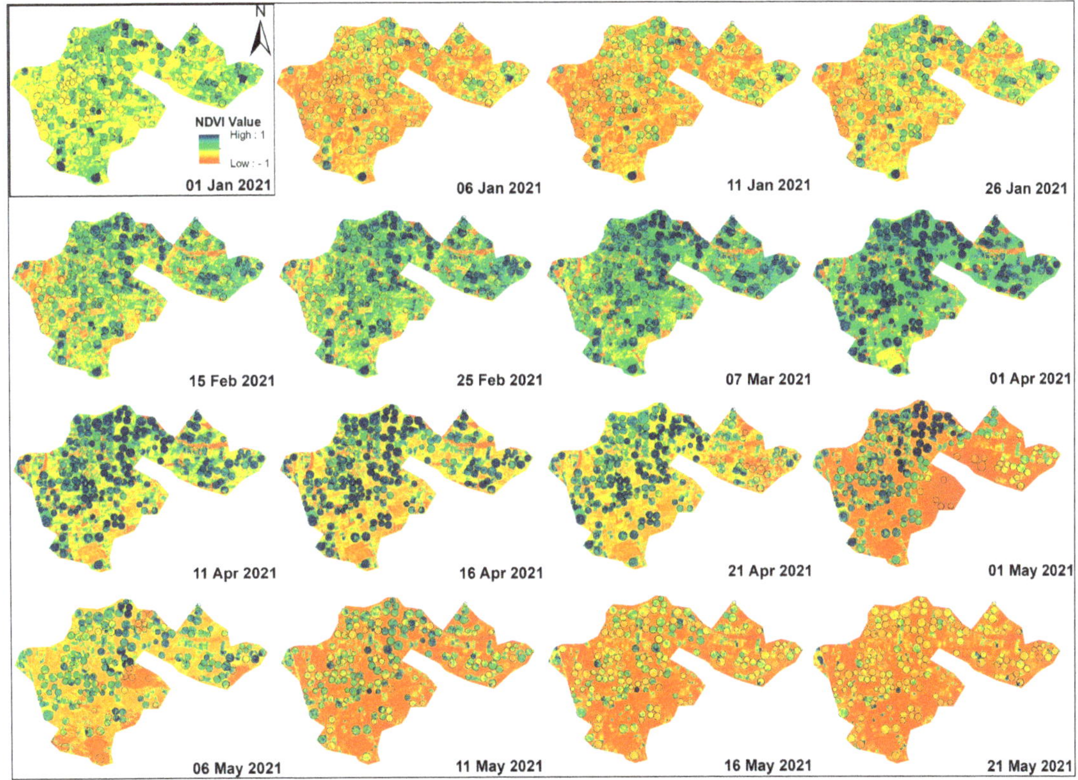

Figure 5. Temporal Variation of the NDVI Value-Based Vegetation Density Classes of Wheat Field in 2021.

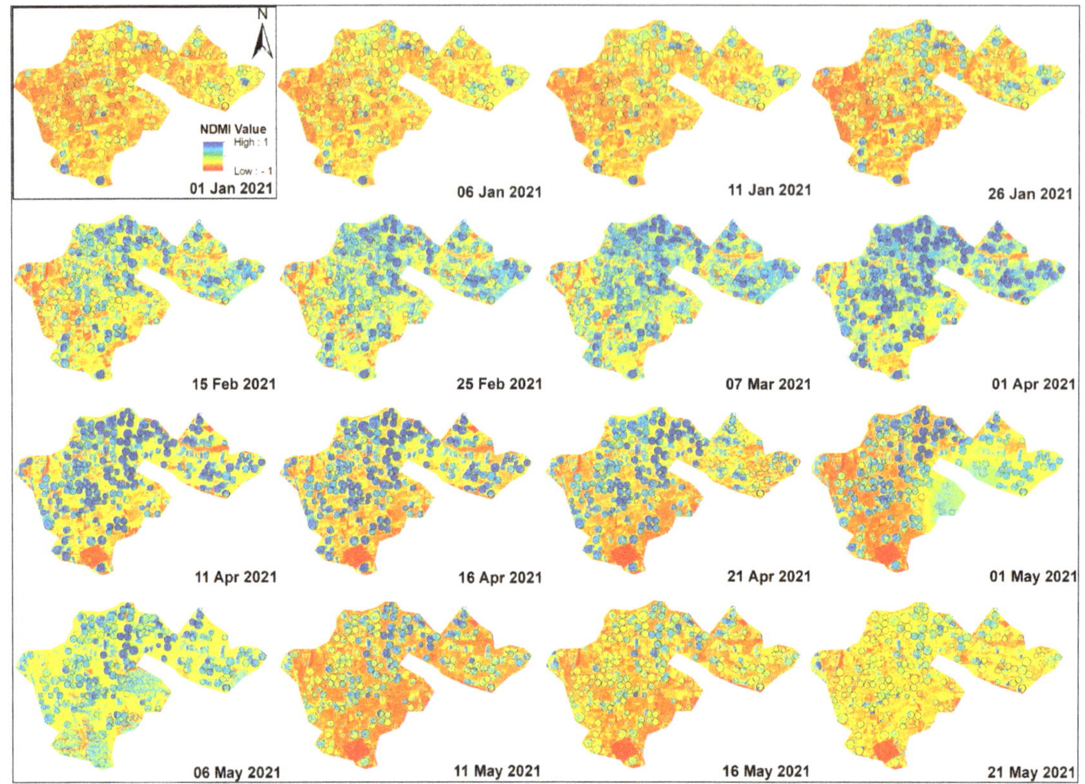

Figure 6. Temporal Variation of the NDMI Value-Based Vegetation Density Classes of Wheat Field in 2021.

Table 2. Statistical indices of measured NDMI Value, and area of each Classes.

NDMI Classes	1-Jan-21	6-Jan-21	11-Jan-21	26-Jan-21	15-Feb-21	25-Feb-21	7-Mar-21	1-Apr-21
HIGH	0.50–1.00	0.50–1.00	0.50–1.00	0.50–1.00	0.50–1.00	0.50–1.00	0.50–1.00	0.50–1.00
Area/ha	0.28	1.68	3.72	0.04	65.96	115.88	239.64	337.28
MODERATE	0.30–0.50	0.30–0.50	0.30–0.50	0.30–0.50	0.30–0.50	0.30–0.50	0.30–0.50	0.30–0.50
Area/ha	63.12	90.56	95.48	184.76	1618.2	2329.76	2961.12	3515.56
LOW	0.10–0.30	0.10–0.30	0.10–0.30	0.10–0.30	0.10–0.30	0.10–0.30	0.10–0.30	0.10–0.30
Area/ha	354.76	667.24	779.72	1700.92	3576.84	4301.56	4486.52	4859.68
OPEN SOIL	0.00––1.00	0.00––1.00	0.00––1.00	0.00––1.00	0.00––1.00	0.00––1.00	0.00––1.00	0.00––1.00
Area/ha	9563.54	9250.72	7202.48	3567.1	4781.4	3066.52	2331.81	921.62

NDMI Classes	11-Apr-21	16-Apr-21	21-Apr-21	1-May-21	6-May-21	11-May-21	16-May-21	21-May-21
HIGH	0.50–1.00	0.50–1.00	0.50–1.00	0.50–1.00	0.50–1.00	0.50–1.00	0.50–1.00	0.50–1.00
Area/ha	877.52	979.24	5.12	367.44	37.7	5.48	2.52	1.84
MODERATE	0.30–0.50	0.30–0.50	0.30–0.50	0.30–0.50	0.30–0.50	0.30–0.50	0.30–0.50	0.30–0.50
Area/ha	2616.08	2125.84	2313.64	1187.68	1005.96	594.6	86.76	40.36
LOW	0.10–0.30	0.10–0.30	0.10–0.30	0.10–0.30	0.10–0.30	0.10–0.30	0.10–0.30	0.10–0.30
Area/ha	3554.6	1700.84	1850.08	1024.64	1647.48	1840.68	361.7	489.92
OPEN SOIL	0.00––1.00	0.00––1.00	0.00––1.00	0.00––1.00	0.00––1.00	0.00––1.00	0.00––1.00	0.00––1.00
Area/ha	2968.74	3588.14	4579.58	671.2	5759.16	2529.62	8867.86	9504.2

Table 1 shows the correlation between drought and reduced vegetation area/NDVI mean values, further supporting the importance of water availability for vegetation growth. It is also possible that other factors, such as temperature and nutrient availability, may contribute to the observed differences in vegetation indices between the two fields. There is a clear difference in the vegetation indices (NDVI and NDMI) between the irrigated and rain-fed fields during the growing season months of March and April. The irrigated field shows a significant change in NDVI values and NDMI, which may indicate increased vegetation growth due to water availability. On the other hand, the rain-fed field experiences drought during these months, which is correlated with reduced vegetation area and lower NDVI mean values. It is worth noting that NDVI is a commonly used index for evaluating the health and growth of vegetation, as it quantifies the chlorophyll content in plant leaves. On the other hand, NDMI is an indicator of vegetation water content and can offer insights into soil moisture conditions [7].

Consequently, the observed dissimilarities in NDVI and NDMI values between irrigated and rain-fed fields imply that water availability is a crucial determinant of vegetation growth and productivity. Nonetheless, it was observed that NDMI values increased significantly during the months of March and April, especially in the irrigated field, when the crops were subjected to 72 to 96 h of irrigation per month. In contrast, the NDVI value showed very little vegetation, particularly in the non-irrigated section of the study area (see Figures 5 and 6).

These findings are consistent with those of a previous study by [43]. The temporal fluctuations in NDVI and NDMI demonstrated that the drought resulted in a decline in the value and extent of spectral indices from normal to extreme levels, with the non-irrigated area being the driest. Conversely, the field where supplementary irrigation was applied during the specified period had high NDVI coverage. The NDVI values revealed that the non-irrigated area in the study region was affected by drought (see Figures 5 and 6). However, the severity and spatial extent of the drought varied. Low values indicated a dry season, whereas high values indicated a wet season [44]. From January to May, the NDVI values exhibited temporal and spatial variations across the region (Figures 5 and 6), primarily attributed to differences in precipitation amount, frequency, and intensity [45]. Moreover, meteorological factors such as precipitation, temperature, and relative humidity were crucial in influencing NDVI variability [46]. The spatial variability of NDVI values was determined by several factors, including climate, soil, temperature, ETo, and ETc fluctuations (refer to Table 3).

Table 3. The fourteen fields were monitored during five months of irrigation, with irrigation hours used for each center pivot by month.

| | | | | | | | | | Irrigation/h/mm | | |
Location	Longitude	Latitude	Yield Kg/ Hectare	January	February	March	April	May	Total/ Hours	Mm/ Hours	Mm/ Season
1	43.96674	36.35929	3120	24	48	72	72	18	234	2.1	491.4
2	43.96334	36.36667	3400	36	48	86	72	18	260	2.1	546.0
3	43.96235	36.37074	3320	24	36	72	72	12	216	2.1	453.6
4	43.96983	36.37265	2960	24	48	72	72	12	228	2.1	478.8
5	43.97896	36.37460	3520	36	48	96	96	6	282	2.1	592.2
6	43.98112	36.36880	3360	24	48	72	80	6	230	2.1	483.0
7	43.98278	36.36585	3400	12	48	72	72	12	216	2.1	453.6
8	43.97472	36.37033	3400	12	36	72	80	12	212	2.1	445.2
9	43.97036	36.36933	3800	24	48	72	72	12	228	2.1	478.8
10	43.97635	36.36438	4800	24	48	96	96	12	276	2.1	579.6
11	43.97226	36.36283	3520	18	36	72	80	6	212	2.1	445.2
12	43.97428	36.36111	4400	24	48	72	96	24	264	2.1	554.4
13	43.95905	36.28151	5000	24	48	76	96	12	256	2.1	537.6
14	43.97158	36.28487	4800	36	48	76	96	18	274	2.1	575.4

The NDMI is a remote sensing metric used to measure how much water is in vegetation. Higher values mean that the vegetation has more water. Such areas may be irrigated or have access to other sources of water. The NDMI values for certain months and areas. Larger values mean that plants are especially healthy because there is a lot of water. Specifically,

NDMI values for the High Values category in the months of January through May. The values for this category are 0.28, 1.68, 3.72, 0.04, 65.96, 115.88, 239.64, 337.28, 877.52, 979.24, 5.12, 367.44, 37.7, 5.48, 2.52, and 1.84, respectively. High Values in the table indicate areas with abundant water in the vegetation, which signifies particularly healthy vegetation. Both the NDVI and the NDMI are commonly used as remote sensing indices to measure how healthy plants are and how much water they have. This information could be used to improve irrigation efficiency and determine how much water plants need based on satellite data. NDVI is typically used to estimate crop yield, vegetation cover, and photosynthetic activity. In this study, NDVI was used to find plants that were water stressed, which is a common problem in agriculture that uses water. Farmers can improve irrigation schedules, increase crop yields, and waste less water by keeping an eye on NDVI values.

The NDMI index is utilized for the estimation of soil moisture and water stress in crops, and it can aid farmers in adjusting their irrigation schedules to optimize water usage and improve crop yield. Changes in NDVI and NDMI values prior to and following irrigation can assist farmers in gauging the amount of water required to attain the desired vegetation growth or moisture content. The NDMI values were observed to be at their minimum during January and May, indicating reduced vegetation and lower NDVI values (Figures 5 and 6), which could be attributed to decreased rainfall during those months. However, during February, March, and April, NDMI values increased, which was positively correlated with the application of supplementary irrigation, suggesting that irrigation had a beneficial impact on vegetation growth during these months. The decline in NDMI values during the growth season could be attributable to the adverse effects of high temperatures on vegetation growth. The study area witnessed a significant decrease in annual rainfall averages, which could have impacted overall vegetation growth. The 14 fields received 445.2 mm and 579.6 mm of irrigation water, respectively.

During the 2021 monitoring season, irrigation amounts were measured in Fields C1 to C14 using a water meter, which recorded a total of 350 mm of water. One or two irrigation events were performed during the early growing stage to ensure the establishment of robust young plants. Subsequently, irrigation for Fields C1 to C14 was scheduled based on NDMI sensor measurements to irrigate when the depletion level within the rooting zone was ready to drop 60% of the total available below. An algorithm was executed with a crop file created based on Fields measurements and observations and dynamically modified during the cultivation season by adjusting weather parameters. Field-specific constraints, such as not allowing irrigation events to occur more often than once every four days, were integrated into the scheduling process. In cases where the generated irrigation schedule was not followed (e.g., due to an electrical failure in the pumping station or damage to the irrigation system), the algorithm was re-executed, and the irrigation schedule was adjusted accordingly. Despite these adjustments, consistent violations of the first depletion threshold occurred, as irrigation was applied either before the threshold was reached or after the water content had fallen below it.

In dryland areas, supplemental irrigation is often necessary to support crop growth, especially during the winter months from January to May. The region employs various irrigation technologies, including Center pivot irrigation and homogeneous irrigation systems. Monitoring vegetation variability using central pivots is feasible, and the NDMI is a useful input variable for irrigation prediction models, as it correlates with the NDVI, which is considered moderate in Center Pivot techniques (as demonstrated in Figures 6 and 7). It is crucial to validate modeling results to ensure their reliability. Figure 8 shows low variability, indicating that the modeling approaches used in this study were robust. The study utilized a combination of remote sensing techniques and meteorological data to evaluate the efficacy of irrigation systems, and the results indicated that this approach was reliable and robust. Remote sensing was employed to collect data on the water supply to crops and its impact on crop growth, while meteorological data provided information on weather patterns, including precipitation, temperature, humidity, ETo, and ETc, which affected crop growth and irrigation needs. The integration of these two data types enabled the development of a comprehensive model for assessing irrigation system performance. The study found that this approach was effective in providing accurate information about the system's performance, which could help optimize irrigation methods and enhance crop yields.

Notably, all methodologies employed in this study were linear based. Ideally, a well-designed irrigation system would ensure uniform water distribution, which would result in homogeneous water consumption patterns across fields. However, several factors, including deficient infrastructure, inadequate management practices, soil type, water quality, and fertilization, can lead to non-uniform water use across fields and zones. The amount of irrigation water applied in each event was measured through the use of rain gauges positioned above the crop canopy (see Figures 7 and 8). This methodology was proposed and implemented by María [47].

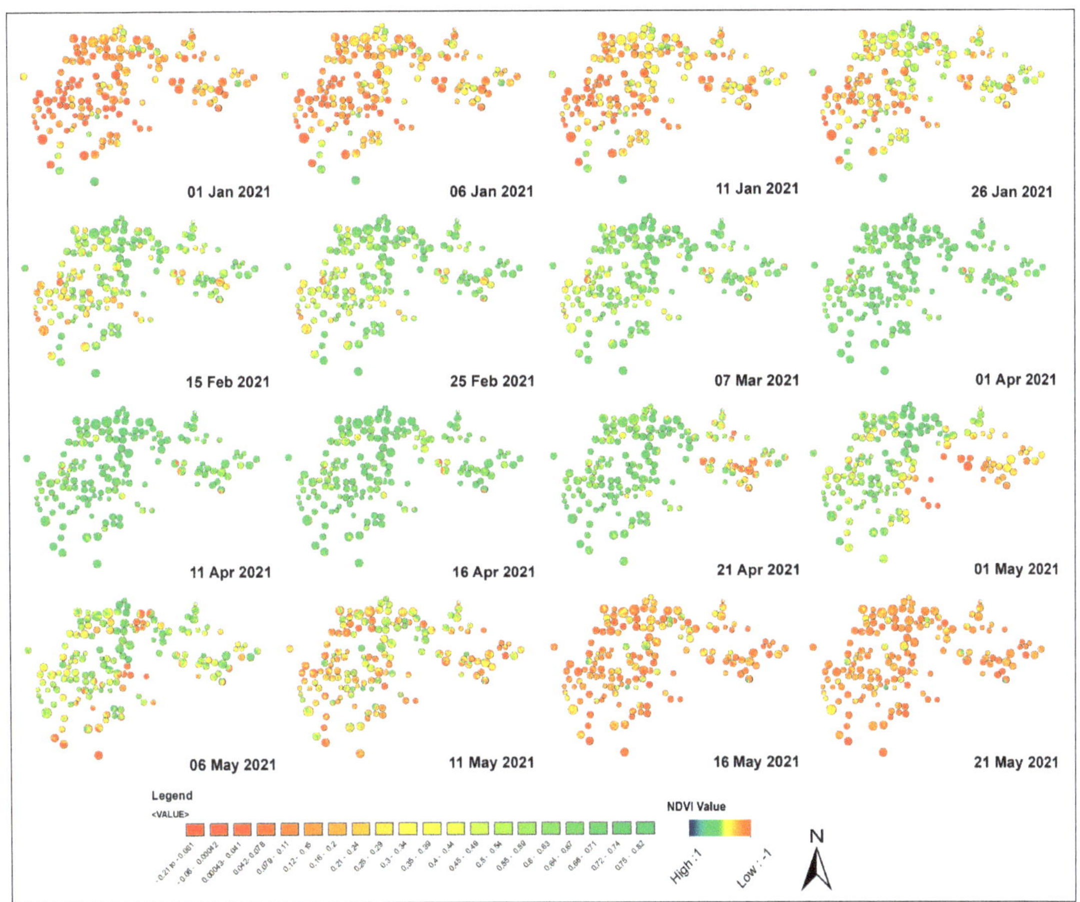

Figure 7. Temporal Variation of the NDVI Value of Centre pivot fields in 2021.

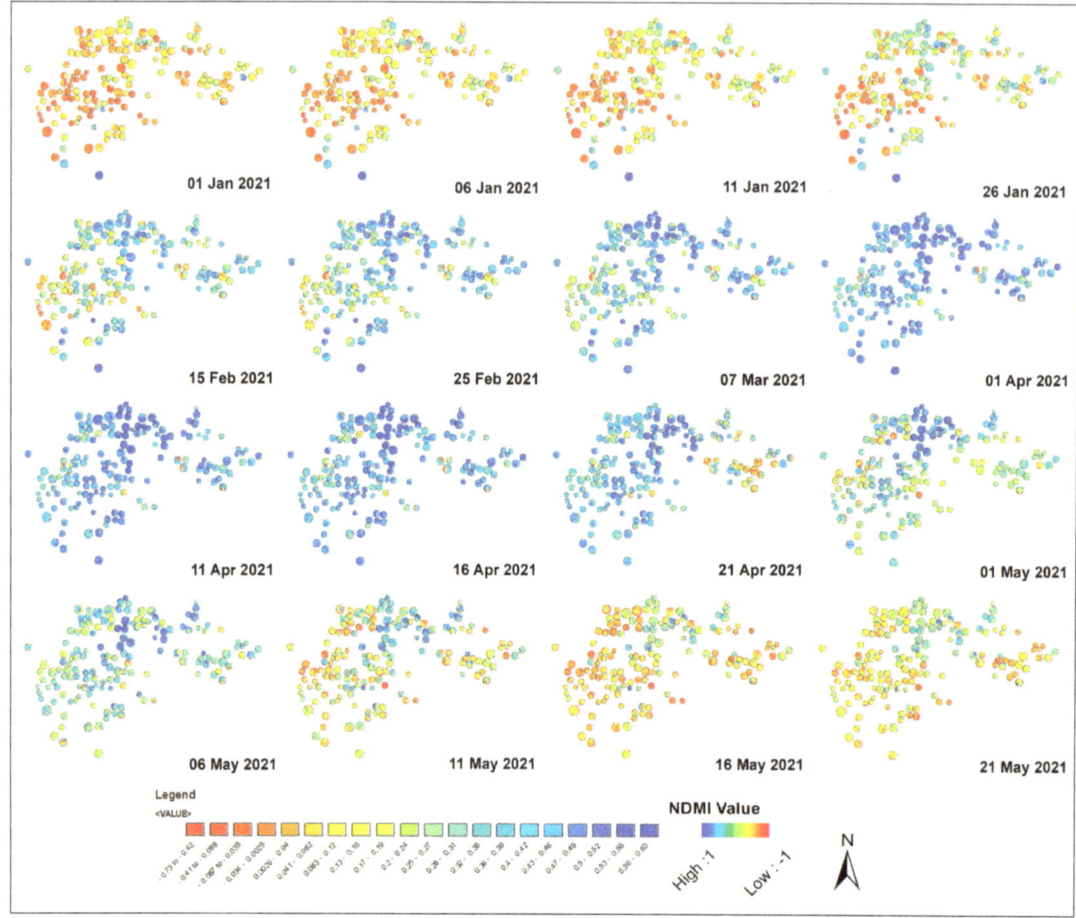

Figure 8. Temporal Variation in the NDMI Value of Center pivot fields in 2021.

3.2. Crop Water Use and Water Use Efficiency

During the period from January 2021 to May 2021, a total of 14 Center pivot (C) fields were subjected to rigorous monitoring to ascertain their irrigation schedules, crop growth, and soil moisture profiles, enabling the assessment of water productivity. The irrigation schedules were established by the farmers, who relied on weather forecasts to optimize their plans. In this regard, the irrigation plan was adjusted to compensate for the lack of spring precipitation, thereby ensuring a reliable supply of water for crop development in the long term. However, this approach was unable to account for the non-linear effect of water shortages during specific phonological stages on crop yields. Irrigation was carried out using the Center pivot technique, and the monthly amount of water was distributed evenly over two to three irrigation events, spaced two weeks apart. The first two rounds of watering were conducted in January, following a principle that was empirically established by the farmers as optimal for maintaining the moisture content of Vertisol, while minimizing water losses due to frequent irrigation. The remaining fields at each site served to verify the measurements obtained from the intensively monitored fields. Groundwater served as the primary source of irrigation water in the wheat fields. While direct measurements are typically more accurate, transpiration and evapotranspiration are challenging to measure and must be derived from the Cropwat table).

In Table 3, the validation of irrigation measurements was conducted for 14 wheat fields (C1, C2, C3, … , C14) from January to May. The validation included a comparison of the weekly evolution of irrigation and observed actual ET, along with the amount of irrigation and rainfall for each of the 14 wheat fields. The validation fields exhibited a high level of coherence, indicating the effectiveness of the irrigation hours in mitigating crop water stress through supplementary irrigation. In April, the center pivot of C12, C13, and C14 with regular irrigation had higher vegetation density and yield, as evidenced by NDVI, whereas fields in other center pivots had lower density. Integrating remote sensing techniques and meteorological data can provide valuable insights into the performance of an ideal irrigation system, which can enhance crop productivity. The study results suggest that remote sensing and meteorological data integration can facilitate the identification of optimal irrigation schedules and strategies for specific crops and regions. By monitoring vegetation health using remote sensing techniques, farmers and irrigation managers can adjust irrigation schedules to match crop water requirements, thereby reducing water waste and improving crop productivity. More frequent irrigation may be necessary in hot and dry conditions, whereas irrigation may not be required for a certain period.

3.3. NDVI and NDMI Time Series of Wheat Crop

In the period from January to May 2021, some center-pivot fields were planted with a single crop type, while others had alternating crop types, resulting in variations in NDVI and NDMI time series behavior. To examine this variation, NDVI and NDMI were compared for the Wheat Crop. Specifically, NDVI and NDMI were identified from 143 fields where only Wheat Crop was planted throughout the study period. The fields, each with a diameter of approximately 800 m, were divided into 2500 pixels, and the mean value for each field was determined by averaging the NDVI and NDMI of these pixels. Crop statistics were derived from the NDVI and NDMI time series data for the wheat crop, including maximum (Max) and minimum (Min) values, amplitude variation (Amp), and standard deviation (Std), to establish differences in time series behavior (see Appendix B Tables A2 and A3). In March and April, NDVI and NDMI values fluctuated significantly more in irrigated lands than in non-irrigated lands, possibly due to the influence of ETo and temperature on vegetation growth. To assess the changes in vegetation, the changes in classes (1, 2, 3, and 4) for NDVI and Dense, Moderate, Sparse, and Open soil classes were calculated. The category with the greatest change relative to the other four was then identified as the dominant category.

To demonstrate changes over time, the relative dominance of each class category was computed for each period. The NDVI and NDMI indices revealed that the non-irrigated portions of the field experienced drought conditions in February, March, and April, as indicated by the low NDVI values. The severity and extent of the drought varied over time and space. The values for January to May displayed spatial and temporal variability across the region, which can be attributed to the prevailing temperature and humidity conditions. The fluctuation in NDVI values is influenced by weather-related parameters such as rainfall, temperature, and humidity levels. Temporal variations in climate, soil, and temperature cause changes in NDVI values.

3.4. Evolutions of Observed NDVI and NDMI

While Figures 7 and 8 focused on the relationship between NDMI-based soil moisture and NDVI-based vegetation health in Wheat Crop, we aim to assess the possibility of detecting irrigation events using Sentinel-2 data by analyzing their association with real-world data. As illustrated in Figure 8, the NDMI in the center pivot area that was irrigated was notably higher than that in the nearby bare soil area.

In the Center pivot field, soil moisture levels were high due to irrigation, while the other field showed significant variability in soil moisture levels before June 2021. The wheat crop was grown and harvested frequently from January to June 2021, with irrigation durations ranging from 12 to 96 h, as per farmer records. The field was planted with maize from August 2020 to December 2020, and irrigation maintained high soil moisture levels during that period. After the maize was picked, irrigation stopped until January 2021. This led to a sharp drop in the amount of water in the soil. However, soil moisture increased in mid-February 2021 after an 18 mm precipitation event from 11–15 February, which was captured by the Meteorological Station and resulted in an increase in both NDVI and NDMI (Figure 8). High temperatures are believed to impact vegetation strength and cause plant stress. The mean NDMI values increased in all center pivot irrigated fields, consistent with the trend of NDVI variation (see Figures 7 and 8). Although a lack of precipitation is typically the primary cause of drought, elevated potential evapotranspiration is associated with temperature and relative humidity. The evidence presented indicates that soil moisture plays a limiting role in the actual evapotranspiration process, especially during drought conditions, as it prevents excessive drying. Temperature and other factors that influence vegetation development and phenology also have an impact on actual evapotranspiration [48]. Thus, NDVI is a reliable sign of drought, and it is important to use empirical NDVI-NDMI-based indices to improve the performance of irrigation systems and increase crop yields. The average NDVI, which reflects the greenness of vegetation, is strongly linked to seasonal rainfall, indicating the potential use of NDVI as a drought predictor.

3.5. Correlation Matrix between NDVI and NDMI

Figure 9 displays the average NDVI time series profile for 2021, which shows a significant correlation between NDVI and NDMI. The projected green cover increased from about 0.40 to 0.80 between February and April, likely due to irrigation, before dropping to 0.6 by the end of April as irrigation decreased in preparation for the harvest season. The average NDVI increased due to an increase in NDMI but then dropped to 0.20 in May after the maturation and harvesting phase. Sentinel2 was able to detect changes in vegetation or physiological density accurately. All monthly index values were significantly correlated with meteorological parameters from various locations with a 95% confidence level. NDVI increased at several locations in the 143 center pivot and rain-fed area due to increased rainfall, irrigation, and temperature. The combination of moisture content and temperature played a significant role in reducing NDVI values. Ongoing land and crop degradation and yield losses occurred due to a lack of vegetation cover and relative humidity caused by insufficient rainfall during the growing season (Table A1). Increased temperatures can have severe natural consequences, increasing ETo and ETc. This suggests that crop development was not uniformly distributed across the region, even in times of water shortage, and some farmers were able to achieve acceptable crop growth by effectively using water through supplementary irrigation and implementing good agricultural practices. Remote-sensing-based NDVI and NDMI can provide valuable insights into the spatial and temporal distribution of crop irrigation water requirements at the field level, which can be compared to in situ monitoring of observed irrigation water supplied at the irrigation district scale. Acceptable estimates of ET0 and ETc for wheat (73.7, 98.4, 125.1, 229.4, and 319.2 mm/month) and (42.7, 57.1, 93.8, 229.4, and 383.0 mm/month) were obtained under various weather and water management conditions with irrigation periods of 12, 24, 48, 72, and 96 h, respectively, according to farmer records for the months from January to May. Figure 9 provide further details on this.

The significant rise in temperature can be attributed to the absence of rainfall, which caused a lack of moisture and resulted in higher ETo and ETc. One of the consequences of the increased ETc in April and May in the study area is the elevated temperature and decreased NDVI (Figure 9). This indicates the adverse effect of high temperature on the vegetation growth environment, resulting in lower vegetation values and area (NDVI) in the region. Only a few Center pivot areas in the study saw an increase in NDMI values, which was reflected in a vegetation increase in NDVI. Due to its location, Iraq's climate has transformed into a semi-arid climate, with a heavy influence from the Mediterranean climate characterized by hot, dry summers and warm, wet winters. Rainfall, temperature, sunshine, and wind speed have a negative impact on wheat yield and chlorophyll content in semi-arid environments. The study suggests that the irrigation schedule followed by farmers compensated for the insufficient spring rainfall, resulting in sustainable crop growth. However, this approach did not consider the non-linear effects of water scarcity during specific phenological stages on crop yield. The newly proposed approach helps prevent harmful stress on wheat and ensures the proper distribution of water based on agronomic principles in a center pivot system. These results align with the findings of Polinova et al. (2019) [25].

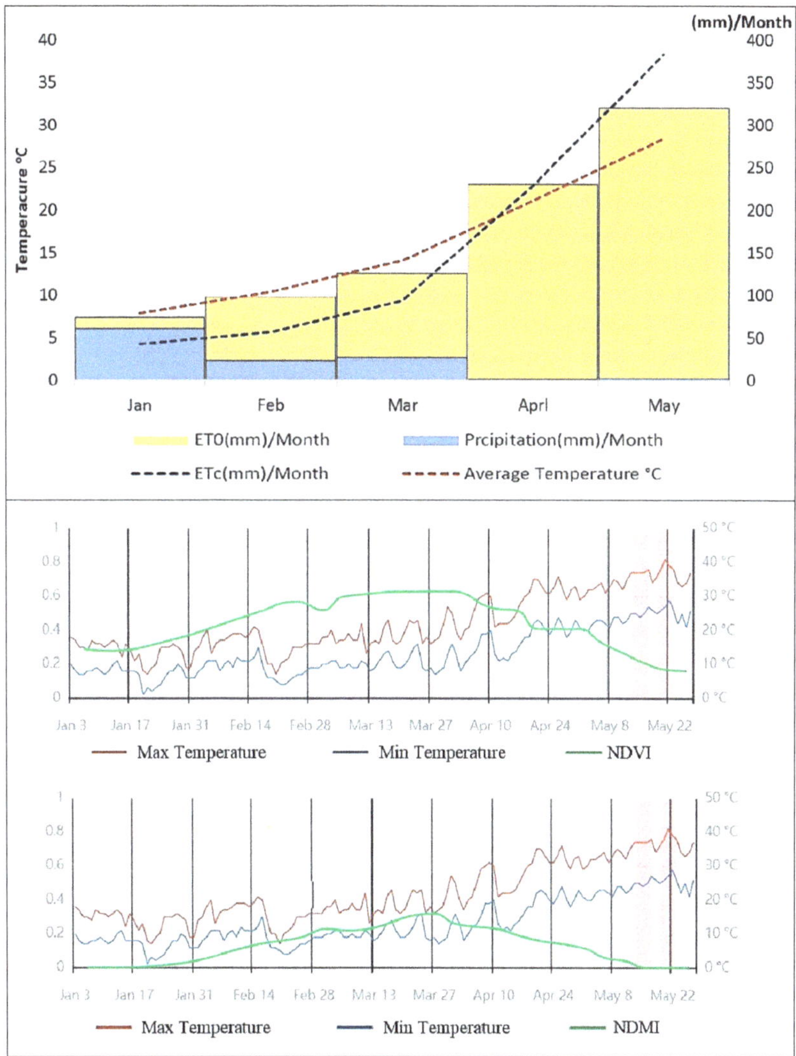

Figure 9. Monthly Precipitation, ETo, ETc, NDVI, NDMI, Average Temperature in study area recorded in 2021.

The NDVI has been associated with numerous vegetative properties, including photosynthetic capacity. Sufficient irrigation must be provided for a larger wheat yield to encourage strong output intensity. Figures 7 and 8 demonstrate the observed irrigation water supply distribution, showcasing clear shifts from one month to another in both the applied and required irrigation water. The different shapes of the irrigation distribution reveal that farmers' use of irrigation water varies significantly from month to month, which may be influenced by both farmers' irrigation decision making and the small portion of crop water requirements met by rainfall, which was higher in the dry months of January and February 2021. During the research period (March), there was a slight increase in the average NDVI, although the increase was more noticeable. The average NDVI remained steady throughout the growing seasons (March to April). Even though 90% of the precipitation occurred during the growing seasons, annual precipitation varied greatly from 60.8 mm in January to 2.6 mm in May 2021 (Figures 9 and 10). These significant changes seemed to impact crop development, suggesting that sufficient irrigation water was available to support wheat growth throughout the growing season. This study defines water consumption as the amount of water that evaporates and seeps from an agricultural zone rather than the amount used for crop irrigation or diverted for that purpose. It was shown that by combining remote sensing and GIS techniques, a comprehensive evaluation of irrigation efficiency can be achieved for different irrigation systems [49].

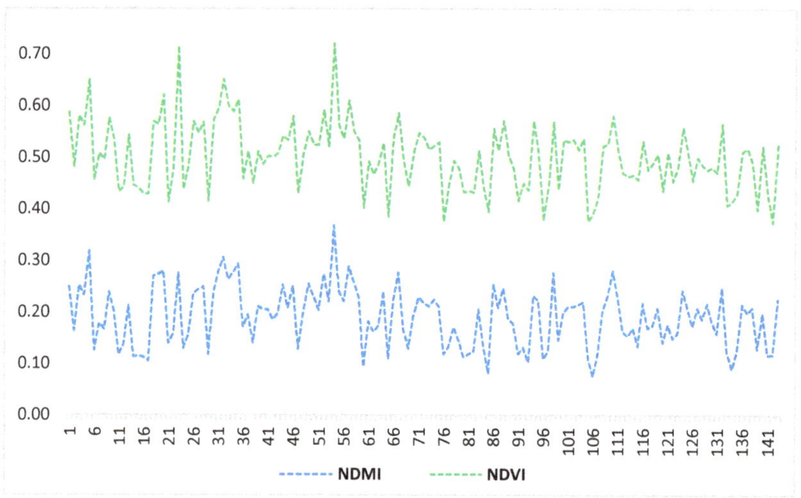

Figure 10. Temporal Variation of the NDVI and NDMI Value Based Vegetation Density Classes of 143 Center pivot wheat Field in 2021.

3.6. Crop Growing Period (NDVI)

Figure 11 displays the vegetation cover trend over a period of five months, indicating an upward trend in January, reaching its highest point in March and April, and declining to the lowest level at the end of May, which is consistent with the reported rates. The fluctuations in NDVI values during these months are heavily influenced by the precipitation and irrigation water requirements, as evidenced by the strong correlation between NDVI and NDMI, which varied in unison with an R2 of 0.904 (Figure 11). Remote sensing provides a holistic perspective of ground conditions. This study utilized available data to explore the correlation between irrigation schedules, crop yield, and remote sensing data. The final stage of the research involved a comparison of the findings obtained through the analysis of remote sensing data. By considering all the environmental factors that impact crop growth, and by evaluating the performance of irrigation, the study aimed to determine the root causes of variations in NDVI and NDMI values [50]. Nonetheless, it is worth mentioning that certain factors may not be pertinent or suitable in all situations [51,52]. The study period covered five months from January to May, with one growing season per year for wheat (Figure 4). The variability of climatic influences on drought, temporal changes, and their relationship with precipitation, irrigation hours, ETo, ETc, and temperature is evident (Figure 12).

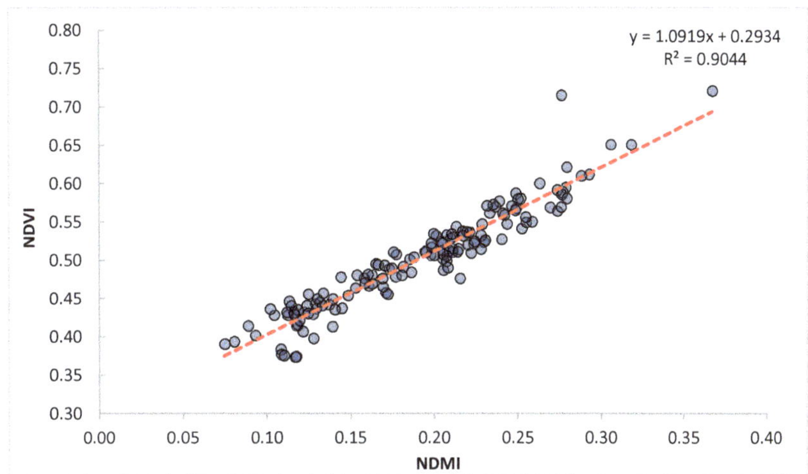

Figure 11. NDMI Correlation with NDVI-wheat in 143 Center pivot wheat Field -2021.

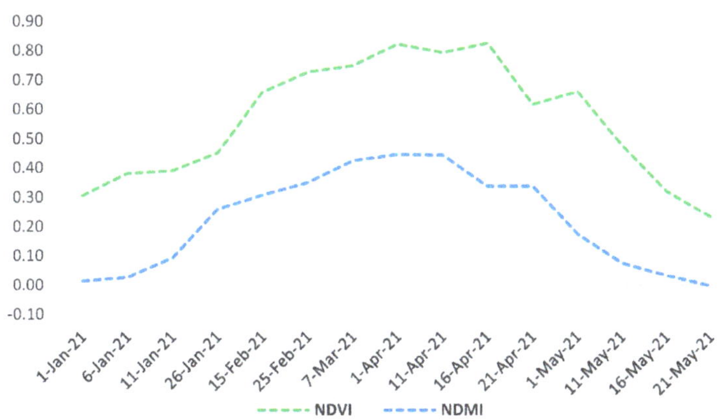

Figure 12. Temporal variation of the NDVI and NDMI value-based vegetation density classes of Center pivot wheat field from 1 January to 21 May 2021.

The NDVI is influenced by ETo, soil temperature, and irrigation, resulting in correlation values of 0.49, 0.55, and 0.76, respectively. Despite this, the growth rate of the crop is significantly affected by the NDMI. Although there is a strong relationship between NDMI and irrigation, with a correlation of approximately 0.91, it is not statistically significant, as shown in Figure 11 and Table A1. By analyzing 143 selected fields, it was found that the correlation between NDVI and NDMI increased the yield output when sufficient irrigation was provided in February, March, and April (Figure 11). Using EXLSTAT, the correlation coefficients for NDVI, NDMI, Irrigation, ETo, and ETc were calculated during the growing season spanning from January to May, an average of five months. These correlation coefficients are illustrated in Figures 12–14.

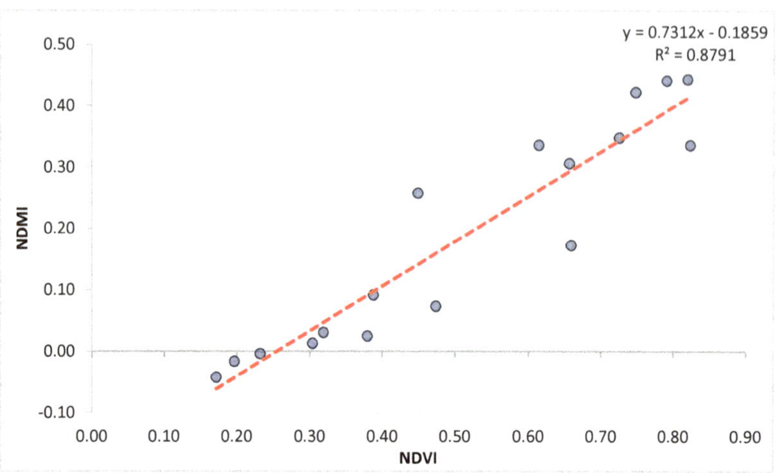

Figure 13. NDMI Correlation with NDVI-wheat in 143 Center pivot wheat field products with a 5-day temporal resolution.

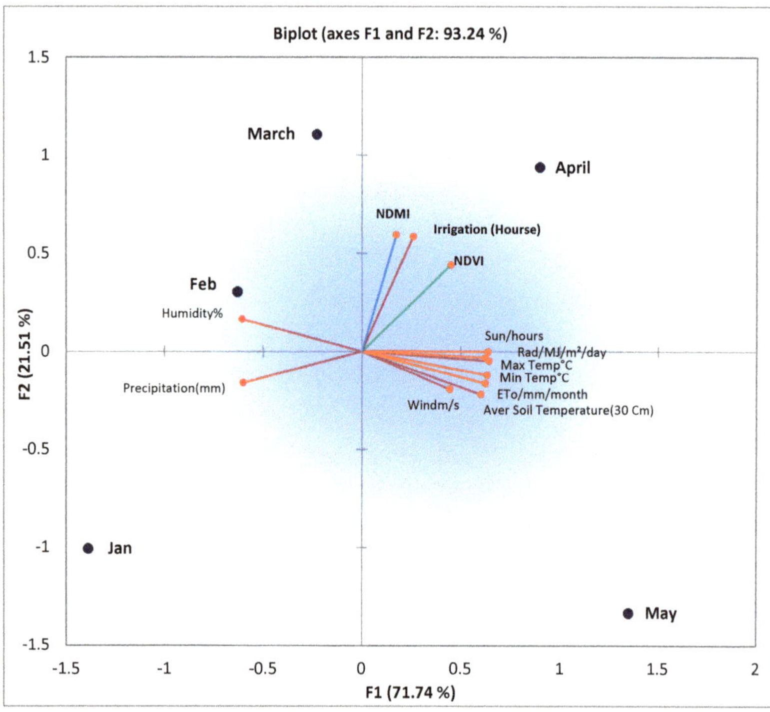

Figure 14. Principal Component Analysis of temporal Pattern changes of NDVI and NDMI with Meteorological parameters.

The findings revealed a strong and positive association between ETc and both NDVI and NDMI. Similarly, there was a positive correlation observed between NDVI and NDMI. These correlations between remote sensing-derived spectral indices and meteorological parameters were statistically significant. After 2020, the planting of maize ceased in the field, and the surface was bare until September 2021. Both time series of NDVI and NDMI present fluctuations in vegetation, with peaks up to 1 and 0.5 for NDVI and NDMI, respectively, occurring during growing seasons. High values with a maximum NDVI of 1 and NDWI of 0.5 were observed in March 2021, which may have been caused by the presence of supplementary irrigation. The study also noticed through the trend of increasing the value of NDVI that the effect of supplementary irrigation during the months of the study was different, and this change mainly depended on the amount of precipitation falling for each month in the study area, and other environmental factors that also play a prominent role in the growth of the wheat crop (Figure 12). The results are consistent with multiple investigations [53–55].

Temperature is a key factor affecting the transpiration process of crops, as demonstrated by the significant fluctuation in ET0 values between January and May, corresponding to the growth and harvest stages of the crop (Figure 13). Additionally, the wheat fields studied through farmer interviews were located in 14 Center pivot fields.

The fields had similar irrigation schedules, with an average of 48 irrigation hours distributed over two periods in January due to sufficient rainfall. However, in February, with a decrease in the precipitation rate and the application of fertilizers, the irrigation requirements of the fields increased to one hour, resulting in the need for three irrigation periods. PCA was utilized to determine the primary modes of variability in a dataset that examined the temporal changes in NDVI and NDMI in relation to meteorological parameters. NDVI and NDMI are remote sensing indices that are commonly used to monitor vegetation and moisture content, respectively (Figure 14). By examining the changes in these indices over time and their connection to temperature and precipitation, PCA identified the crucial factors influencing the changes in vegetation and moisture patterns. To carry out the PCA analysis, a covariance matrix of the dataset was constructed, and the matrix was eigendecomposed to obtain the principal components, which are linear combinations of the original variables that represent unique patterns of variability in the data. Typically, the first few principal components account for most of the variance in the dataset and are used to summarize the dominant modes of variability. The PCA of the temporal pattern changes in NDVI and NDMI in relation to meteorological parameters yielded valuable insights into the underlying factors responsible for changes in vegetation and moisture patterns over time and can aid in the identification of critical variables for monitoring and predicting these changes.

4. Conclusions

The NDMI is a valuable tool for detecting moisture deficiencies in crops and identifying under-irrigated areas. This information can be used to divide fields into zones with different water needs and schedule precision irrigation events as needed. Crop Monitoring provides historical and current precipitation data, NDVI index graphs, and precipitation monitoring graphs. By analyzing seasonal weather and precipitation patterns, it is possible to plan precision irrigation strategies for different fields. In addition, Crop Monitoring offers a 5-day weather forecast for each field, which helps farmers decide on the need for watering activities to ensure proper soil moisture for crops. Precision irrigation is cost effective, providing sufficient water supply to crops in areas with limited rainfall. Crop Monitoring tracks changes in the NDVI for individual fields throughout the season. This helps farmers identify areas of weak and strong productivity across the field and create special maps for variable-rate applications of seeds and fertilizers. NDVI values can vary throughout the growing season, indicating water stress or waterlogging, and can be visualized through maps and graphs. Crop Monitoring also provides current and historical soil moisture data and NDVI index graphs, which help farmers track the correlation between rainfall and moisture levels in the field. NDMI values also vary throughout the growing season and can be visualized through maps and graphs, indicating water stress or waterlogging. A decrease in NDMI values suggests water stress, while abnormally high values could signal waterlogging. Visualization of NDMI through maps and graphs helps farmers detect problem areas in the field and save time and resources. Water supply is a critical factor for plant growth, along with sunlight, nutrients, and soil temperature. While fields in areas with frequent rainfall receive sufficient water for crop growth, additional watering is necessary to maximize yields in semi-arid regions [56–58]

For a crop to progress from germination to the reproductive growth stage, where it can begin to produce grain, it requires at least 100 mm of water. It is worth mentioning that, according to the local meteorological station data, the 2021 crop season had the least precipitation and was classified as a drought season. In this study, the focus was on quantifying the loss of water due to evaporation from soil and transpiration from plants, which is collectively known as evapotranspiration. The most commonly used method for calculating evapotranspiration is to determine the reference evapotranspiration (ETo), which is the amount of water that would be lost from a standardized grass surface under specific conditions [59].

The Penman–Monteith equation is often used to calculate ETo, taking into account weather variables such as temperature, humidity, wind speed, and solar radiation. To adjust ETo for different crop types, a crop coefficient (Kc) is used, which represents the ratio of the actual evapotranspiration of a crop to the reference evapotranspiration. The Kc value is dependent on the crop growth stage, canopy cover, and other factors, and can be determined from tables or estimated using empirical formulas. Crop evapotranspiration (ETc) is calculated by multiplying ETo with Kc, and it represents the amount of water required to meet the crop's water needs. ETc is calculated for each crop growth stage and added up over the entire growing season to estimate irrigation water requirements, which is the amount of water needed to supplement natural precipitation to meet the crop's water requirements. Irrigation water requirements can be calculated for different time periods, such as monthly, depending on the capacity of the irrigation system. This study engaged in collaboration with farmers and implemented a systematic approach to determine optimal irrigation water quantity and timing, aiming to satisfy crop water requirements, improve yield, reduce water usage, and address environmental concerns. The outcomes of this research align with findings from prior investigations. Furthermore, it was observed that the elevated temperatures in the growing season and spring led to heightened evaporation rates, necessitating longer irrigation periods [56–59].

This suggests a considerable need for additional irrigation during the drought year 2021. This study investigated the use of remote sensing techniques and meteorological data to assess the ideal irrigation system performance scenarios for improving crop productivity in the Erbil province of Iraq. The study period covered five months of the wheat growing season. The results demonstrated that vegetation coverage was significantly affected by climatic factors such as precipitation, irrigation hours, and temperature, particularly in the months of March, April, and May. The study also showed that the use of supplementary irrigation in rain-fed areas is essential to mitigate the impact of drought. The correlation between NDMI and NDVI was significant, likely due to the delayed effect of insufficient precipitation on vegetation. The Sentinel2 satellite provided accurate predictions of irrigation conditions, and the developed model was suitable for different time scales, ensuring the reliability and efficiency of water requirement predictions. Severe water stress was observed throughout the rain-fed area in March, April, and May, despite the widespread drought conditions in Iraq during that year. However, north Erbil's center pivot irrigation areas were unaffected by the lack of precipitation and water stress [56,57,60].

This study employed meteorological data and remote sensing-based indices to identify optimal irrigation schedules and water use anomalies. Including agricultural production and surface evaporation data could further improve the assessment of these factors during drought periods. Integrating remote sensing and evapotranspiration models can support irrigation managers in optimizing irrigation schedules and detecting areas with irrigation problems, leading to more efficient and sustainable irrigation practices. However, remote sensing techniques have limitations, such as inadequate spatial and temporal resolutions and inaccurate calibration. While continuous monitoring of the same field over multiple years is impractical due to the study area's crop rotation method, precision irrigation systems that depend on integrated remote sensing and meteorological data can significantly impact agriculture and water resource management. Such systems can provide up-to-date information on crop health and moisture levels, which can be used to design and optimize precision irrigation systems that deliver water and nutrients where and when needed [56–58,60]. This approach can increase crop yields and quality, reduce water usage, and mitigate adverse environmental impacts such as soil erosion, nutrient leaching, and water runoff. Adopting precision irrigation systems based on remote sensing can also help address water scarcity issues at both the global and local levels, given that agriculture accounts for most freshwater usage worldwide. The development and adoption of such systems represent a significant advancement in agriculture and water resource management, which has the potential to enhance food security and environmental sustainability. Despite some moderate research linking wheat crops with NDMI and NDVI, there is still little investigation of wheat crop monitoring based on these indices in the KRI.

Author Contributions: Conceptualization, A.M.F.A.-Q., H.A.A.G. and K.M.; Data curation, H.A.A.G., A.M.F.A.-Q., H.A.S.R., K.K.H. and S.H.Z.; Formal analysis, H.A.A.G. and A.M.F.A.-Q.; Investigation, A.M.F.A.-Q., K.M., M.R. and H.A.A.G.; Methodology, H.A.A.G., K.K.H. and A.M.F.A.-Q.; Resources, A.M.F.A.-Q., K.M., H.A.S.R., S.H.Z. and H.A.A.G.; Supervision, A.M.F.A.-Q.; Validation, H.A.A.G. and A.M.F.A.-Q.; Visualization, A.M.F.A.-Q., P.H.A., M.R., K.K.H. and H.A.A.G.; Writing—original draft, A.M.F.A.-Q. and H.A.A.G.; Review, editing, improving, A.M.F.A.-Q., K.K.H., M.R., P.H.A. and K.K.H. All authors have read and agreed to the published version of the manuscript.

Funding: This research received no external funding.

Data Availability Statement: Some data in this manuscript were obtained from the Ministry of Agriculture and Water Resources, Kurdistan Region, Iraq, and the other data from the United States Geological Service (USGS) which provided the Landsat images freely on its website, and Statistical analysis of the parameters.

Acknowledgments: The authors extend their appreciation to the United States Geological Service (USGS) for making the Landsat images available on their website at no cost. Additionally, we express our deep gratitude to Ayad M. Fadhil Al-Quraishi of Tishk International University, Erbil, for funding the publication of this paper. We also want to thank Salahaddin University and Tishk International University in Erbil, Kurdistan Region, Iraq, for their valuable support. We are also thankful to Nuffic, the Orange Knowledge Programme, through the OKP-IRA-104278, Wageningen University & Research, The Netherlands, the Ministry of Agriculture and Water Resources/Water Resources Department.

Conflicts of Interest: The authors declare no conflict of interest.

Appendix A

Table A1. Correlation matrix (Pearson (n)).

Variables	NDVI	NDMI	Precipitation (mm)	Min Temp °C	Max Temp °C	Humidity %	ETo/mm/ Month	ETc/mm/ Month	Soil Temperature (30 cm)	Irrigation (Hours)
NDVI	1.00	0.93	−0.86	0.42	0.63	−0.44	0.49	0.80	0.55	0.67
NDMI	0.93	1.00	−0.62	0.07	0.32	−0.12	0.15	0.59	0.22	0.89
Precipitation (mm)	−0.86	−0.62	1.00	−0.77	−0.89	0.78	−0.83	−0.95	−0.85	−0.26
Min Temp °C	0.42	0.07	−0.77	1.00	0.96	−0.96	0.99	0.79	0.98	−0.27
Max Temp °C	0.63	0.32	−0.89	0.96	1.00	−0.97	0.98	0.93	0.99	0.01
Humidity %	−0.44	−0.12	0.78	−0.96	−0.97	1.00	−0.98	−0.87	−0.98	0.14
ET0/mm/month	0.49	0.15	−0.83	0.99	0.98	−0.98	1.00	0.86	1.00	−0.17
ETc/mm/month	0.80	0.59	−0.95	0.79	0.93	−0.87	0.86	1.00	0.89	0.32
Soil Temperture	0.55	0.22	−0.85	0.98	0.99	−0.98	1.00	0.89	1.00	−0.11
Irrigation (h)	0.67	0.89	−0.26	−0.27	0.01	0.14	−0.17	0.32	−0.11	1.00

Note: Values in bold are different from 0 with a significance level alpha = 0.05.

Appendix B

Table A2. Monthly Average NDVI in 143 Center Pivot Values Zonal Statistics.

Field	NDVI-Jan	NDVI-Febf	NDVI-Mar	NDVI-Apr	NDVI-May	Count	Sum	Mean	St. Dev	Variance	Mean
1	0.71	0.81	0.81	0.77	0.31	2634.0	2209.8	0.84	0.07	0.00	0.84
2	0.20	0.56	0.72	0.83	0.53	2568.0	2198.4	0.86	0.04	0.00	0.86
3	0.65	0.85	0.84	0.80	0.31	1639.0	1425.4	0.87	0.05	0.00	0.87
4	0.57	0.85	0.83	0.80	0.35	1944.0	1678.4	0.86	0.04	0.00	0.86
5	0.84	0.88	0.85	0.80	0.36	2876.0	2507.8	0.87	0.04	0.00	0.87
6	0.18	0.43	0.61	0.80	0.53	3793.0	3089.1	0.81	0.08	0.01	0.81
7	0.41	0.61	0.68	0.79	0.44	1512.0	1237.9	0.82	0.03	0.00	0.82
8	0.40	0.61	0.62	0.76	0.43	864.0	693.0	0.80	0.06	0.00	0.80
9	0.57	0.76	0.81	0.83	0.42	865.0	765.2	0.88	0.04	0.00	0.88
10	0.40	0.70	0.77	0.80	0.47	1720.0	1417.8	0.82	0.06	0.00	0.82
11	0.19	0.53	0.65	0.73	0.45	835.0	632.5	0.76	0.06	0.00	0.76
12	0.25	0.54	0.69	0.79	0.38	1787.0	1479.5	0.83	0.07	0.00	0.83
13	0.42	0.76	0.81	0.81	0.46	1506.0	1269.7	0.84	0.04	0.00	0.84
14	0.21	0.25	0.37	0.72	0.66	1861.0	1229.1	0.66	0.18	0.03	0.66
15	0.25	0.62	0.67	0.65	0.44	829.0	558.1	0.67	0.12	0.01	0.67
16	0.23	0.56	0.65	0.74	0.38	1727.0	1332.9	0.77	0.06	0.00	0.77
17	0.18	0.43	0.62	0.74	0.47	1703.0	1363.0	0.80	0.06	0.00	0.80
18	0.40	0.78	0.79	0.84	0.58	2014.0	1732.6	0.86	0.04	0.00	0.86
19	0.44	0.82	0.81	0.80	0.53	2076.0	1773.0	0.85	0.04	0.00	0.85
20	0.72	0.89	0.88	0.81	0.37	985.0	854.9	0.87	0.05	0.00	0.87
21	0.29	0.59	0.65	0.60	0.33	1988.0	1420.8	0.71	0.23	0.05	0.71
22	0.24	0.56	0.66	0.77	0.55	844.0	662.4	0.78	0.05	0.00	0.78
23	0.73	0.79	0.66	0.74	0.77	874.0	730.8	0.84	0.10	0.01	0.84
24	0.22	0.51	0.59	0.78	0.44	1810.0	1451.4	0.80	0.06	0.00	0.80
25	0.25	0.70	0.78	0.74	0.48	1838.0	1412.2	0.77	0.05	0.00	0.77
26	0.45	0.77	0.80	0.77	0.56	1997.0	1627.5	0.81	0.14	0.02	0.81
27	0.38	0.77	0.78	0.81	0.55	2117.0	1778.2	0.84	0.04	0.00	0.84
28	0.44	0.79	0.83	0.82	0.52	1510.0	1300.7	0.86	0.04	0.00	0.86
29	0.17	0.49	0.60	0.76	0.40	1014.0	808.1	0.80	0.09	0.01	0.80
30	0.55	0.78	0.80	0.82	0.43	1986.0	1690.8	0.85	0.06	0.00	0.85
31	0.52	0.81	0.86	0.79	0.54	2013.0	1720.3	0.85	0.03	0.00	0.85
32	0.69	0.87	0.88	0.82	0.53	1699.0	1448.5	0.85	0.04	0.00	0.85
33	0.50	0.80	0.85	0.80	0.58	1097.0	938.9	0.86	0.05	0.00	0.86
34	0.50	0.86	0.85	0.82	0.52	3061.0	2688.2	0.88	0.04	0.00	0.88
35	0.63	0.86	0.85	0.83	0.44	1897.0	1628.4	0.86	0.04	0.00	0.86
36	0.27	0.72	0.81	0.75	0.35	1943.0	1653.0	0.85	0.05	0.00	0.85
37	0.32	0.69	0.74	0.81	0.49	1807.0	1531.8	0.85	0.03	0.00	0.85
38	0.29	0.61	0.71	0.74	0.33	2107.0	1663.0	0.79	0.05	0.00	0.79
39	0.56	0.83	0.81	0.63	0.31	1910.0	1511.9	0.79	0.04	0.00	0.79
40	0.52	0.81	0.80	0.65	0.26	1606.0	1311.1	0.82	0.03	0.00	0.82
41	0.37	0.75	0.81	0.75	0.41	2374.0	2018.5	0.85	0.09	0.01	0.85
42	0.42	0.71	0.75	0.77	0.36	2020.0	1614.6	0.80	0.07	0.00	0.80
43	0.22	0.60	0.72	0.80	0.64	3040.0	2545.3	0.84	0.06	0.00	0.84
44	0.41	0.81	0.81	0.80	0.46	2252.0	1917.3	0.85	0.05	0.00	0.85
45	0.45	0.79	0.81	0.77	0.41	4415.0	3704.2	0.84	0.04	0.00	0.84
46	0.48	0.82	0.85	0.81	0.51	4122.0	3596.2	0.87	0.03	0.00	0.87
47	0.35	0.61	0.66	0.67	0.30	1145.0	893.7	0.78	0.06	0.00	0.78
48	0.47	0.74	0.79	0.74	0.34	1445.0	1168.7	0.81	0.10	0.01	0.81
49	0.52	0.83	0.83	0.82	0.33	1663.0	1469.9	0.88	0.03	0.00	0.88
50	0.33	0.73	0.77	0.77	0.55	1074.0	900.4	0.84	0.05	0.00	0.84
51	0.44	0.77	0.81	0.79	0.37	1924.0	1611.9	0.84	0.04	0.00	0.84
52	0.46	0.84	0.87	0.83	0.56	3039.0	2651.7	0.87	0.03	0.00	0.87
53	0.25	0.67	0.79	0.83	0.59	1864.0	1625.3	0.87	0.04	0.00	0.87
54	0.73	0.89	0.90	0.84	0.69	1139.0	997.6	0.88	0.02	0.00	0.88
55	0.51	0.78	0.83	0.78	0.44	1167.0	972.6	0.83	0.04	0.00	0.83

Table A2. *Cont.*

Field	NDVI-Jan	NDVI-Febf	NDVI-Mar	NDVI-Apr	NDVI-May	Count	Sum	Mean	St. Dev	Variance	Mean
56	0.43	0.78	0.81	0.77	0.46	612.0	508.9	0.83	0.04	0.00	0.83
57	0.53	0.86	0.89	0.84	0.52	2074.0	1802.1	0.87	0.02	0.00	0.87
58	0.44	0.83	0.86	0.78	0.46	2331.0	2037.8	0.87	0.05	0.00	0.87
59	0.44	0.80	0.83	0.79	0.40	2475.0	2078.2	0.84	0.04	0.00	0.84
60	0.20	0.46	0.59	0.76	0.35	2504.0	1993.8	0.80	0.08	0.01	0.80
61	0.41	0.72	0.75	0.77	0.33	1957.0	1614.6	0.83	0.04	0.00	0.83
62	0.33	0.66	0.77	0.77	0.33	3045.0	2469.6	0.81	0.06	0.00	0.81
63	0.24	0.70	0.72	0.81	0.46	1678.0	1405.9	0.84	0.07	0.00	0.84
64	0.47	0.82	0.84	0.77	0.34	1829.0	1515.3	0.83	0.08	0.01	0.83
65	0.25	0.43	0.53	0.59	0.39	3014.0	2110.2	0.70	0.11	0.01	0.70
66	0.61	0.79	0.81	0.71	0.28	2074.0	1670.3	0.81	0.13	0.02	0.81
67	0.45	0.86	0.88	0.81	0.53	2009.0	1781.4	0.89	0.04	0.00	0.89
68	0.23	0.68	0.75	0.81	0.50	2264.0	1911.2	0.84	0.04	0.00	0.84
69	0.18	0.39	0.62	0.81	0.51	1767.0	1487.3	0.84	0.10	0.01	0.84
70	0.29	0.72	0.75	0.83	0.50	1618.0	1383.9	0.86	0.08	0.01	0.86
71	0.42	0.83	0.84	0.81	0.44	1870.0	1594.9	0.85	0.06	0.00	0.85
72	0.37	0.75	0.79	0.83	0.50	1966.0	1695.6	0.86	0.07	0.00	0.86
73	0.53	0.83	0.84	0.74	0.25	1208.0	1004.4	0.83	0.04	0.00	0.83
74	0.54	0.86	0.85	0.75	0.25	1013.0	859.8	0.85	0.03	0.00	0.85
75	0.74	0.80	0.77	0.61	0.20	1938.0	1482.1	0.76	0.06	0.00	0.76
76	0.26	0.53	0.58	0.50	0.33	2413.0	1588.0	0.66	0.29	0.08	0.66
77	0.16	0.30	0.51	0.82	0.58	1046.0	850.2	0.81	0.05	0.00	0.81
78	0.40	0.74	0.77	0.78	0.32	1906.0	1594.4	0.84	0.08	0.01	0.84
79	0.41	0.62	0.63	0.72	0.38	1521.0	1114.7	0.73	0.07	0.01	0.73
80	0.18	0.49	0.71	0.72	0.46	1378.0	1023.4	0.74	0.15	0.02	0.74
81	0.30	0.63	0.63	0.66	0.38	1226.0	858.2	0.70	0.06	0.00	0.70
82	0.32	0.63	0.73	0.68	0.30	2679.0	2022.2	0.75	0.07	0.00	0.75
83	0.47	0.82	0.81	0.75	0.33	1704.0	1401.2	0.82	0.06	0.00	0.82
84	0.27	0.67	0.68	0.73	0.36	954.0	740.3	0.78	0.06	0.00	0.78
85	0.31	0.56	0.60	0.62	0.27	2026.0	1393.7	0.69	0.18	0.03	0.69
86	0.47	0.84	0.85	0.82	0.41	1806.0	1568.4	0.87	0.06	0.00	0.87
87	0.40	0.77	0.77	0.77	0.40	1482.0	1202.8	0.81	0.06	0.00	0.81
88	0.41	0.80	0.85	0.84	0.53	1929.0	1682.8	0.87	0.03	0.00	0.87
89	0.39	0.65	0.70	0.78	0.44	1354.0	1115.4	0.82	0.03	0.00	0.82
90	0.28	0.61	0.72	0.77	0.47	3457.0	2815.9	0.81	0.09	0.01	0.81
91	0.22	0.47	0.55	0.73	0.42	1128.0	865.4	0.77	0.06	0.00	0.77
92	0.19	0.48	0.62	0.78	0.51	1553.0	1253.3	0.81	0.11	0.01	0.81
93	0.18	0.39	0.55	0.76	0.54	1193.0	910.4	0.76	0.10	0.01	0.76
94	0.58	0.80	0.81	0.77	0.43	1696.0	1386.2	0.82	0.05	0.00	0.82
95	0.42	0.83	0.85	0.82	0.29	1732.0	1499.9	0.87	0.04	0.00	0.87
96	0.16	0.39	0.55	0.76	0.32	1258.0	998.8	0.79	0.09	0.01	0.79
97	0.19	0.55	0.73	0.74	0.45	1606.0	1306.9	0.81	0.07	0.01	0.81
98	0.47	0.84	0.87	0.81	0.47	1800.0	1531.6	0.85	0.04	0.00	0.85
99	0.21	0.56	0.64	0.75	0.43	1882.0	1486.4	0.79	0.12	0.01	0.79
100	0.35	0.71	0.76	0.79	0.55	1706.0	1401.2	0.82	0.03	0.00	0.82
101	0.41	0.75	0.78	0.82	0.41	1430.0	1233.9	0.86	0.04	0.00	0.86
102	0.51	0.81	0.81	0.80	0.31	1954.0	1654.8	0.85	0.04	0.00	0.85
103	0.42	0.79	0.84	0.83	0.33	1796.0	1571.3	0.87	0.03	0.00	0.87
104	0.37	0.76	0.80	0.83	0.48	1864.0	1587.3	0.85	0.06	0.00	0.85
105	0.26	0.56	0.61	0.69	0.19	1693.0	1216.8	0.72	0.08	0.01	0.72
106	0.23	0.40	0.50	0.73	0.34	1807.0	1346.3	0.75	0.09	0.01	0.75
107	0.27	0.58	0.62	0.73	0.33	2854.0	2193.8	0.77	0.12	0.01	0.77
108	0.23	0.64	0.69	0.81	0.67	1813.0	1480.6	0.82	0.06	0.00	0.82
109	0.39	0.71	0.79	0.73	0.52	1619.0	1358.1	0.84	0.04	0.00	0.84
110	0.56	0.84	0.84	0.73	0.48	1872.0	1557.7	0.83	0.09	0.01	0.83
111	0.33	0.73	0.83	0.77	0.49	2219.0	1893.3	0.85	0.04	0.00	0.85
112	0.38	0.69	0.78	0.63	0.40	1124.0	930.3	0.83	0.06	0.00	0.83

Table A2. *Cont.*

Field	NDVI-Jan	NDVI-Febf	NDVI-Mar	NDVI-Apr	NDVI-May	Count	Sum	Mean	St. Dev	Variance	Mean
113	0.43	0.77	0.79	0.68	0.25	1222.0	955.1	0.78	0.09	0.01	0.78
114	0.31	0.58	0.70	0.80	0.38	1625.0	1365.7	0.84	0.07	0.00	0.84
115	0.23	0.56	0.72	0.77	0.47	3150.0	2604.8	0.83	0.07	0.00	0.83
116	0.40	0.75	0.80	0.80	0.45	2733.0	2322.8	0.85	0.10	0.01	0.85
117	0.41	0.75	0.77	0.71	0.31	1380.0	1075.2	0.78	0.07	0.00	0.78
118	0.16	0.54	0.76	0.83	0.60	908.0	789.5	0.87	0.03	0.00	0.87
119	0.42	0.71	0.77	0.76	0.39	1422.0	1153.8	0.81	0.09	0.01	0.81
120	0.22	0.65	0.73	0.75	0.33	1188.0	978.2	0.82	0.04	0.00	0.82
121	0.27	0.71	0.73	0.79	0.55	551.0	453.8	0.82	0.05	0.00	0.82
122	0.31	0.64	0.73	0.76	0.32	1740.0	1416.5	0.81	0.05	0.00	0.81
123	0.34	0.53	0.55	0.74	0.53	936.0	726.9	0.78	0.05	0.00	0.78
124	0.43	0.79	0.84	0.80	0.50	2891.0	2422.4	0.84	0.11	0.01	0.84
125	0.42	0.74	0.80	0.77	0.36	445.0	372.6	0.84	0.06	0.00	0.84
126	0.32	0.73	0.80	0.68	0.34	2584.0	2171.8	0.84	0.03	0.00	0.84
127	0.42	0.77	0.81	0.76	0.33	761.0	649.7	0.85	0.03	0.00	0.85
128	0.39	0.75	0.79	0.75	0.32	875.0	700.9	0.80	0.04	0.00	0.80
129	0.36	0.79	0.68	0.74	0.34	1960.0	1659.0	0.85	0.09	0.01	0.85
130	0.43	0.74	0.80	0.74	0.28	2010.0	1754.8	0.87	0.03	0.00	0.87
131	0.30	0.73	0.78	0.75	0.37	805.0	651.1	0.81	0.15	0.02	0.81
132	0.50	0.80	0.85	0.74	0.49	1836.0	1583.5	0.86	0.05	0.00	0.86
133	0.26	0.52	0.64	0.69	0.32	922.0	701.8	0.76	0.08	0.01	0.76
134	0.19	0.51	0.66	0.70	0.40	1071.0	821.4	0.77	0.07	0.00	0.77
135	0.19	0.56	0.73	0.78	0.36	941.0	801.6	0.85	0.03	0.00	0.85
136	0.50	0.82	0.83	0.81	0.21	1294.0	1131.7	0.87	0.03	0.00	0.87
137	0.47	0.83	0.85	0.74	0.32	1419.0	1214.4	0.86	0.02	0.00	0.86
138	0.33	0.64	0.74	0.74	0.48	935.0	808.2	0.86	0.02	0.00	0.86
139	0.29	0.64	0.71	0.71	0.17	1201.0	958.4	0.80	0.16	0.02	0.80
140	0.50	0.75	0.78	0.75	0.35	1148.0	923.8	0.80	0.16	0.03	0.80
141	0.19	0.37	0.54	0.75	0.50	957.0	706.4	0.74	0.12	0.01	0.74
142	0.26	0.55	0.62	0.61	0.23	3077.0	2179.4	0.71	0.18	0.03	0.71
143	0.43	0.82	0.84	0.81	0.33	1062.0	946.9	0.89	0.02	0.00	0.89

Table A3. Monthly Average NDMI in 143 Center Pivot Values Zonal Statistics.

Field	NDMI-Jan	NDMI-Feb	NDMI-Mar	NDMI-April	NDMI-May	Count	Sum	Mean	St. Dev	Variance	Mean
1	0.27	0.36	0.41	0.44	0.08	658.0	181.6	0.28	0.06	0.00	0.46
2	−0.10	0.17	0.32	0.45	0.18	643.0	−69.3	−0.11	0.03	0.00	0.44
3	0.23	0.45	0.47	0.43	0.07	410.0	85.4	0.21	0.09	0.01	0.47
4	0.15	0.44	0.44	0.43	0.10	484.0	61.0	0.13	0.08	0.01	0.45
5	0.40	0.46	0.45	0.47	0.15	722.0	296.8	0.41	0.10	0.01	0.50
6	−0.12	0.06	0.21	0.41	0.14	961.0	−113.2	−0.12	0.03	0.00	0.40
7	0.04	0.19	0.28	0.42	0.14	380.0	19.1	0.05	0.06	0.00	0.42
8	0.03	0.20	0.23	0.38	0.12	213.0	5.0	0.02	0.07	0.00	0.37
9	0.14	0.33	0.41	0.47	0.16	218.0	28.8	0.13	0.07	0.00	0.49
10	0.02	0.28	0.36	0.44	0.17	427.0	8.4	0.02	0.06	0.00	0.43
11	−0.09	0.12	0.22	0.35	0.12	210.0	−20.5	−0.10	0.02	0.00	0.34
12	−0.07	0.15	0.30	0.43	0.10	447.0	−38.0	−0.08	0.04	0.00	0.43
13	0.04	0.33	0.40	0.44	0.17	375.0	9.0	0.02	0.05	0.00	0.44
14	−0.09	−0.05	0.02	0.34	0.24	465.0	−19.2	−0.04	0.08	0.01	0.27
15	−0.05	0.21	0.28	0.28	0.09	207.0	−13.3	−0.06	0.03	0.00	0.29
16	−0.08	0.16	0.24	0.36	0.08	430.0	−40.3	−0.09	0.03	0.00	0.36
17	−0.10	0.07	0.20	0.37	0.11	425.0	−44.3	−0.10	0.03	0.00	0.39
18	0.03	0.36	0.39	0.50	0.31	508.0	2.8	0.01	0.03	0.00	0.47

Table A3. *Cont.*

Field	NDMI-Jan	NDMI-Feb	NDMI-Mar	NDMI-April	NDMI-May	Count	Sum	Mean	St. Dev	Variance	Mean
19	0.06	0.41	0.42	0.48	0.29	516.0	12.1	0.02	0.03	0.00	0.48
20	0.27	0.48	0.48	0.45	0.12	245.0	64.6	0.26	0.10	0.01	0.48
21	-0.04	0.20	0.27	0.29	0.15	501.0	-24.2	-0.05	0.06	0.00	0.35
22	-0.07	0.16	0.26	0.40	0.21	216.0	-20.4	-0.09	0.04	0.00	0.37
23	0.27	0.34	0.22	0.35	0.29	219.0	63.2	0.29	0.09	0.01	0.41
24	-0.08	0.14	0.20	0.40	0.12	450.0	-33.5	-0.07	0.03	0.00	0.38
25	-0.06	0.28	0.38	0.37	0.13	457.0	-37.8	-0.08	0.04	0.00	0.38
26	0.07	0.36	0.41	0.42	0.22	498.0	27.4	0.06	0.07	0.01	0.44
27	0.02	0.35	0.38	0.47	0.26	530.0	-1.5	0.00	0.03	0.00	0.45
28	0.06	0.38	0.44	0.47	0.23	374.0	15.2	0.04	0.06	0.00	0.47
29	-0.11	0.11	0.22	0.38	0.11	254.0	-29.5	-0.12	0.03	0.00	0.37
30	0.14	0.38	0.41	0.45	0.13	495.0	63.9	0.13	0.10	0.01	0.45
31	0.12	0.39	0.47	0.47	0.26	504.0	54.9	0.11	0.07	0.00	0.48
32	0.24	0.45	0.50	0.47	0.24	423.0	99.4	0.24	0.06	0.00	0.46
33	0.09	0.37	0.46	0.45	0.27	280.0	21.6	0.08	0.06	0.00	0.47
34	0.10	0.43	0.46	0.49	0.25	768.0	56.0	0.07	0.04	0.00	0.48
35	0.20	0.46	0.48	0.50	0.19	474.0	86.9	0.18	0.07	0.01	0.50
36	-0.07	0.27	0.39	0.42	0.15	488.0	-38.5	-0.08	0.05	0.00	0.45
37	-0.02	0.28	0.35	0.45	0.17	450.0	-15.4	-0.03	0.08	0.01	0.46
38	-0.03	0.20	0.31	0.38	0.08	530.0	-22.5	-0.04	0.04	0.00	0.40
39	0.15	0.42	0.42	0.34	0.11	479.0	68.8	0.14	0.09	0.01	0.42
40	0.13	0.40	0.40	0.36	0.10	405.0	48.0	0.12	0.09	0.01	0.44
41	0.02	0.31	0.40	0.43	0.18	591.0	1.9	0.00	0.04	0.00	0.44
42	0.04	0.30	0.34	0.40	0.11	502.0	11.7	0.02	0.06	0.00	0.40
43	-0.09	0.19	0.31	0.45	0.30	763.0	-82.3	-0.11	0.03	0.00	0.42
44	0.05	0.38	0.41	0.47	0.23	564.0	9.5	0.02	0.04	0.00	0.47
45	0.06	0.38	0.43	0.43	0.11	1099.0	36.7	0.03	0.06	0.00	0.45
46	0.08	0.41	0.46	0.47	0.20	1036.0	62.0	0.06	0.06	0.00	0.49
47	-0.01	0.20	0.22	0.33	0.08	287.0	-3.4	-0.01	0.06	0.00	0.35
48	0.09	0.31	0.38	0.40	0.12	357.0	28.8	0.08	0.08	0.01	0.43
49	0.11	0.40	0.43	0.46	0.18	415.0	38.8	0.09	0.06	0.00	0.47
50	0.00	0.32	0.36	0.44	0.26	267.0	-7.0	-0.03	0.02	0.00	0.43
51	0.06	0.33	0.39	0.43	0.13	482.0	25.4	0.05	0.07	0.00	0.44
52	0.07	0.41	0.48	0.49	0.27	761.0	29.2	0.04	0.06	0.00	0.49
53	-0.07	0.24	0.37	0.49	0.30	464.0	-37.1	-0.08	0.05	0.00	0.46
54	0.28	0.48	0.54	0.52	0.35	284.0	80.6	0.28	0.09	0.01	0.53
55	0.11	0.35	0.43	0.46	0.18	292.0	28.9	0.10	0.08	0.01	0.48
56	0.05	0.35	0.41	0.44	0.18	150.0	5.0	0.03	0.06	0.00	0.45
57	0.13	0.43	0.49	0.50	0.25	518.0	56.5	0.11	0.07	0.00	0.49
58	0.06	0.39	0.45	0.46	0.24	588.0	26.0	0.04	0.07	0.00	0.47
59	0.08	0.37	0.41	0.43	0.19	618.0	33.3	0.05	0.04	0.00	0.43
60	-0.10	0.09	0.18	0.36	0.05	627.0	-66.5	-0.11	0.02	0.00	0.38
61	0.04	0.32	0.36	0.41	0.08	486.0	4.2	0.01	0.04	0.00	0.43
62	-0.02	0.26	0.37	0.41	0.09	761.0	-17.2	-0.02	0.05	0.00	0.42
63	-0.09	0.28	0.31	0.43	0.18	421.0	-46.1	-0.11	0.03	0.00	0.41
64	0.11	0.40	0.46	0.44	0.17	458.0	53.2	0.12	0.04	0.00	0.45
65	-0.05	0.09	0.16	0.28	0.15	754.0	-40.9	-0.05	0.05	0.00	0.31
66	0.19	0.37	0.42	0.41	0.07	519.0	99.6	0.19	0.11	0.01	0.46
67	0.07	0.42	0.48	0.48	0.27	503.0	21.5	0.04	0.05	0.00	0.48
68	-0.09	0.25	0.33	0.43	0.16	566.0	-57.9	-0.10	0.05	0.00	0.43
69	-0.11	0.05	0.21	0.43	0.17	442.0	-51.1	-0.12	0.09	0.01	0.43
70	-0.04	0.30	0.35	0.46	0.20	407.0	-12.6	-0.03	0.07	0.01	0.45
71	0.05	0.42	0.48	0.45	0.15	466.0	11.1	0.02	0.04	0.00	0.46
72	0.00	0.33	0.39	0.47	0.21	491.0	-17.1	-0.03	0.04	0.00	0.46
73	0.13	0.39	0.42	0.40	0.09	306.0	39.1	0.13	0.06	0.00	0.44
74	0.13	0.42	0.45	0.42	0.10	253.0	32.3	0.13	0.06	0.00	0.47
75	0.31	0.37	0.36	0.29	0.04	487.0	156.2	0.32	0.12	0.02	0.39

Table A3. *Cont.*

Field	NDMI-Jan	NDMI-Feb	NDMI-Mar	NDMI-April	NDMI-May	Count	Sum	Mean	St. Dev	Variance	Mean
76	−0.05	0.15	0.22	0.24	0.16	605.0	−37.9	−0.06	0.05	0.00	0.30
77	−0.12	−0.01	0.14	0.44	0.19	256.0	−31.5	−0.12	0.03	0.00	0.41
78	0.02	0.30	0.36	0.41	0.07	476.0	9.3	0.02	0.05	0.00	0.43
79	0.04	0.22	0.23	0.33	0.08	378.0	17.1	0.05	0.08	0.01	0.32
80	−0.12	0.11	0.30	0.35	0.12	347.0	−46.1	−0.13	0.05	0.00	0.37
81	−0.02	0.21	0.23	0.29	0.09	307.0	−10.9	−0.04	0.04	0.00	0.30
82	−0.02	0.22	0.32	0.31	0.04	671.0	−19.3	−0.03	0.04	0.00	0.36
83	0.09	0.41	0.43	0.39	0.09	426.0	26.9	0.06	0.07	0.00	0.43
84	−0.04	0.27	0.29	0.35	0.06	239.0	−11.0	−0.05	0.03	0.00	0.36
85	−0.03	0.16	0.20	0.26	0.00	506.0	−17.7	−0.04	0.09	0.01	0.30
86	0.08	0.42	0.45	0.49	0.20	450.0	24.1	0.05	0.06	0.00	0.48
87	0.04	0.35	0.37	0.42	0.16	368.0	5.1	0.01	0.05	0.00	0.40
88	0.04	0.37	0.44	0.48	0.23	485.0	0.8	0.00	0.05	0.00	0.47
89	0.03	0.26	0.30	0.43	0.15	343.0	3.0	0.01	0.04	0.00	0.43
90	−0.04	0.22	0.32	0.43	0.18	863.0	−49.0	−0.06	0.06	0.00	0.42
91	−0.08	0.11	0.17	0.38	0.12	282.0	−28.0	−0.10	0.04	0.00	0.37
92	−0.10	0.11	0.22	0.41	0.14	387.0	−38.6	−0.10	0.04	0.00	0.40
93	−0.11	0.04	0.14	0.36	0.15	302.0	−32.1	−0.11	0.04	0.00	0.34
94	0.15	0.37	0.40	0.41	0.14	432.0	60.1	0.14	0.09	0.01	0.42
95	0.05	0.41	0.44	0.49	0.10	430.0	10.8	0.03	0.06	0.00	0.50
96	−0.11	0.06	0.16	0.37	0.15	316.0	−39.2	−0.12	0.05	0.00	0.36
97	−0.09	0.14	0.30	0.37	0.10	402.0	−36.6	−0.09	0.03	0.00	0.40
98	0.08	0.42	0.48	0.49	0.24	438.0	25.5	0.06	0.06	0.00	0.49
99	−0.08	0.17	0.24	0.39	0.14	469.0	−42.2	−0.09	0.05	0.00	0.39
100	−0.02	0.29	0.37	0.43	0.19	428.0	−11.7	−0.03	0.05	0.00	0.44
101	0.05	0.34	0.39	0.47	0.13	361.0	18.1	0.05	0.08	0.01	0.47
102	0.11	0.40	0.42	0.43	0.07	485.0	45.2	0.09	0.08	0.01	0.45
103	0.04	0.36	0.43	0.47	0.12	450.0	11.2	0.02	0.05	0.00	0.48
104	0.01	0.34	0.41	0.46	0.20	465.0	−1.8	0.00	0.06	0.00	0.44
105	−0.04	0.17	0.23	0.31	0.07	424.0	−22.3	−0.05	0.03	0.00	0.32
106	−0.09	0.05	0.12	0.34	0.04	453.0	−42.9	−0.09	0.04	0.00	0.32
107	−0.05	0.19	0.23	0.36	0.07	711.0	−49.6	−0.07	0.03	0.00	0.37
108	−0.07	0.22	0.30	0.44	0.30	453.0	−31.7	−0.07	0.03	0.00	0.42
109	0.04	0.29	0.38	0.42	0.27	405.0	9.5	0.02	0.07	0.00	0.44
110	0.16	0.42	0.45	0.43	0.27	467.0	70.1	0.15	0.07	0.01	0.45
111	0.02	0.31	0.42	0.45	0.25	554.0	−2.8	0.00	0.07	0.00	0.47
112	0.02	0.26	0.36	0.33	0.14	284.0	4.8	0.02	0.06	0.00	0.41
113	0.06	0.35	0.37	0.32	0.02	305.0	17.6	0.06	0.07	0.01	0.37
114	−0.02	0.19	0.30	0.45	0.14	407.0	−13.5	−0.03	0.05	0.00	0.45
115	−0.08	0.16	0.31	0.38	0.12	785.0	−67.0	−0.09	0.03	0.00	0.41
116	0.04	0.33	0.40	0.44	0.18	684.0	13.2	0.02	0.06	0.00	0.44
117	0.04	0.33	0.37	0.35	0.08	345.0	9.5	0.03	0.06	0.00	0.38
118	−0.14	0.14	0.34	0.48	0.25	228.0	−35.0	−0.15	0.02	0.00	0.46
119	0.07	0.30	0.36	0.41	0.16	353.0	20.7	0.06	0.05	0.00	0.42
120	−0.08	0.23	0.32	0.39	0.08	298.0	−26.5	−0.09	0.04	0.00	0.42
121	−0.04	0.28	0.33	0.43	0.15	138.0	−8.9	−0.06	0.02	0.00	0.43
122	−0.02	0.23	0.32	0.39	0.07	432.0	−14.7	−0.03	0.04	0.00	0.41
123	0.00	0.14	0.17	0.38	0.21	236.0	−2.4	−0.01	0.05	0.00	0.35
124	0.05	0.35	0.43	0.46	0.24	717.0	14.7	0.02	0.05	0.00	0.45
125	0.05	0.31	0.39	0.42	0.16	119.0	3.8	0.03	0.09	0.01	0.46
126	0.00	0.31	0.40	0.37	0.13	646.0	−10.5	−0.02	0.03	0.00	0.45
127	0.06	0.35	0.40	0.43	0.14	191.0	7.3	0.04	0.04	0.00	0.46
128	0.06	0.34	0.39	0.39	0.10	213.0	7.5	0.04	0.05	0.00	0.39
129	0.04	0.37	0.35	0.42	0.20	493.0	3.2	0.01	0.05	0.00	0.46
130	0.05	0.31	0.39	0.39	0.10	500.0	15.0	0.03	0.06	0.00	0.46
131	−0.05	0.32	0.40	0.39	0.08	199.0	−14.2	−0.07	0.05	0.00	0.41
132	0.08	0.35	0.46	0.43	0.25	458.0	35.8	0.08	0.13	0.02	0.47

Table A3. *Cont.*

Field	NDMI-Jan	NDMI-Feb	NDMI-Mar	NDMI-April	NDMI-May	Count	Sum	Mean	St. Dev	Variance	Mean
133	−0.05	0.14	0.25	0.35	0.09	230.0	−9.5	−0.04	0.04	0.00	0.36
134	−0.10	0.12	0.24	0.33	0.06	269.0	−26.7	−0.10	0.03	0.00	0.35
135	−0.10	0.15	0.29	0.41	0.08	236.0	−24.2	−0.10	0.02	0.00	0.44
136	0.08	0.39	0.42	0.44	0.10	324.0	16.6	0.05	0.08	0.01	0.47
137	0.09	0.39	0.45	0.42	0.06	354.0	20.4	0.06	0.05	0.00	0.48
138	0.00	0.24	0.36	0.42	0.25	233.0	−4.0	−0.02	0.04	0.00	0.48
139	−0.03	0.24	0.30	0.34	0.05	297.0	−9.8	−0.03	0.06	0.00	0.38
140	0.09	0.33	0.37	0.40	0.10	288.0	23.7	0.08	0.07	0.00	0.41
141	−0.10	0.04	0.16	0.38	0.15	240.0	−28.2	−0.12	0.07	0.00	0.34
142	−0.04	0.16	0.24	0.29	0.10	770.0	−36.6	−0.05	0.05	0.00	0.32
143	0.06	0.37	0.43	0.45	0.16	268.0	11.4	0.04	0.04	0.00	0.46

References

1.	Ghazaryan, G.; König, S.; Rezaei, E.E.; Siebert, S.; Dubovyk, O. Analysis of Drought Impact on Croplands from Global to Regional Scale: A Remote Sensing Approach. *Remote Sens.* **2020**, *12*, 4030. [CrossRef]
2.	Ministry of Agriculture and Water Resources. *Review of the Agricultural Sector in the Kurdistan Region of Iraq: Analysis on Crops, Water Resources And Irrigation, and Selected Value Chains*; Annual Report; Ministry of Agriculture and Water Resources: Erbil, Iraq, 2019.
3.	Yousuf, M.A.; Rapantova, N.; Younis, J.H. Sustainable Water Management in Iraq (Kurdistan) as a Challenge for Governmental Responsibility. *Water* **2018**, *10*, 1651. [CrossRef]
4.	Dong, J.; Liu, W.; Han, W.; Xiang, K.; Lei, T.; Yuan, W. A Phenology-Based Method for Identifying the Planting Fraction of Winter Wheat Using Moderate-Resolution Satellite Data. *Int. J. Remote Sens.* **2020**, *41*, 6892–6913. [CrossRef]
5.	Jin, X.; Kumar, L.; Li, Z.; Xu, X.; Yang, G.; Wang, J. Estimation of Winter Wheat Biomass and Yield by Combining the Aquacrop Model and Field Hyperspectral Data. *Remote Sens.* **2016**, *8*, 972. [CrossRef]
6.	Van Dam, J.C.; Singh, R.; Bessembinder, J.J.E.; Leffelaar, P.A.; Bastiaanssen, W.G.M.; Jhorar, R.K.; Kroes, J.G.; Droogers, P. Assessing Options to Increase Water Productivity in Irrigated River Basins Using Remote Sensing and Modelling Tools. *Int. J. Water Resour. Dev.* **2006**, *22*, 115–133. [CrossRef]
7.	Kamal, K.; Hakzi, A. Spatiotemporal Variation of Wheat Yield and Water Productivity in Centre Pivot Irrigation Systems. Master's Thesis, IHE Delft Institute for Water Education, Delft, The Netherlands, 2022.
8.	FAO; WFP. *Dates, Grapes, Tomatoes and Wheat*; Food and Agriculture Organization: Rome, Italy, 2021; ISBN 9789251336342.
9.	Huang, J.; Sedano, F.; Huang, Y.; Ma, H.; Li, X.; Liang, S.; Tian, L.; Zhang, X.; Fan, J.; Wu, W. Assimilating a Synthetic Kalman Filter Leaf Area Index Series into the WOFOST Model to Improve Regional Winter Wheat Yield Estimation. *Agric. For. Meteorol.* **2016**, *216*, 188–202. [CrossRef]
10.	Morel, J.; Bégué, A.; Todoroff, P.; Martiné, J.-F.; Lebourgeois, V.; Petit, M. Coupling a Sugarcane Crop Model with the Remotely Sensed Time Series of FIPAR to Optimise the Yield Estimation. *Eur. J. Agron.* **2014**, *61*, 60–68. [CrossRef]
11.	FAO; IHE Delft. *Water Accounting in the Jordan River Basin, Remote Sensing for Water Productivity*; Food and Agriculture Organization: Rome, Italy, 2020; ISBN 9789251326619.
12.	Nichols, P.D.; Murphy, M.K.; Kenney, D.S. *Water and Growth in Colorado*; Natural Resources Law Center: Boulder, CO, USA, 2021; ISBN 1045021091.
13.	International Organization for Migration. *All Small Scale Irrigation Infrastructure Development in Iraq: A Feasibility Review*; International Organization for Migration: Grand-Saconnex, Switzerland, 2022.
14.	Schnepf, R. Iraq Agriculture and Food Supply: Background and Issues. *CRS Rep. Congr.* **2004**, *57*, 62.
15.	Bastiaanssen, W.G.M.; Molden, D.J.; Makin, I.W. Remote Sensing for Irrigated Agriculture: Examples from Research and Possible Applications. *Agric. Water Manag.* **2000**, *46*, 137–155. [CrossRef]
16.	FAO; IHE Delft. *WaPOR Quality Assessment*; Food and Agriculture Organization: Rome, Italy, 2019; ISBN 9789251315354.
17.	Jaradat, A. Agriculture in Iraq: Resources, Potentials, Constraints, Research Needs and Priorities. *Agriculture* **2003**, *1*, 160–166.
18.	Cooper, B.P.J.M.; Gregoryf, P.J.; Tully, D.; Harris, H.C.; August, A. Improving water use efficiency of annual crops in the rainfed farming systems of west asia and north africa. *Exp. Agric.* **1987**, *23*, 113–158. [CrossRef]
19.	Jin, X.; Yang, G.; Li, Z.; Xu, X.; Wang, J.; Lan, Y. Estimation of Water Productivity in Winter Wheat Using the AquaCrop Model with Field Hyperspectral Data. *Precis. Agric.* **2018**, *19*, 1–17. [CrossRef]
20.	Awchi, T.A.; Jasim, A.I. Rainfall Data Analysis and Study of Meteorological Draught in Iraq for the Period 1970–2010. *Tikrit J. Eng. Sci.* **2017**, *24*, 110–121. [CrossRef]
21.	Gaznayee, H.A.A.; Al-Quraishi, A.M.F.; Mahdi, K.; Ritsema, C. A Geospatial Approach for Analysis of Drought Impacts on Vegetation Cover and Land Surface Temperature in the Kurdistan Region of Iraq. *Water* **2022**, *14*, 927. [CrossRef]

22. Al-Quraishi, A.M.F.; Gaznayee, H.A.; Crespi, M. Drought Trend Analysis in a Semi-Arid Area of Iraq Based on Normalized Difference Vegetation Index, Normalized Difference Water Index and Standardized Precipitation Index. *J. Arid Land* **2021**, *13*, 413–430. [CrossRef]

23. Namdar, R.; Karami, E.; Keshavarz, M. Climate Change and Vulnerability: The Case of Mena Countries. *ISPRS Int. J. Geo-Inf.* **2021**, *10*, 794. [CrossRef]

24. Ahmad, M.D.; Turral, H.; Nazeer, A. Diagnosing Irrigation Performance and Water Productivity through Satellite Remote Sensing and Secondary Data in a Large Irrigation System of Pakistan. *Agric. Water Manag.* **2009**, *96*, 551–564. [CrossRef]

25. Polinova, M.; Salinas, K.; Bonfante, A.; Brook, A. Irrigation Optimization under a Limited Water Supply by the Integration of Modern Approaches into Traditional Water Management on the Cotton Fields. *Remote Sens.* **2019**, *11*, 2127. [CrossRef]

26. Mustafa, M.T.; Hassoon, K.I.; Hussain, H.M.; Modher, H. Using Water Indices (Ndwi, Mndwi, Ndmi, Wri and Awei) to Detect Physical and Chemical Parameters by Apply. *Int. J. Res.-Granthaalayah* **2017**, *5*, 117–128. [CrossRef]

27. Elbeltagi, A.; Aslam, M.R.; Mokhtar, A.; Deb, P.; Abubakar, G.A.; Kushwaha, N.L.; Venancio, L.P.; Malik, A.; Kumar, N.; Deng, J. Spatial and Temporal Variability Analysis of Green and Blue Evapotranspiration of Wheat in the Egyptian Nile Delta from 1997 to 2017. *J. Hydrol.* **2021**, *594*, 125662. [CrossRef]

28. Gaznayee, H.A.; Al-Quraishi, A.M.F. Analysis of Agricultural Drought's Severity and Impacts in Erbil Province, the Iraqi Kurdistan Region Based on Time Series NDVI and TCI Indices for 1998 through 2017. *J. Adv. Res. Dyn. Control Syst.* **2019**, *11*, 287–297. [CrossRef]

29. Wan, Z.; Wang, P. Using MODIS Land Surface Temperature and Normalized Difference Vegetation Index Products for Monitoring Drought in the Southern Great Plains, USA. *Int. J. Remote Sens.* **2004**, *25*, 61–72. [CrossRef]

30. Heydari, H.; Zoej, M.J.V.; Maghsoudi, Y.; Dehnavi, S. An Investigation of Drought Prediction Using Various Remote-Sensing Vegetation Indices for Different Time Spans. *Int. J. Remote Sens.* **2018**, *39*, 1871–1889. [CrossRef]

31. Qader, S.H.; Dash, J.; Atkinson, P.M. Forecasting Wheat and Barley Crop Production in Arid and Semi-Arid Regions Using Remotely Sensed Primary Productivity and Crop Phenology: A Case Study in Iraq. *Sci. Total Environ.* **2018**, *613–614*, 250–262. [CrossRef] [PubMed]

32. Jaafar, H.H.; Ahmad, F.A. Crop Yield Prediction from Remotely Sensed Vegetation Indices and Primary Productivity in Arid and Semi-Arid Lands. *Int. J. Remote Sens.* **2015**, *36*, 4570–4589. [CrossRef]

33. Bolton, D.K.; Friedl, M.A. Forecasting Crop Yield Using Remotely Sensed Vegetation Indices and Crop Phenology Metrics. *Agric. For. Meteorol.* **2013**, *173*, 74–84. [CrossRef]

34. Ghazaryan, G.; Dubovyk, O.; Graw, V.; Kussul, N.; Schellberg, J. Local-Scale Agricultural Drought Monitoring with Satellite-Based Multi-Sensor Time-Series. *GIScience Remote Sens.* **2020**, *57*, 704–718. [CrossRef]

35. Das, A.C.; Noguchi, R.; Ahamed, T. An Assessment of Drought Stress in Tea Estates Using Optical and Thermal Remote Sensing. *Remote Sens.* **2021**, *13*, 2730. [CrossRef]

36. AlJawa, S.B.; Al-Ansari, N. Open Access Online Journal of the International Association for Environmental Hydrology assessment of groundwater quality using. *Groundw. Qual. Using* **2017**, *26*, 1–14.

37. Al-quraishi, A.M.F.; Negm, A.M. *Environmental Remote Sensing and GIS in Iraq*; Springer: Berlin/Heidelberg, Germany, 2019; ISBN 9783030213435.

38. Phiri, D.; Simwanda, M.; Salekin, S.; Nyirenda, V.R.; Murayama, Y.; Ranagalage, M. Sentinel-2 Data for Land Cover/Use Mapping: A Review. *Remote Sens.* **2020**, *12*, 2291. [CrossRef]

39. Mahdianpari, M.; Salehi, B.; Mohammadimanesh, F.; Homayouni, S.; Gill, E. The First Wetland Inventory Map of Newfoundland at a Spatial Resolution of 10 m Using Sentinel-1 and Sentinel-2 Data on the Google Earth Engine Cloud Computing Platform. *Remote Sens.* **2019**, *11*, 43. [CrossRef]

40. Yengoh, G.T.; Dent, D.; Olsson, L.; Tengberg, A.E.; Tucker, C.J. *Use of the Normalized Difference Vegetation Index (NDVI) to Assess Land Degradation at Multiple Scaless: A Review of the Current Status, Future Trends, and Practical Considerations*; Springer: Berlin/Heidelberg, Germany, 2014; ISBN 978-3-319-24110-4.

41. Hunt, E.R.; Rock, B.N. Detection of Changes in Leaf Water Content Using Near- and Middle-Infrared Reflectances. *Remote Sens. Environ.* **1989**, *30*, 43–54. [CrossRef]

42. Anderson, R.G.; French, A.N. Crop Evapotranspiration. *Agronomy* **2019**, *9*, 614. [CrossRef]

43. Kharrou, M.H.; Simonneaux, V.; Er-raki, S.; Page, M.L.; Khabba, S.; Chehbouni, A. Assessing Irrigation Water Use with Remote Sensing-based Soil Water Balance at an Irrigation Scheme Level in a Semi-arid Region of Morocco. *Remote Sens.* **2021**, *13*, 1133. [CrossRef]

44. Anyamba, A.; Tucker, C.J. Historical Perspectives on AVHRR NDVI and Vegetation Drought Monitoring. *Remote Sens. Drought Innov. Monit. Approaches* **2012**, *23*, 20. [CrossRef]

45. Mukherjee, A.; Wang, S.Y.S.; Promchote, P. Examination of the Climate Factors That Reduced Wheat Yield in Northwest India during the 2000s. *Water* **2019**, *11*, 343. [CrossRef]

46. Aquino, D.d.N.; Neto, O.C.d.R.; Moreira, M.A.; Teixeira, A.d.S.; de Andrade, E.M. Use of Remote Sensing to Identify Areas at Risk of Degradation in the Semi-Arid Region. *Rev. Cienc. Agron.* **2018**, *49*, 420–429. [CrossRef]

47. Pedro-Monzonís, M.; Solera, A.; Ferrer, J.; Estrela, T.; Paredes-Arquiola, J. A Review of Water Scarcity and Drought Indexes in Water Resources Planning and Management. *J. Hydrol.* **2015**, *527*, 482–493. [CrossRef]

48. Maimaitijiang, M.; Sagan, V.; Sidike, P.; Hartling, S.; Esposito, F.; Fritschi, F.B. Soybean Yield Prediction from UAV Using Multimodal Data Fusion and Deep Learning. *Remote Sens. Environ.* **2020**, *237*, 111599. [CrossRef]

49. Taghvaeian, S.; Neale, C.M.U.; Osterberg, J.C.; Sritharan, S.I.; Watts, D.R. Remote Sensing and GIS Techniques for Assessing Irrigation Performance: Case Study in Southern California. *J. Irrig. Drain. Eng.* **2018**, *144*, 05018002. [CrossRef]

50. Elmetwalli, A.H.; Mazrou, Y.S.A.; Tyler, A.N.; Hunter, P.D.; Elsherbiny, O.; Yaseen, Z.M.; Elsayed, S. Assessing the Efficiency of Remote Sensing and Machine Learning Algorithms to Quantify Wheat Characteristics in the Nile Delta Region of Egypt. *Agriculture* **2022**, *12*, 332. [CrossRef]

51. Ali, M.H. *Practices of Irrigation & On-Farm Water Management: Volume 2*; Springer Science & Business Media: Berlin/Heidelberg, Germany, 2011; ISBN 9781441976369.

52. Abuzar, M.; Whitfield, D.; McAllister, A.; Sheffield, K. Application of ET-NDVI-Relationship Approach and Soil-Water-Balance Modelling for the Monitoring of Irrigation Performance of Treed Horticulture Crops in a Key Fruit-Growing District of Australia. *Int. J. Remote Sens.* **2019**, *40*, 4724–4742. [CrossRef]

53. Mahmoud, S.H.; Gan, T.Y. Irrigation Water Management in Arid Regions of Middle East: Assessing Spatio-Temporal Variation of Actual Evapotranspiration through Remote Sensing Techniques and Meteorological Data. *Agric. Water Manag.* **2019**, *212*, 35–47. [CrossRef]

54. Gadédjisso-Tossou, A.; Avellán, T.; Schütze, N. Potential of Deficit and Supplemental Irrigation under Climate Variability in Northern Togo, West Africa. *Water* **2018**, *10*, 1803. [CrossRef]

55. Ma, C.; Johansen, K.; McCabe, M.F. Monitoring Irrigation Events and Crop Dynamics Using Sentinel-1 and Sentinel-2 Time Series. *Remote Sens.* **2022**, *14*, 1205. [CrossRef]

56. Karamage, F.; Zhang, C.; Fang, X.; Liu, T.; Ndayisaba, F.; Nahayo, L.; Kayiranga, A.; Nsengiyumva, J.B. Modeling Rainfall-Runoffresponse to Land Use and Land Cover Change in Rwanda (1990–2016). *Water* **2017**, *9*, 147. [CrossRef]

57. Armitage, F.B. *Irrigated Forestry in Arid and Semi-Arid Lands: A Synthesis*; IDRC: Ottawa, ON, USA, 1986; Volume 61, ISBN 088936432X.

58. Nouri, H.; Beecham, S.; Anderson, S.; Nagler, P. High Spatial Resolution WorldView-2 Imagery for Mapping NDVI and Its Relationship to Temporal Urban Landscape Evapotranspiration Factors. *Remote Sens.* **2013**, *6*, 580–602. [CrossRef]

59. Moran, M.S.; Clarke, T.R.; Inoue, Y.; Vidal, A. Estimating Crop Water Deficit Using the Relation between Surface-Air Temperature and Spectral Vegetation Index. *Remote Sens. Environ.* **1994**, *49*, 246–263. [CrossRef]

60. Bryla, D.R. Crop evapotranspiration and irrigation scheduling in blueberry. In *Evapotranspiration—From Measurements to Agricultural and Environmental Applications*; Intech: Rijeka, Croatia, 2011; pp. 167–186.

water

Article

Automatic Water Control System and Environment Sensors in a Greenhouse

Yousif Yakoub Hilal [1,*], Montaser Khairie Khessro [1], Jos van Dam [2] and Karrar Mahdi [2]

[1] Department of Agricultural Machines and Equipment, College of Agriculture and Forestry, University of Mosul, Mosul 41002, Iraq; montaser.hussain@uomosul.edu.iq
[2] Soil Physics and Land Management Group, Wageningen University & Research, 6708 PB Wageningen, The Netherlands; jos.vandam@wur.nl (J.v.D.); karrar.mahdi@wur.nl (K.M.)
* Correspondence: yousif.yakoub@uomosul.edu.iq; Tel.: +964-77-3101-6244

Abstract: Iraqi greenhouses require an active microcontroller system to ensure a suitable microclimate for crop production. At the same time, reliable and timely Water Consumption Rate (WCR) forecasts provide an essential means to reduce the amount of water loss and maintain the environmental conditions inside the greenhouses. The Arduino micro-controller system is tested to determine its effectiveness in controlling the WCR, Temperature (T), Relative Humidity (RH), and Irrigation Time (IT) levels and improving plant growth rates. The Arduino micro-controller system measurements are compared with the traditional methods to determine the quality of the work of the new control system. The development of mathematical models relies on T, RH, and IT indicators. Based on the results, the new system proves to reliably identify the amount of WCR, IT, T, and RH necessary for plant growth. A t-test for the values from the Arduino microcontroller system and traditional devices for both conditions show no significant difference. This means that there is solid evidence that the WCR, IT, T, and RH levels for these two groups are no different. In addition, the linear, two-factor interaction (2FI), and quadratic models display acceptable performance very well since multiple coefficients of determination (R^2) reached 0.962, 0.969, and 0.977% with IT, T, and RH as the predictor variables. This implies that 96.9% of the variability in the WCR is explained by the model. Therefore, it is possible to predict weekly WCR 14 weeks in advance with reasonable accuracy.

Keywords: Arduino microcontroller; environment; sensors; models; greenhouses

Citation: Hilal, Y.Y.; Khessro, M.K.; van Dam, J.; Mahdi, K. Automatic Water Control System and Environment Sensors in a Greenhouse. *Water* **2022**, *14*, 1166. https://doi.org/10.3390/w14071166

Academic Editor: José Alberto Herrera-Melián

Received: 1 March 2022
Accepted: 21 March 2022
Published: 6 April 2022

Publisher's Note: MDPI stays neutral with regard to jurisdictional claims in published maps and institutional affiliations.

1. Introduction

Agriculture is a pillar of the economic lifeline, but traditional, broad forms of agriculture are no longer able to meet the development requirements of modem agriculture; therefore, developing precision agriculture has become an inevitable trend. Water represents a natural extension of the agricultural concept. Water is vital for farm production, and more water can increase crop production [1]. Irrigation allows farmers to apply nutrients more precisely and uniformly to the wetted root volume, where the active roots are concentrated. When environments of plants are consistently maintained and kept within their comfort zone, plants are more photosynthetically efficient and can grow stress-free [2].

In arid and semi-arid zones, the main constraints limiting crop production in open fields are the scarcity and disparity in rainfall, high temperatures, extreme solar radiation, and the spread of weeds and diseases [3,4]. Since the beginning of this century, agriculture in Iraq has undergone many changes [5,6]. Agriculture was making valuable contributions to the Iraqi economy until production costs rose and farmers lacked any real support from the government. Neighbouring countries began to produce more at lower prices, and Iraqi farmers struggled to compete. Luckily, in the past few years, agriculture has become the one sector that has contributed the most to national food security, economic growth, and employment [7–9]. Greenhouses can provide high-quality products year-round with efficient production resources, including fertilizer, water, pesticides, and labour. Countries like the United States, Europe, China, and countless others have used greenhouses to increase crop production [10–12]. The practice of producing crops in protected environments has developed rapidly, and by 2019, there were nearly 5,630,000 ha covered by greenhouses, high tunnels, low tunnels, or direct covers worldwide [13]. In Iraq, the cultivated area under greenhouses was 41,776 ha in 2019 [14].

A controlled greenhouse environment can provide suitable conditions for the optimal growth of vegetables and flowers in Iraq. Furthermore, greenhouses can be used to minimize the annual importation of vegetables. Greenhouses protect against high winds, unstable air temperature, insects, and airborne diseases. In addition to crop protection, the humidity of the air in these closed environments is considerably increased. Water productivity is increased and freshwater resources are used more efficiently [15,16].

Automated greenhouse monitoring has been addressed on a large scale from multiple perspectives and has been mostly focused on specific applications such as precision farming, irrigation, environmental control, yield prediction, and weed detection, to name a few [17]. Reference [18] explained that the interest in automated greenhouse monitoring has now reached an impressive level and noted that the interest in cultivation applications is growing exponentially. This increasing interest is also reflected in the significant advances in relevant technology such as various sensors as well as the development of cloud computing and machine learning techniques [19]. New technological improvements should allow farmers to realize their long-term expectations when using sensors in greenhouses [20]. For example, the water monitoring system improved water sustainability and reduced the daily water use of a beverage factory by 11% [21].

Unfortunately, due to poor management, loss of climate control, and overuse of water, there can be heavy losses of up to 40% for crops grown in greenhouses [13,14]. One of the major environmental stressors affecting plant growth and productivity is temperature. High-temperature stress frequently causes physiological disturbance and reduced yield and negatively affects the primary functions of the root systems [22,23].

Farmers took the agricultural traditions normally practiced in open fields in Iraq, such as using large quantities of water, and transferred them to greenhouses. Because of these traditions, water control systems and environmental sensors are not widely used, and farmers do not think that they are useful.

Studies carried out in countries neighbouring Iraq have presented many findings related to the use of automated monitoring in greenhouses under environmental conditions that differ from those in Iraq. The aims of these studies were to provide specific and non-exhaustive information on what automated monitoring should provide for greenhouse applications in their respective countries. Following the same vein, the present study aims to complement previous research by closely examining automated monitoring for greenhouse applications. The main objective of this research is to improve the methods used for water control systems and environmental sensors. Sensors monitor information related to the water levels and the general environments in greenhouses. Additionally, this paper focuses on how automated greenhouse monitoring could help meet the specific requirements of different stakeholders for several major greenhouse applications. Providing an overview of the emerging opportunities that could enhance the role of automated monitoring in Iraq by providing operational and efficient greenhouse application services will hopefully help reduce crop imports. Another goal of this study is to develop a simplified mathematical model of the water system, specifically concerning water consumption in greenhouses.

2. Materials and Methods

2.1. Test Site and Climate

The experiments were conducted in greenhouses affiliated with the College of Agriculture and Forestry, Mosul University, in the Mosul Governorate (36°23′24.1″ N 43°07′55.1″ E longitude, at an altitude of 234 m), Northern Iraq, as depicted in Figure 1. The soil was silt loam. It consisted of 17.4% Sand, 56.7% Silt, and 25.9% Clay. Soil EC and PH were 1.607 dS/m and 7.7%, respectively.

Figure 1. Greenhouses affiliated with the College of Agriculture and Forestry, Mosul University. https://goo.gl/maps/SuGqvhba4uQ3Tw6s5, accessed on 6 September 2021.

The study area was similar to a Mediterranean climate (a hot semi-arid climate) with a sweltering, prolonged, dry summer, brief and mild autumn, spring, and moderately wet, relatively cool winter. The ranges of rainfall, temperature, relative humidity, average cloud, and wind speed were 50–150 mm, 14–45 °C, 14–53%, 1–39%, and 14–22 km/h, respectively. Table 1 presents the average seasonal rainfall during the experimental period along with some climate characteristics.

Table 1. Average monthly climate at the test site.

Months	Climate			
	Temperatures (°C)	Precipitation (mm)	Average Cloud (%)	Average Humidity (%)
January	14	156.7	32	49
February	16	71.2	29	47
March	18	179.6	36	49
April	24	62.1	27	40
May	35	10.3	14	20
June	42	0	2	14
July	45	0	1	14
August	45	0	0	15
September	38	2.3	2	14
October	32	64.4	28	24
November	21	91.9	34	47
December	17	139.7	39	53

2.2. Specifications of the Plants

Cucumber F1 Bahar was one of the more suitable plants chosen largely due to high returns and a short growth period. The desired temperature for plant growth is between 15 °C and 32 °C with a growth period between 50 and 70 days. Humidity can range from 50 to 90% according to the growth stage [13].

2.3. Description of the Greenhouse and Control System

The greenhouse was made of galvanized steel tubes and covered with a polyethylene material. It was also equipped with a fan and a ventilation cooling system that was used in the study. The dimensions of the greenhouse were 44 m in length, 9 m in width, and 4 m in height. The greenhouse was constructed at an elevation of 234 m. The fans were distributed inside the greenhouse according to the method described by [24] and the standard guideline of the American Society of Agricultural Engineering [25]. The polyethylene used as covering material (200 µm thick) had a low transmission coefficient compared to glass. The light transmittance was 88%, with 80% U.V. and 77% infrared.

The Arduino microcontroller system (Figure 2) was developed to control all electrical components involved in measuring temperature, humidity, and irrigation processes. The designed control board consisted of main components such as wireless communication, multiple voltage regulator circuitry (5 V, 6 V, and 9 V), and control modules integrated onto a single Printed Circuit Board. The 9 Amp/hour rechargeable battery was used as a backup when the main power supply was cut off. The Arduino was used primarily in the microcontroller system to create an interactive electronic project that could include different environmental sensors to measure temperature, humidity, and irrigation.

Figure 2. Actual prototype of the Arduino microcontroller system installed inside the greenhouse.

The Arduino microcontroller system observed daily WCR, IT, T, and RH values obtained using WCR, IT, T, and RH sensors and readings on an LCD panel. The values could be downloaded onto a USB flash drive. Four inputs and six outputs were used to control the module temperature, humidity, and irrigation processes, as described in Figure 3.

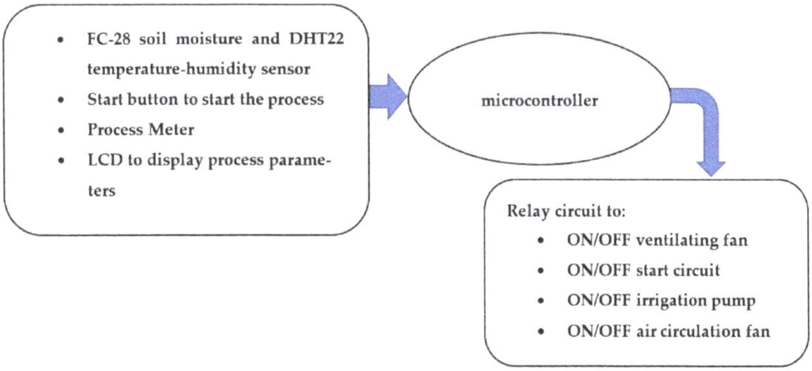

Figure 3. Input and output pin connections for the Arduino microcontroller system.

2.4. Execution of the Study and Development of Mathematical Models

The drip irrigation system used inside the greenhouse was one of the best irrigation systems suitable for greenhouse agriculture. Four irrigation pipes extended along the greenhouse. The irrigation pipe had drippers with a discharge of 3.6 L per hour, and the distance between the drippers was 20 cm. Each pipe was 40 m long and contained 200 drippers. The irrigation system was operated for two days before planting to carry out the calibration process and ensure that the drippers worked.

The traditional method used to grow this type of cucumber was followed to verify the accuracy of measuring and controlling the temperature, humidity, and irrigation processes (water consumption and irrigation time) of a new control unit (an Arduino microcontroller system). The system monitored the temperature, humidity, and irrigation processes using DHT22 and FC-28 soil moisture sensors. The system had two functions, the first one was to turn on and off the ventilating fan, air circulation fan, and the pump when the temperature, humidity, and soil moisture fell below a certain reference value. The secondary function of the system was to display the status of the temperature, humidity, and soil using an LCD. The FC-28 soil moisture sensor read soil moisture, and the DHT22 sensor read the value for temperature and humidity throughout the greenhouse as input from the system that the Arduino microcontroller system could process. Temperature, humidity, and soil moisture content were measured in three locations (at the entrance, in the middle, and at the end of the greenhouse).

The HTC-2 Digital Thermometer Hygrometer Electronic was used to measure humidity and temperature using three devices placed throughout the greenhouse. An Extech -Mo750 device was used to measure the moisture content of the soil. The details of the devices are shown in Table 2. For the purpose of verifying the accuracy of the control, the measurements obtained from the conventional devices (HTC-2 Digital Thermometer Hygrometer Electronic and Extech -Mo750) were compared with the measurements obtained from the developed control unit. Fertilization, pest control, seed quantity, greenhouse preparation, and irrigation systems were carried out as were typical for the study area [13].

Table 2. Specifications and range of the HTC-2 Digital Thermometer Hygrometer Electronic and Extech -Mo750.

Extech -Mo750	HTC-2 Digital Thermometer Hygrometer Electronic
Sensor Type Integrated Contact Probe	Material: ABS Size: $10.5 \times 9.8 \times 2.4$ cm/$4.13 \times 3.86 \times 0.94''$ Power supply: 1.5 V $\times$ 1(AAA battery)
Moisture Content 0 to 50% Accuracy $\pm$(5% + 5 digits) FS @23 $\pm$ 5 °C	Temperature measurement range: -10 °C ~ $+50$ °C Temperature measurement accuracy: ±1 °C Temperature resolution: 0.1 °C
Operating Temperature (0 to 50 °C) Operating Humidity < 80% RH	Humidity measurement range: 10% RH–99% RH Humidity measurement accuracy: $\pm$5% RH Humidity resolution: 1%
Max Resolution 0.1% Dimensions $14.7 \times 1.6 \times 1.6''$ ($374 \times 40 \times 40$ mm) Weight 9.4 oz (267 g)	

The Independent Samples t-test was used to test the research question. The research question was: is there a difference in WCR, IT, T, and RH measurements between an Arduino microcontroller system and traditional devices? Therefore, the Hypotheses was:

The null hypothesis (H0): *There is no difference in mean WCR, IT, T, and RH measurements between an Arduino microcontroller system and traditional devices.*

The alternative hypothesis (H1): *There is a difference in mean WCR, IT, T, and RH measurements between an Arduino microcontroller system and traditional devices.*

The data analysis and building multiple mathematical models to predict the WRC were performed using the Design-Expert Version 13 software. The software was from Stat-Ease Inc., Minneapolis, MN, USA. The model followed the steps to building multiple regression models described in [26].

2.5. Simulation Models

A simulation model was necessary to measure the power and efficiency of any developed forecasting algorithm. It allowed the researchers to verify the robustness of the different algorithms by considering the component combination of the model to obtain the best test scenario. The model's performance was evaluated by the correctness of its estimation, its ability to reproduce the actual return in the simulation, and its stability. We wanted to regularly forecast the weather 14 weeks in advance and produce a new forecast each week. Hence, the estimated model in this study was validated and evaluated based on its forecasting power by using the mean square error (MSE), mean absolute percentage error (MAPE), and average accuracy percentage (AAP). The following performance measure functions were employed [27,28]:

1-The form of MSE can be written as follows:

$$MSE = \frac{1}{N} \sum_{1}^{N} (\text{Actual yield} - \text{Forecasted yield})^2$$

2-Mean absolute percentage error MAPE

$$MAPE = \frac{1}{N} \sum_{1}^{N} \frac{/(\text{Actual yield} - \text{Forecasted yield})/}{\text{Actual yield}} \times 100$$

3-Average accuracy percentage AAP

$$AAP\% = 100\% - MAPE$$

where: N is the number of data points for $I = 1, 2, \ldots, N$.

3. Results and Discussion

3.1. Comparison of the Arduino Microcontroller System and Traditional Devices

WCR, IT, T, and RH levels were measured during both sunny and cloudy days by the Arduino microcontroller system and traditional devices in the greenhouse. The results were compared using an Independent Samples *t*-test as described in Table 3. Levene's test checked the null hypothesis that the variances of the two groups were equal. In this study, the p-value for WCR, IT, T and RH levels was 0.778, 0.589, 0.651, and 0.985, respectively. The assumption of equal variances was not violated so we can look at the top row of Table 3.

The values of the t statistic are 1.669, 1.251, 1.298, and 1.355, and the *p*-value is displayed as 0.107, 0.222, 0.206, and 0.187. This means that there is a very small probability of these results occurring by chance under the alternative hypothesis of difference between the two groups. The alternative hypothesis is formally rejected when accepting the null hypothesis. There is no difference in mean measurements between an Arduino microcontroller system and traditional devices. This means that there is very strong evidence that the WCR, IT, T, and RH levels for these two groups are no different, which can be seen clearly in Figures 4–7.

The Arduino microcontroller system was tested and compared with traditional measuring methods commonly used in greenhouses. Observed values of daily WCR, IT, T, and RH were obtained by means of WCR, IT, T, and RH sensors and readings on an LCD panel. The values could be downloaded onto a USB flash drive.

The daily WCR, IT, T, and RH were recorded for 100 days (14 weeks) and the weekly averages of WCR, IT, T, and RH were calculated. Figures 4–7 show the comparisons between weekly measurements of the WCR, IT, T, and RH from the Arduino microcontroller system

3.2. Modelling and Relationships between WCR, IT, T, and RH

In Iraqi greenhouses, measurements of temperature, humidity, and irrigation duration that are collected by most commercial plantations are minimal since normally, only data concerning monthly whole rainy days and daily rainfall days are available [13]. Therefore, it is critical to provide a simplified methodology for weekly WCR using the limited temperature, humidity, and irrigation time data. The mathematical models used the T, RH, and IT data from the Arduino microcontroller system.

The resulting linear, 2FI, and quadratic equations and the corresponding values of R2 for the testing period of 14 weeks are presented in Table 4. The model equation with WCR, IT, T, and RH in weeks as the independent variables showed that R2 = 0.962, 0.969, and 0.977, implying that the model explained well 96.2, 96.9, and 97.7% of the variability in WCR. A linear relationship between actual and predicted weekly WCR showed a strong positive association with high significance at $p < 0.0001$.

Table 4. Mathematical models under an Arduino microcontroller system in a greenhouse.

Model	F-Value	p-Value	R^2	Adj. R^2	Pred. R^2	Std. Dev.
WCR = $-0.0173451.0.001595$T + 0.004261RH + 0.05329IT	320.91	<0.0001	0.962	0.959	0.952	0.093
WCR = $8.12464 - 0.489734$T $- 0.119744$RH + 0.052929IT $+ 0.007382$T $\times$ RH $- 0.002221$T $\times$ IT $+ 0.000465$RH $\times$ IT	184.02	<0.0001	0.969	0.964	0.954	0.087
WCR = $20.54747 - 1.17731$T $- 0.275757$RH $- 0.012634$IT $+ 0.012368$T $\times$ RH $- 0.000944$T $\times$ IT $+ 0.001470$RH $\times$ IT $+ 0.008004$T^2 + 0.000328RH^2 $- 0.000588$IT^2	155.26	<0.0001	0.977	0.971	0.959	0.077

The F-test of overall significance indicates whether a model provides a better fit to the data than a model that contains no independent variables. Model F-values of 320.91, 184.02, and 155.26 and p-values less than 0.0500 indicate that the model terms are significant. The F-values with a probability of 0.0001 indicate that the regression coefficients are nonzero. There is only a 0.01% chance that F-values these large could occur due to noise. The predicted R^2 of 0.9528, 0.9542, and 0.9598 are in reasonable agreement with the Adjusted R^2 of 0.9590, 0.9640, and 0.9713, i.e., the difference is less than 0.2 for linear, 2FI, and quadratic models, respectively.

Figure 8 was a scatter plot that showed the correlation between predicted values and actual values for linear, 2FI, and quadratic models to see how the models performed with the mean and best values of the variables in every prediction. Figures graphically represent an association of the actual and predicted WCR for all models. The magnitude of error expressed as the difference between predicted R^2 and adjusted R^2 values is also shown in the same Table 4.

The practical predictive ability of the linear, 2FI, and quadratic models was visual, and the representation strongly suggests a goodness of fit for models in predicting WCR; the magnitude of error was very small. The mean error is preferred for the whole population rather than a single sample [29]. According to Figure 8A, the linear predicted model fully fit (without deviations) to the actual values and the R^2 value is 0.962. Reference [26] explained that the best model has an R-square value above 80% and it uses the smallest number of parameters.

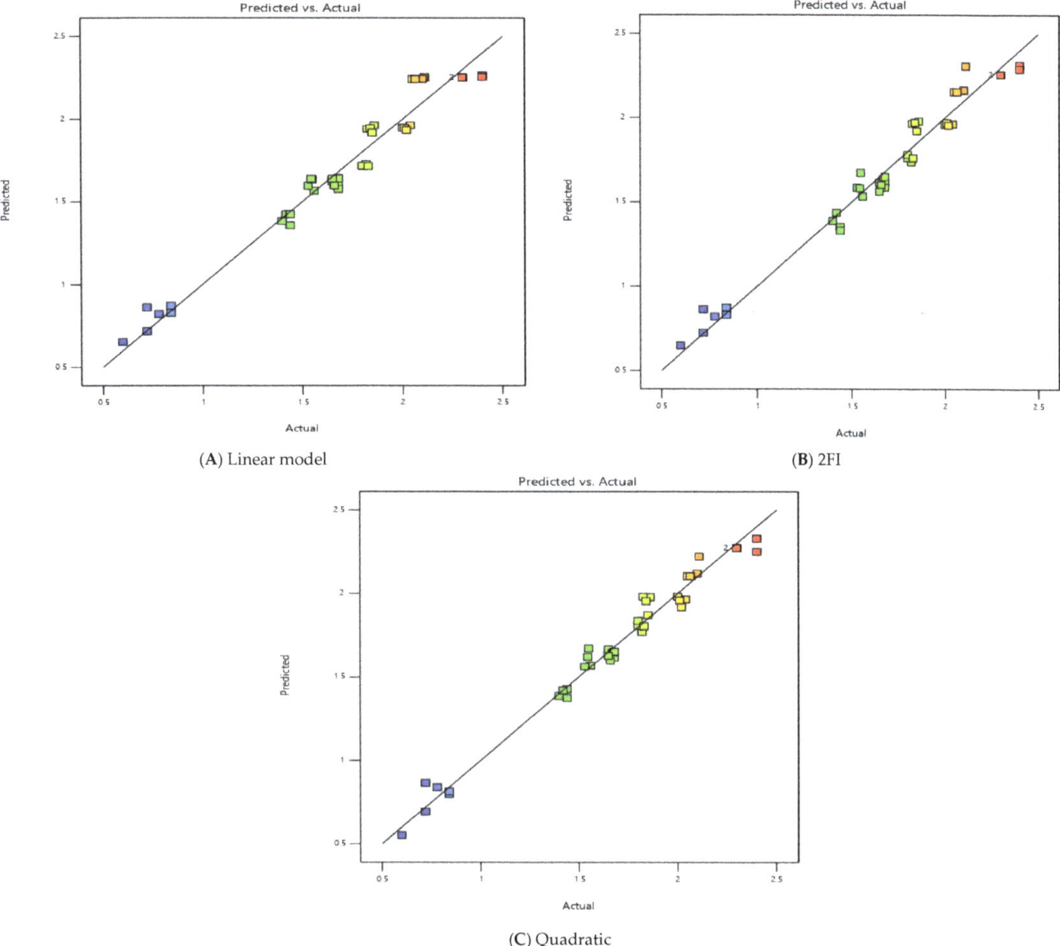

(A) Linear model

(B) 2FI

(C) Quadratic

Figure 8. The scatter plot of the predicted vs. actual values for linear, 2FI, and quadratic models. (**A**) Linear model (**B**) 2FI model (**C**) Quadratic model.

The effect of WCR varies with the degree of temperature, humidity, and irrigation duration for the cucumber plants. Water deficit stress during pre-flowering and grain filling stages massively affects plant performance due to the imprecise traits function. Water stress increased the flowering days and days to maturity while it decreased the leaf number and led to a loss of normal root architecture which further led to a reduction in yield [30]. The changes in WCR due to temperature, humidity, and irrigation duration variations usually contribute to increasing yield [17,31]. Overall, our present study indicated that WCR has a strong relationship with temperature, humidity, and irrigation time. These factors accounted for increasing the R^2 of the models as WCR is dependent on temperature, humidity, and irrigation duration.

Table 5 provides a comparison between the linear forecasts and the actual values based on the MSE, MAPE, and AAP within the studied states. The results were obtained for the forecasted WCR for 14 weeks. The outputs of the linear model were found to be closer to the actual values for the WCR. The finding indicated that the average accuracy percentage for the forecasts by the linear models had a record-high value of 92.038. In contrast, the importance of the MSE and MAPE for the model was lower. Therefore, the linear model was appropriate for forecasting the data and can be used as an alternative model to predict the WCR.

Table 5. Simulation models of WCR in greenhouses.

Weeks	Actual WCR m^3	Predicted WCR m^3
1	0.6	0.589
2	0.72	0.623
3	1.44	1.285
4	1.4	1.304
5	1.42	1.357
6	1.44	1.391
7	1.68	1.592
8	1.65	1.580
9	1.67	1.564
10	1.68	1.534
11	1.82	1.677
12	1.8	1.574
13	1.8	1.576
14	1.83	1.576
MSE %	0.019	
MAPE %	7.961	
AAP %	92.038	

4. Conclusions

The main objective of an optimal environment and evaluation is to control critical parameters such as water consumption, irrigation time, temperature, and humidity, based on the design of a microcontroller system in a greenhouse which is a purely sensor-based system. The results show that the developed system maintained WCR, IT, T, and RH levels and reduced water consumption by approximately 12.5%. We hope that the new system will be used as a prototype for Iraqi greenhouses to automatically control and monitor WCR, IT, T, and RH levels.

The developed models are simple, efficient, and easily applied. The models indicated that the regression equation well represented 96.9% of the variability in weekly WCR. A linear relationship between actual and predicted weekly WCR shows that the correlation = 0.984, which is a strong positive association. Additionally, the linear model was chosen as the best model to use in greenhouses, with the average accuracy percentage simulation and mean square error being 92.038 and 0.01988%, respectively. The researchers suggest conducting experiments with other crops under the same conditions to verify the accuracy of the proposed equations. Future research could develop a raspberry pi microcontroller system and compare it with the Arduino microcontroller system. Additionally, the Arduino microcontroller system used in this study could be applied to other types of crops, thus increasing the general applicability of the method presented in this article.

Author Contributions: Conceptualization, Y.Y.H., M.K.K., J.v.D. and K.M.; Methodology, Y.Y.H. and M.K.K.; Software, Y.Y.H.; Validation, Y.Y.H., M.K.K., J.v.D. and K.M.; Formal analysis, Y.Y.H.; Investigation, Y.Y.H.; Resources and data curation, Y.Y.H. and M.K.K.; Writing—original draft preparation, Y.Y.H.; Writing—review and editing, J.v.D.; Visualization, supervision; project administration and funding acquisition, K.M. All authors have read and agreed to the published version of the manuscript.

Funding: This study was funded by Nuffic, the Orange Knowledge Programme, through the OKP-IRA-104278 project titled "Efficient water management in Iraq switching to climate smart agriculture: capacity building and knowledge development" coordinated by Wageningen University & Research, The Netherlands.

Institutional Review Board Statement: Not applicable.

Informed Consent Statement: Not applicable.

Data Availability Statement: Not applicable.

Acknowledgments: The authors wish to express their appreciation to Mosul University, Nuffic and Wageningen University for funding this research under the OKP-IRA-104278 project.

Conflicts of Interest: The authors declare no conflict of interest. The funders had no role in the design of the study; in the collection, analyses, or interpretation of data; in the writing of the manuscript, or in the decision to publish the results.

References

1. Sarker, K.K.; Hossain, A.; Timsina, J.; Biswas, S.K.; Malone, S.L.; Alam, M.K.; Bazzaz, M. Alternate furrow irrigation can maintain grain yield and nutrient content, and increase crop water productivity in dry season maize in sub-tropical climate of South Asia. *Agric. Water Manag.* **2020**, *238*, 106229. [CrossRef]
2. Wang, Y.P.; Zhang, L.S.; Mu, Y.; Liu, W.H.; Guo, F.X.; Chang, T.R. Effect of A Root-Zone Injection Irrigation Method on Water Productivity and Apple Production in A Semi-Arid Region in North-Western China. *Irrig. Drain.* **2020**, *69*, 74–85. [CrossRef]
3. Sahu, N.; Reddy, G.; Dash, B.; Kumar, N.; Singh, S. Assessment on spatial extent of arid and semi-arid climatic zones of India using GIS. *J. Agrometeorol.* **2021**, *23*, 189–193. [CrossRef]
4. Tsafaras, I.; Campen, J.B.; Stanghellini, C.; de Zwart, H.F.; Voogt, W.; Scheffers, K.; Al Assaf, K. Intelligent greenhouse design decreases water use for evaporative cooling in arid regions. *Agric. Water Manag.* **2021**, *250*, 106807. [CrossRef]
5. International Organization for Migration (IOM), Iraq, "Rural Areas in Ninewa Legacies of Conflict on Rural Economies and Communities in Sinjar and Ninewa Plains," August 2019. Available online: https://reliefweb.int/sites/reliefweb.int/files/resources/IOM%20Rural%20areas%20in%20Ninewa.pdf (accessed on 6 September 2021).
6. Jongerden, J.; Wolters, W.; Dijkxhoorn, Y.; Gür, F.; Öztürk, M. The politics of agricultural development in Iraq and the Kurdistan Region in Iraq (KRI). *Sustainability* **2019**, *11*, 5874. [CrossRef]
7. Drebee, H.A.; Abdul-Razak, N.A. The Impact of Corruption on Agriculture Sector in Iraq: Econometrics Approach. In Proceedings of the IOP Conference Series: Earth and Environmental Science, University of Al-Qadisiyah, Al Diwaniyah, Iraq, 31 May–1 June 2020.
8. Hilmi, M. Entrepreneurship in Farming: Small-Scale Farming and Agricultural Mechanization Hire Service Enterprises in Iraq. *Middle East J. Agric. Res.* **2021**, *10*, 10–52.
9. Al-Ansari, N.; Abed, S.A.; Ewaid, S.H. Agriculture in Iraq. *J. Earth Sci. Geotech. Eng.* **2021**, *11*, 223–241. [CrossRef]
10. Shirokov, Y.; Tikhnenko, V. Analysis of environmental problems of crop production and ways to solve them. In Proceedings of the E3S Web of Conferences, Rostov on Don, Russia, 24–26 February 2021.
11. Narimani, M.; Hajiahmad, A.; Moghimi, A.; Alimardani, R.; Rafiee, S.; Mirzabe, A.H. Developing an aeroponic smart experimental greenhouse for controlling irrigation and plant disease detection using deep learning and IoT. In Proceedings of the 2021 ASABE Annual International Virtual Meeting, American Society of Agricultural and Biological Engineers, St. Joseph, MI, USA, 12–16 July 2021.
12. Zhou, D.; Meinke, H.; Wilson, M.; Marcelis, L.F.; Heuvelink, E. Towards delivering on the sustainable development goals in greenhouse production systems. *Resour. Conserv. Recycl.* **2021**, *169*, 105379. [CrossRef]
13. Al-Qaissy, M.M.J. A Study of Some Environmental and Economic Indicators of an Automated Greenhouse in Comparison with the Traditional Ones When Growing Cucumbers (*Cucumis sativus* L.). Master's Thesis, Mosul University, Mosul, Iraq, 2020.
14. Central Statistical Organization CSO. Wheat and Barley Production in 2018 (In Arabic), 2019. Available online: https://www.cosit.gov.iq/ar/ (accessed on 6 September 2021).
15. Hamza, A.A.; Almasraf, S.A. Evaluation of the yield and water use efficiency of the cucumber inside greenhouses. *J. Babylon Univ. Eng. Sci.* **2016**, *24*, 95–106.
16. Aljubury, I.M.A.; Ridha, H.D.A. Enhancement of evaporative cooling system in a greenhouse using geothermal energy. *Renew. Energy* **2017**, *111*, 321–331. [CrossRef]
17. Nath, S.D.; Hossain, M.S.; Chowdhury, I.A.; Tasneem, S.; Hasan, M.; Chakma, R. Design and Implementation of an IoT Based Greenhouse Monitoring and Controlling System. *J. Comput. Sci. Technol. Stud.* **2021**, *3*, 01–06. [CrossRef]
18. Achour, Y.; Ouammi, A.; Zejli, D. Technological progresses in modern sustainable greenhouses cultivation as the path towards precision agriculture. *Renew. Sustain. Energy Rev.* **2021**, *147*, 111251. [CrossRef]

19. Sivagami, A.; Kandavalli, M.A.; Yakkala, B. Design and Evaluation of an Automated Monitoring and Control System for Greenhouse Crop Production. In *Next-Generation Greenhouses for Food Security*; Shamshiri, R.R., Ed.; Intechopen: London, UK, 2021; p. 164.
20. Oo, Z.Z.; Phyu, S. Greenhouse environment monitoring and controlling system based on IoT technology. In Proceedings of the AIP Conference Proceedings, Walt Whitman, MI, USA, 3 May 2021.
21. Jagtap, S.; Skouteris, G.; Choudhari, V.; Rahimifard, S.; Duong, L.N.K. An Internet of Things Approach for Water Efficiency: A Case Study of the Beverage Factory. *Sustainability* **2021**, *13*, 3343. [CrossRef]
22. Hmiz, D.J.; Ithbayyib, I.J. Effect of the Root Zone Temperature and Salt Stress on Plant Growth, Main Branches and some other Chemical Characteristics of Tomato Fruit *Solanum lycopersicum* L. cv. memory. *Basrah J. Agric. Sci.* **2021**, *34*, 156–170. [CrossRef]
23. Xu, C.; Wang, M.T.; Yang, Z.Q.; Han, W.; Zheng, S.H. Effects of high temperature on photosynthetic physiological characteristics of strawberry seedlings in greenhouse and construction of stress level. *Ying Yong Sheng Tai Xue Bao J. Appl. Ecol.* **2021**, *32*, 231–240. [CrossRef]
24. Lee, T.S.; Kang, G.C.; Paek, Y.; Moon, J.P.; Oh, S.S.; Kwon, J.K. Analysis of temperature and humidity distributions according to arrangements of air circulation fans in single-span tomato greenhouse. *Prot. Hortic. Plant Fact.* **2016**, *25*, 277–282. [CrossRef]
25. ASAE. Heating, Ventilating and Cooling Greenhouses. ANSI/ASAE EP406.4 JAN03, ASAE Standard 2003. Available online: http://ceac.arizona.edu/sites/default/files/asae_-_heating_ventilating_and_cooling_greenhouses.pdf (accessed on 6 September 2021).
26. Hilal, Y.Y.; Ishak, W.; Yahya, A.; Asha'ari, Z.H. Development of genetic algorithm for optimization of yield models in oil palm production. *Chil. J. Agric. Res.* **2018**, *78*, 228–237. [CrossRef]
27. Khamis, A.; Wahab, A. Comparative study on predicting crude palm oil prices using regression and neural network models. *Int. J. Sci. Technol.* **2016**, *5*, 1–6.
28. Ranjit, K.P.; Sinha, K. Forecasting crop yield: A comparative assessment of arimax and narx model. *RASHI* **2016**, *1*, 72–87.
29. Sun, X.D.; Zhou, M.X.; Sun, Y.Z. Variables selection for quantitative determination of cotton content in textile blends by near infrared spectroscopy. *Infrared Phys. Technol.* **2016**, *77*, 65–72. [CrossRef]
30. Sah, R.P.; Chakraborty, M.; Prasad, K.; Pandit, M.; Tudu, V.K.; Chakravarty, M.K.; Moharana, D. Impact of water deficit stress in maize: Phenology and yield components. *Sci. Rep.* **2020**, *10*, 1–15. [CrossRef] [PubMed]
31. Aubert, L.; Konrádová, D.; Barris, S.; Quinet, M. Different drought resistance mechanisms between two buckwheat species Fagopyrum esculentum and Fagopyrum tataricum. *Physiol. Plant.* **2021**, *172*, 577–586. [CrossRef] [PubMed]

Article

Assessing Suitable Techniques for Rainwater Harvesting Using Analytical Hierarchy Process (AHP) Methods and GIS Techniques

Ammar Adham [1,2], Michel Riksen [1,*], Rasha Abed [1,3], Sameer Shadeed [4] and Coen Ritsema [1]

[1] Soil Physics and Land Management Group, Wageningen University, 6700 AA Wageningen, The Netherlands; engammar2000@uoanbar.edu.iq (A.A.); rashahameed@mtu.edu.iq (R.A.); coen.ritsema@wur.nl (C.R.)

[2] Dams and Water Resources Engineering Department, College of Engineering, University of Anbar, Baghdad 55431, Iraq

[3] Anbar Technical Institute, Middle Technical University, Baghdad 10074, Iraq

[4] Water and Environmental Studies Institute, An-Najah National University, Nablus 62451, Palestine; sshadeed@najah.edu

* Correspondence: michel.riksen@wur.nl

Citation: Adham, A.; Riksen, M.; Abed, R.; Shadeed, S.; Ritsema, C. Assessing Suitable Techniques for Rainwater Harvesting Using Analytical Hierarchy Process (AHP) Methods and GIS Techniques. *Water* 2022, 14, 2110. https://doi.org/10.3390/w14132110

Academic Editor: Aizhong Ye

Received: 28 May 2022
Accepted: 27 June 2022
Published: 1 July 2022

Publisher's Note: MDPI stays neutral with regard to jurisdictional claims in published maps and institutional affiliations.

Abstract: The objective of this study is to produce suitability maps for potential rainwater harvesting techniques (RWHT) in the West Bank (WB), Palestine. These techniques aim to reduce water scarcity, which is a major problem for the conservation of water resources in the area. Based on literature reviews and expert recommendations, seven RWHts were selected (runoff basin system, contour ridges, cisterns, eyebrow terrace, check dam, on-farm pond, and bench terraces). Analysis methods performed in the Arc GIS environment include spatial analysis and data reclassification. Other calculations include multi-criteria analysis for assigning suitability. Five criteria (rainfall, runoff, land use, slope, and soil texture) for RWHt were analyzed to produce a suitability map for each technique. The results show that runoff basin systems in the northeast and southwest of WB are the most suitable, with about 50% of the area of WB moderately suitable for this technique, while 70% of the area of WB is very suitable for the contour ridge technique. Furthermore, this analysis shows that almost 50% of the WB is very suitable for cisterns. Sixty percent of the area is very suitable for on-farm puddling, especially in the north and southwest of WB. The areas with high suitability for the different techniques comprehensively cover the WB, as shown in the RWHt suitability maps and the integrated map. Nevertheless, this approach can help decision makers in making an initial selection of RWH techniques suitable for their region.

Keywords: rainwater harvesting technique (RWHt); the West Bank (Palestine); analytical hierarchy process method (AHP); GIS

1. Introduction

Irregular rainfall patterns and a lack of precipitation have caused water shortages around the world. People living in many areas with highly variable rainfall and unpredictable periods of drought or flooding are severely affected by water scarcity and often face livelihood insecurity [1]. Those regions, including Palestine, are characterized by arid to semi-arid climatic conditions and have uncertain water supplies. Population growth to approximately 2.9 million [2] and expansion of agriculture activities increase the stress on limited and uncertain water supplies; furthermore, the current political situation poses another accessibility limitation of water resources for Palestinians. The water shortage issues include the domestic and the agriculture sector. For domestic water, Palestinian Water Authority (PWA) statistics showed that, in most of the West Bank (WB) governorates, the average water consumption rate is (72 L/capita/d), which lies below the minimum World Health Organization's standards (150 L/capita/d) [3]. According to the Palestinian

Central Bureau of Statistics (PCBS), the total domestic water supply in WB increased from (85 MCM/year) in 2010 to (120 MCM/year) in 2015 [2]. A recent study showed that the water supply–demand gap in the entire WB will increase from (31.7 MCM/year) in 2015 to (41.2 MCM/year) in 2032 [4].

The main agricultural water source in WB is the groundwater (springs and wells) and the share of water purchased from Mekorot (Israeli Water Company, Israel). In 2015, the WB governorates' agricultural water requirement for all crops was (75 MCM/year) with a water supply–demand gap of (46.5 MCM/year) with almost the same amount in 2032 [4]. As a result, alternative water resources such as rainwater harvesting (RWH) are becoming a common practice in most regions of WB [5].

In 2011, rainwater harvesting techniques (RWHt) contribute about 1.5 MCM/year for agricultural use, whereas cisterns contribute to domestic use at about 4 MCM/year [2]. According to the 2018 PWA water plan, 10 MCM/year may be gathered via the use of various domestic and agricultural RWH approaches [6].

As a result, non-conventional water resources, such as RWH, may be used to alleviate water shortages in the WB, Palestine. The implementation of RWHt is promoted on a small scale by local societies and non-governmental establishments to improve temporal and spatial water shortage for domestic and agricultural uses. The success of RWH systems depends heavily on their technical design and the identification of suitable sites and techniques [6,7].

The identification of appropriate sites for the various RWHt in large areas was a great challenge [8]. Several methodologies have been established for the identification of RWH suitable sites. Some methodologies integrate multi-criteria decision making (MCDM) based on geoinformation and SWAT (Soil and Water Assessment Tool) model [9], while others used the TOPSIS multi-criteria decision analysis [10]. An intensive research effort focused on the development of rainwater harvesting (RWH) site suitability maps in different areas [11–13]. In contrast, less attention has been paid to the development of RWHt suitability maps. Most studies rely on the analysis of site characteristics to determine suitability rather than on the analysis of technical characteristics. Therefore, the preparation of RWHt suitability maps is crucial for determining which RWH technology is suitable for each suitable site. Thus, to successfully plan and implement RWHt, it is important to determine suitability for both the site and the technique.

In their literature study, Ammar et al. [14] evaluated primary criteria that have been used to identify potential RWH locations and procedures in arid and semi-arid areas (ASARs). They classified and contrasted four primary site selection methodologies, indicated three main sets of criteria for choosing RWH sites, and defined the most prevalent RWHt utilized in ASARs. Others used different methods [15], ranging from those based only on biophysical factors to more comprehensive ones, which included the inclusion of socioeconomic criteria, particularly after 2000. Most studies currently employ GIS in conjunction with hydrological models and/or multi-criteria analysis (MCA) to identify RHW suitability sites. Shadeed and Alawna [16] used this method to identify locations for the successful implementation of RWHt for agricultural use in the WB, Palestine. They concluded that 62% of the WB is high to very high suitable for implementing RWHt. However, they did not make a distinction between the different available RWHts. There are several differences among these RWHt from functional, construction, and design requirements. Not all techniques are equally suitable for every location. To select the best RWHt, it is better to make a suitability map for each technique separately.

This study aimed at developing suitability maps for potential RWHt in the WB by employing a GIS-based MCA approach. Moreover, this study aims at preparing an integrated RWHt for all the selected techniques that are of high value for water decision makers to properly identify suitable techniques that can be implemented in the area. This in turn will enhance sustainable water resources in Palestine.

2. Materials and Methods

2.1. Study Area

The West Bank (WB), Palestine, is located in the Middle East (Figure 1) with an area of about 5860 km^2. It has a population of 2.9 million people distributed in 11 administrative governorates [2].

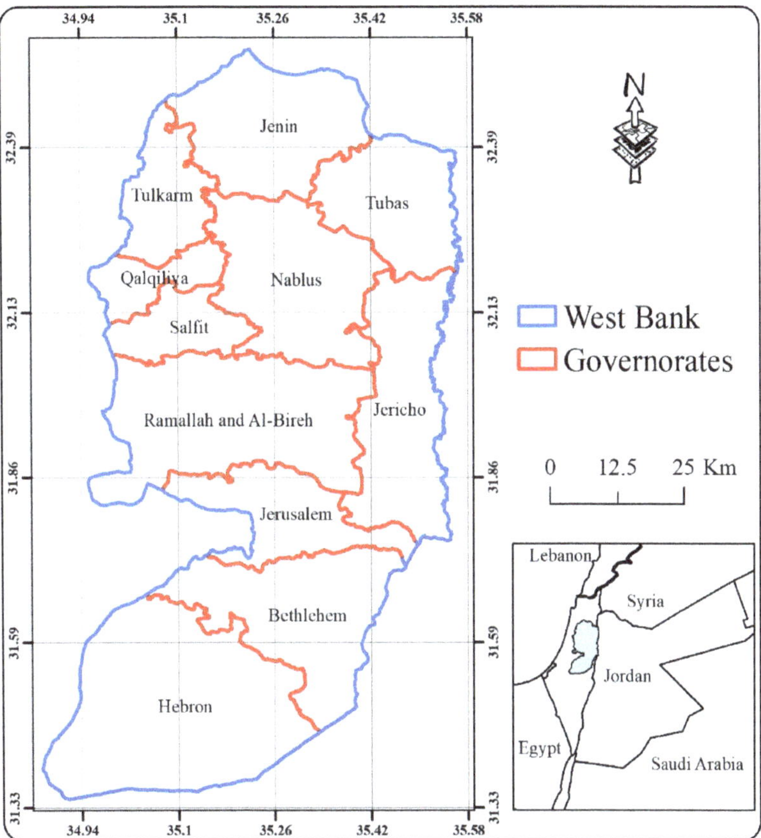

Figure 1. Administrative governates of the West Bank, Palestine.

Geographically, the WB is largely made up of hills (700 and 900 m above sea level) that run north-south and then fall to the Jordan Valley and the Dead Sea on the east side.

In general, the predominant climate is the Mediterranean, with rainy winters and hot, dry summers; the eastern and southern parts are much drier [16]. Surface water is mainly in the Jordan River and ephemeral wadis. However, since 1967, Palestinians do not have the right to access the river; therefore, they mainly rely on groundwater, the discharge from the different springs, and water purchased for domestic and agricultural use.

The average rainfall in the region is about 450 (mm/year). However, the majority of the yearly rainfall (about 80%) falls during the winter [16] with an average runoff curve number of about 50 [17], which indicates the potential for implementing RWH.

Widely variable land-use patterns are determined by the accessibility of water. Comparatively well-watered non-irrigated land within the hills is used for the grazing of sheep and tree crops. Irrigated land within the hills and also the river basin is intensively cultivated for various fruits and vegetables [18].

2.2. Methodology Overview

The identification of rainwater harvesting techniques (RWHt) suitability maps involved four stages:

i.   Selection of RWHt;
ii.   Selection of appropriate criteria for each technique;
iii.   Suitability classification for each criterion;
iv.   GIS application and maps suitability development.

2.2.1. RWHt Selecting

When developing a rainwater harvesting (RWH) system, selecting the appropriate RWHt after determining the RWH site that meets the fundamental technical design requirements for rainwater harvesting is a critical component for assuring long-term implementation. There are numerous strategies and approaches that have been employed to conserve rainwater all over the world.

The design requirements for RWHt as well as the assessment of their site-suitability play a major role in determining technique suitability. The first step in the selection process was to decide which rainwater harvesting techniques could be mapped at the country level, as RWH is a site-specific activity. First, data were collected with respect to the most commonly practiced RWHt in districts with similar climatic conditions and topography.

Next, an overview of the critical values for climate, soil conditions, topography, and other variables affecting each RWHt was created and a pre-selection list for the RWHt was made.

Finally, the pre-selected list was discussed with local experts, taking into account the already implemented RWHt. As a result, seven RWHts were selected: runoff basin system, contour ridges, cisterns, eyebrow terrace, check dam, on-farm pond and bench terraces.

2.2.2. Criteria Selection

This step formulates the set of criteria to choose suitable RWHts based on the primary purpose, expert assessments, literature studies, and, most significantly, data availability. To classify RWHt suitability and influenced by the RWH site suitability criteria, five criteria were selected: annual precipitation (rainfall) as a climate parameter, runoff and curve number (CN) as a hydrology parameter, land use as an agronomy parameter, slope as a topography parameter, and soil texture as a soil parameter.

I.   Rainfall

In any RWH system, the amount and distribution of rainfall are essential factors in determining whether a specific RWHt is suitable or not in a particular location. In ASARs, rainfall is characterized by high temporal and spatial variation [19]. When designing rainwater harvesting systems, the catchment region should receive enough rainfall for storage for future use.

II.   Runoff depth (curve number, CN)

The depth of runoff is used to determine the amount of water available during runoff. The runoff depth was calculated using the curve number (CN) given by the Soil Conservation Service [20]. The effects of soil and land cover on rainfall and runoff predict CN. Land-cover and soil-texture maps were used to estimate CN for each pixel in the research region. The depth of runoff may be stated as follows:

$$Q = \frac{(P - I_a)^2}{(P - I_a) + S} \tag{1}$$

where Q = depth of runoff (mm), P = precipitation (mm), S = potential maximum retention (mm), and I_a = initial abstraction (mm).

$I_a = 0.2\,S$ based on the analysis of rainfall data from small agricultural basins [21]. As a result, Equation (1) may be written as follows.

$$Q = \frac{(P - 0.2S)^2}{(P + 0.8S)} \tag{2}$$

S may be worked out using CN as follows.

$$S = \frac{25400}{\text{CN}} - 254 \tag{3}$$

The runoff reaction to a given rain is represented by CN, which ranges from 0 to 100. The presence of high CNs indicates that a significant percentage of the rainfall will be the surface runoff [22,23]. Shadeed and Almasri [17] created the CN map for the entire WB.

III.   Land use (LU)

At a given location, the runoff generated by rainfall is related to the use of the land. For example, when the land is more densely vegetated, there is less surface runoff and there will be more water infiltration [24]. In the WB and depending on the Ministry of Agriculture (MoA) database, there are seven different types of land uses that have been discovered: built-up, woodland, grazing, irrigated farming, permanent crops, Arab land, and Israeli settlements.

IV.   Slope

Slope plays a direct role in selecting the appropriate RWHt since it interferes with the structure's design. On another hand, it has a significant impact on the runoff generation and, thus, on the sedimentation amount, water flow velocity, and the cost (in terms of time, materials, and effort) required to implement RWH [25]. In Arc GIS 10.2, a 30 m resolution digital elevation model (DEM) was utilized to create the slope map.

V.   Soil texture

In RWHt design, soil texture affects both surface runoff and soil infiltration rates [26,27]. The percentage of sand, silt, and clay in a soil determines its texture class. According to Adham et al. [1], soil with fine and medium texture is usually better suited for RWH as they retain water better. Soil texture is certainly one of the most important factors in deciding where to build an RWHt. This importance varies from one technique to another depending on the design and operation of each technique. In this study, four soil texture classes were selected Sandy loam, Loamy, Clay loam, and Clay [28].

2.2.3. Criterion Suitability Classification

In this step, the different values within the different datasets were converted into a common suitability scale using GIS and MCA. Due to the different measures and weights for the different criteria, each of the five criteria was first classified. The wight for each criterion was determined after assigning scores using the Analytic Hierarchy Process (AHP) and the pairwise comparison matrix [29]. When comparing and scoring two criteria, a continuous 9-point scale is used, with odd numbers 1, 3, 5, 7, and 9 representing the range of suitability (not suitable–very suitable) of the criteria compared to each other. The scores were assigned and adjusted based on active discussions with local experts and engineers, as well as on information from previous scientific work.

2.2.4. GIS Application and Maps Suitability Development

In creating the datasets, the DEM and rainfall station data required additional analysis to derive the input data maps for slope and precipitation. Based on the suitability scale, a new scaled map was created for each input layer. The Spatial Analyst module of Arc GIS 10.2 was used to determine suitability by reclassifying the criteria layers and using the raster calculation tool. The final step is to combine the converted output layers of annual perception, land use, slope, soil texture, and CN. For each RWHt, each criterion was classified as a numerical value and assigned a suitability value. Then, the suitability values were classified into four groups: low (<20), moderate (20–29), high (30–39), and very high (>39). Figure 2 shows the flowchart of the steps taken to derive the RWH suitability map for each RWHt. A Model Builder in ArcGIS10.2.1 was established for generating a suitability model, which included steps for calculating the suitability score for each RWHt.

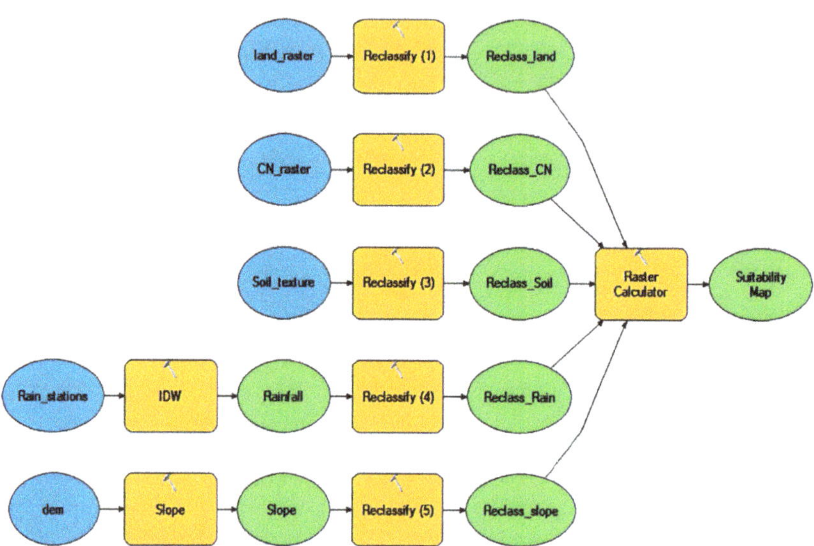

Figure 2. Flow chart for the identification of RWHt suitability map.

3. Results and Discussion

3.1. Input Maps

Figure 3 shows the GIS input maps representing the levels in our model for the suitability analysis. The long-term average annual precipitation for the entire West Bank (WB) is shown in Figure 3a. The amount of precipitation varies greatly within the WB. In particular, the eastern and southern parts of the WB are much drier. Potential runoff, shown in the CN figure in Figure 3b, is low in areas with sandy loam soil in Figure 3f and very high in built-up areas in Figure 3c. Permanent crops, including arable crops, and irrigated farming are found mainly in the regions with higher rainfall. Pasture dominates in the eastern part of the WB. Figure 3d shows the slope driven from the dem in Figure 3e.

3.2. Suitability Score for Each Criterion for the Rwhtt

All seven selected rainwater harvesting techniques (RWHts) were assigned a suitability scale. All layers were reclassified according to their suitability with a specific score, as shown in Table 1. The suitability ratings and criteria selection were the result of several discussions with local experts and engineers with experience in developing RWHt. The ratings were updated and modified several times depending on the previous studies to avoid discrepancies in the allocation of points [25,30–32].

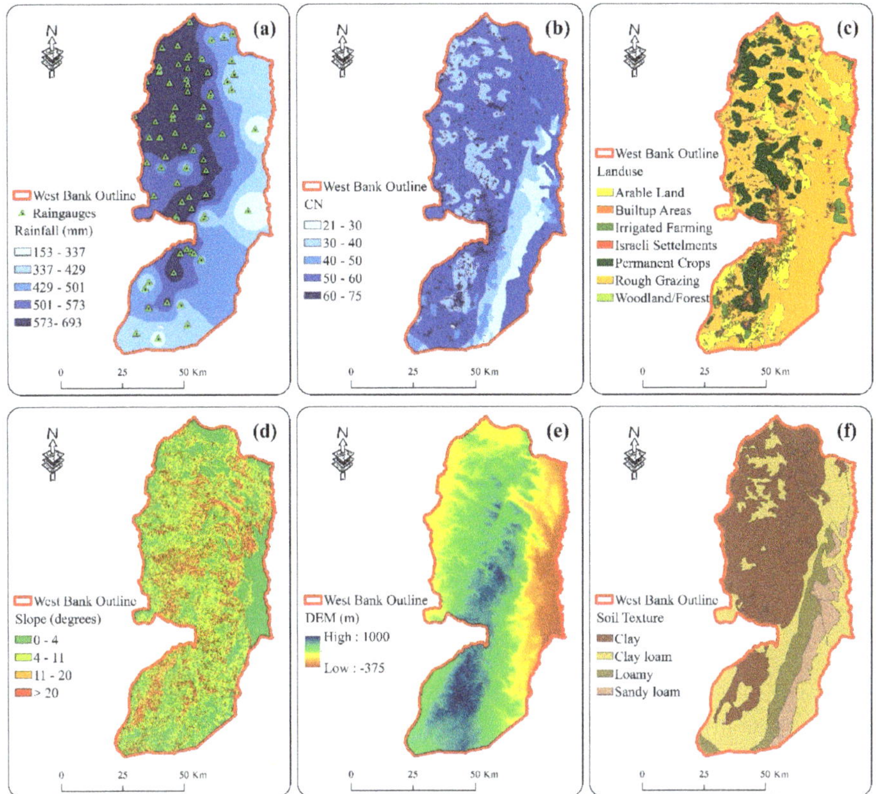

Figure 3. Input maps for the WB: (**a**) annual average rainfall (**b**), runoff curve number, (**c**) land use, (**d**) slope, (**e**) dem, and (**f**) soil texture.

Table 1. The suitability score of the selected rainwater harvesting techniques for the West Bank, Palestine.

#	Criteria	Classes	Runoff Basin	Contour Ridges	Cistern	Eyebrow Terrace	Check Dam	On-Farm Pond	Bench Terrace
							Score		
1	Annual rainfall (mm)	<250	7	3	9	7	5	5	5
		250–500	9	9	7	9	7	7	9
		500–750	3	7	5	5	9	9	7
2	Land use	Arable Land (supporting grains)	7	9	1	7	5	7	7
		Built-up Areas	1	1	5	1	1	1	1
		Woodland/Forest	1	1	1	3	1	1	1
		Rough Grazing/Subsistence Farming	3	5	1	5	5	5	3
		Irrigated Farming	1	3	9	1	9	9	1
		Permanent Crops (Fruits trees)	9	7	7	9	3	3	9
		Israeli Settlements	1	1	1	1	1	1	1

Table 1. *Cont.*

#	Criteria	Classes	Runoff Basin	Contour Ridges	Cistern	Eyebrow Terrace	Check Dam	On-Farm Pond	Bench Terrace
						Score			
3	Slope (%)	flat (0–2)	9	3	3	7	5	3	1
		gentle (2–5)	7	5	9	9	9	7	1
		moderate (5–10)	5	9	7	5	7	9	3
		rolling (10–15)	1	7	5	3	3	5	7
		hilly (15–30)	1	1	1	1	1	1	9
		steep >30	1	1	1	1	1	1	5
4	Soil texture	Sandy loam	3	5	1	5	3	3	7
		Loamy	5	7	3	7	5	5	5
		Clay loam	9	9	5	9	7	7	9
		Clay	7	3	9	3	9	9	3
5	Curve number	≤50	3	5	3	3	3	3	5
		51–60	7	7	5	7	5	5	7
		61–70	9	9	7	9	7	7	9
		>70	5	3	9	5	9	9	3

3.3. The Potential of RWHt

Figure 4 shows potential maps for different types of RWHt. The maps identified by the spatial analyst module show the suitability on a scale from red (not suitable) via yellow (moderate suitable) to green (very high suitable), based on the five selected criteria. Table 2 shows the percentage per suitability class for each RWH technique for the entire WB.

Table 2. Percentage of the WB suitable for different RWH techniques.

Suitability Score	Low	Moderate	High	Very High
	<20.0	20 to 29	30 to 39	>40
On farm pond	3.4	52.9	42.8	1.0
Bench terraces	0.9	29.6	69.5	0.0
Check dam	1.7	40.1	51.7	6.4
Eyebrow terraces	14.8	72.8	12.3	0.0
Cistern	0.0	53.7	46.2	0.1
Contour ridges	3.3	33.0	59.9	3.8
Runoff basin	6.3	64.4	28.7	0.5

To obtain an overview of the suitability assessment for all selected techniques together, Figure 5 shows an integrated suitability map created with specific analysis properties in Arc GIS 10.2. Each RWHt has been assigned a specific color to indicate its high suitability for a specific area in WB. The map shows a large discrepancy in the amount of land suitable for the different RWHts.

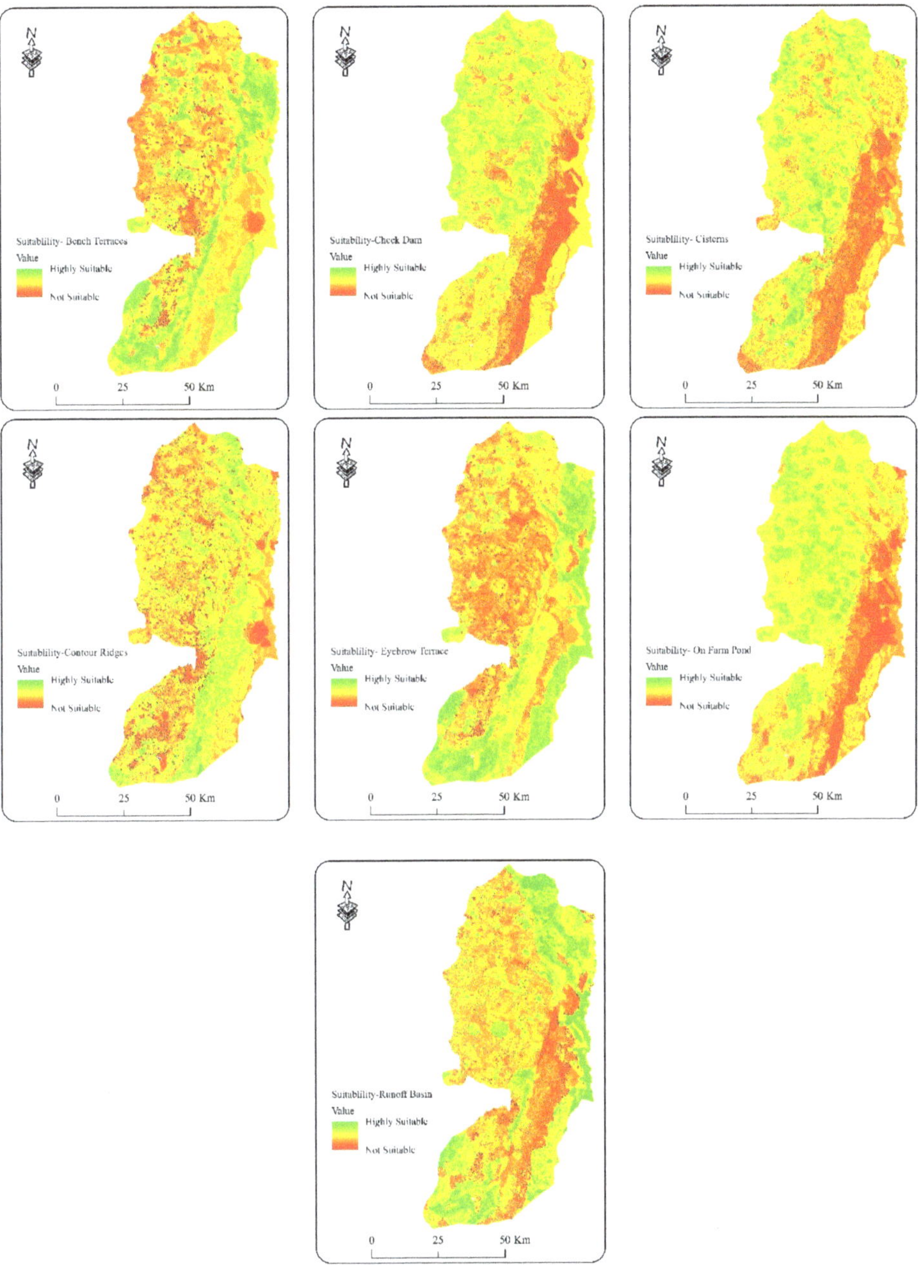

Figure 4. Suitability maps for different types of RWHt in the WB, Palestine, based on soil type, slope, rainfall, land use, and curved number.

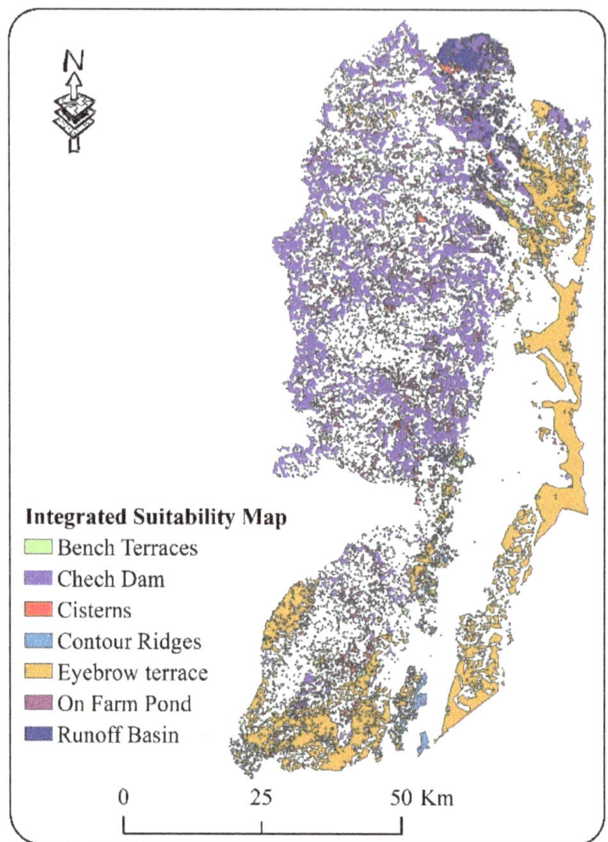

Figure 5. Integrated suitability map of all the seven RWHts for the WB, Palestine.

From the maps in Figure 4 and the integrated map in Figure 5, it is clear that the different RWHts have different suitability in WB. Each RWHt has its own suitability map, which reflects the technical requirements for that technique and is influenced by the criterion. Runoff basins are most suitable on the northern eastern and southern western borders of WB in the relatively shallow parts, and the soil is mostly clay loam. Map statistics show that about 50% of the area of WB is moderately suitable for this technique. Seventy percent of the area of WB is very suitable for the contour ridge technique. Most of the well-suited sites are in the northern east and southern east with cropland and a perceptual range of more than 250 mm. About 50% of the total area of WB is very suitable for the cistern technique. These areas are mainly in the northwest and southwest where the soil texture is mainly classified as clay loam, as the infiltration rate in the catchment area of the cistern is an important factor for suitability. Therefore, the area with high infiltration rate and low CN value shows low suitability for the cistern.

Highly suitable areas for implementing eyebrow terraces are located mainly on the northern and southeastern borders and in the southwestern borders of WB. Statistics showed that about 70% of the area of WB is moderately suitable for this technique. This can be explained by the fact that this technique is less suitable for clay soils. As for the farm pond, the areas with high suitability for this technique are mainly located in the north and southwest of WB, with 60% of the area with moderate slopes and clay soil being very suitable for this technique.

It should be noted that according to this analysis, about 80% of the area of WB, which is mainly in the north and northwest, is highly suitable for the construction of check dams. Figure 3d shows that these areas have a slope of 15 to 30% and a CN of 60 to 70 mm. Areas with high suitability for the use of the bench terrace technique are located in the northeast and southwest of WB where orchard farms are located and these areas have a hilly to steep slopes. Map statistics show that more than 50% of the area of WB is highly suitable for this technique.

A look at the input maps in Figure 3, the scoring system in Table 1, and the result of the individual RWH suitability map in Figure 4 shows how the suitability maps are strongly influenced by the score assigned to each criterion.

The relatively high suitability scores for these RWH techniques do not mean that these are the best solutions for farmers, as socioeconomic, political, and cultural aspects were not considered in this analysis. Bench terraces, for example, are too expensive in most cases due to high labour costs. Eyebrow terraces are less suitable for mechanised cropping operations because of their irregular shape. Nevertheless, this approach can help farmers and decision makers in the initial selection of RWHts suitable for their region.

4. Conclusions

A suitability model based on GIS, created with ModelBuilder in Arc GIS 10.2, was used to identify potential RWHts. A set of criteria (rainfall, runoff, slope, land use and soil texture) were included in the suitability model.

According to the results of this study, this research technique provides an initial meaningful screening of broad areas and is an extremely useful tool for assisting in the development and implementation of a rainwater harvesting (RWH) project, especially in arid and semi-arid environments. Arc GIS 10.2 has proven to be an extremely useful tool in this study for integrating various information to identify ideal locations for different RW. In screening vast regions for the applicability of RWH measures, Arc GIS 10.2 proved to be a versatile, time-saving and cost-effective tool.

Hydrologists, decision makers, and planners will benefit from the suitability map as it will allow them to easily identify which rainwater harvesting technique (RWHt) to use in sites with RWH potential. The quality and accuracy of the data, as well as the way the data were sourced, processed, and produced, were all factors in the quality of the map.

However, to confirm the applicability of the model, it needs to be calibrated and tested in different regions and with different RWHts. In addition, as the suitability ratings have a major impact on the maps of RWHt suitability, a validation study or pilot project is recommended to ensure the margin of error (if any) in determining the preferences for each RWHt. Socio-economic criteria such as investment and maintenance costs and labour input may also be important for water harvesting. Therefore, socio-economic suitability for different RWHts needs to be explored and included in the assessment process. These ideas will improve the realism of the model and broaden the scope of this methodology.

Author Contributions: Conceptualization A.A., M.R. and R.A.; methodology, A.A. and R.A.; software, M.R. and R.A.; formal analysis, S.S. and M.R.; investigation, A.A. and S.S.; resources, A.A. and C.R.; writing—original draft preparation, A.A. and R.A.; writing—review and editing, C.R.; visualization, A.A., S.S. and M.R.; supervision, C.R.; project administration, C.R. and M.R.; funding acquisition, C.R. All authors have read and agreed to the published version of the manuscript.

Funding: This Research is fully funded by NUFIC Orange Knowledge Program, 9OKP-IRA-104278.

Data Availability Statement: Some data in this manuscript were obtained from the Ministry of Agriculture and PWA, Palestine. The other data are from the fieldwork and previous studies.

Conflicts of Interest: The authors declare no conflict of interest.

References

1. Adham, A.; Sayl, K.N.; Abed, R.; Abdeladhim, M.A.; Wesseling, J.G.; Riksen, M.; Ritsema, C.J. A GIS-based approach for identifying potential sites for harvesting rainwater in the Western Desert of Iraq. *Int. Soil Water Conserv. Res.* **2018**, *6*, 297–304. [CrossRef]
2. PCBS. *Manual of Statistical Indicators Provided by Palestinian Central Bureau of Statistics*; Palestinian Central Bureau of Statistics: Ramallah, Palestine, 2016; pp. 1–249. Available online: http://www.pcbs.gov.ps/Downloads/book2196.pdf (accessed on 15 March 2022).
3. PWA. *Status Report of Water Resources in the Occupied State of Palestine*; Palestinian Water Authority: Ramallah, Palestine, 2012; pp. 1–22. Available online: http://www.pwa.ps/page.aspx?id=0PHprpa2520241944a0PHprp (accessed on 15 March 2022).
4. Shadeed, S.; Judeh, T. *Water Supply-Demand Gap Analysis for both Domestic and Agricultural Uses in the West Bank, Palestine [Report]*, 1st ed.; Paduco: Nablus, Palestine, 2018; pp. 1–22. Available online: http://morwater.najah.edu/WaterAttachment//edcb0bd3-d431-4bc0-ab76-4852d6b124e8.pdf (accessed on 15 March 2022).
5. Daoud, A.K.; Swaileh, K.M.; Hussein, R.M.; Matani, M. Quality assessment of roof-harvested rainwater in the West Bank, Palestinian Authority. *J. Water Health* **2011**, *9*, 525–533. [CrossRef]
6. Alawna, S.; Shadeed, S. Rooftop Rainwater Harvesting to Alleviate Domestic Water Shortage in the West Bank, Palestine. *An-Najah Univ. J. Res.-A* **2021**, *35*, 83–108.
7. Adham, A.; Riksen, M.; Ouessar, M.; Abed, R.; Ritsema, C. Development of Methodology for Existing Rainwater Harvesting Assessment in (semi-) Arid Regions. In *Water and Land Security in Drylands: Response to Climate Change*; Ouessar, M., Gabriels, D., Tsunekawa, A., Evett, S., Eds.; Springer International Publishing: Cham, Switzerland, 2017; pp. 171–184, ISBN 978-3-319-54021-4.
8. Prinz, D.; Oweis, T.; Oberle, A. Rainwater harvesting for dry land agriculture-Developing a methodology based on remote sensing and GIS. *Proc. XIII Int. Congr. Agric. Eng.* **1998**, *2*, 2–6.
9. Umugwaneza, A.; Chen, X.; Liu, T.; Mind'je, R.; Uwineza, A.; Kayumba, P.M.; Maniraho, A.P. Integrating a GIS-based approach and a SWAT model to identify potential suitable sites for rainwater harvesting in Rwanda. *AQUA Water Infrastruct. Ecosyst. Soc.* **2022**, *71*, 415–432. [CrossRef]
10. Tahvili, Z.; Khosravi, H.; Malekian, A.; Khalighi Sigaroodi, S.; Pishyar, S.; Singh, V.P.; Ghodsi, M. Locating suitable sites for rainwater harvesting (RWH) in the central arid region of Iran. *Sustain. Water Resour. Manag.* **2021**, *7*, 1–11. [CrossRef]
11. Balkhair, K.S.; Ur Rahman, K. Development and assessment of rainwater harvesting suitability map using analytical hierarchy process, GIS and RS techniques. *Geocarto Int.* **2021**, *36*, 421–448. [CrossRef]
12. Muleta, B.; Seyoum, T.; Assefa, S. GIS-Based Assessment of Suitability Area of Rainwater Harvesting in Daro Labu District, Oromia, Ethiopia. *Am. J. Water Sci. Eng.* **2022**, *8*, 21–35.
13. Ndeketeya, A.; Dundu, M. Application of HEC-HMS Model for Evaluation of Rainwater Harvesting Potential in a Semi-arid City. *Water Resour. Manag.* **2021**, *35*, 4217–4232. [CrossRef]
14. Adham, A.; Riksen, M.; Ouessar, M.; Ritsema, C. Identification of suitable sites for rainwater harvesting structures in arid and semi-arid regions: A review. *Int. Soil Water Conserv. Res.* **2016**, *4*, 108–120.
15. Ziadat, F.; Bruggeman, A.; Oweis, T.; Haddad, N.; Mazahreh, S.; Sartawi, W.; Syuof, M. A participatory GIS approach for assessing land suitability for rainwater harvesting in an arid rangeland environment. *Arid. Land Res. Manag.* **2012**, *26*, 297–311. [CrossRef]
16. Shadeed, S.; Almasri, M. Application of GIS-based SCS-CN method in WB catchments, Palestine. *Water Sci. Eng.* **2010**, *3*, 1–13. [CrossRef]
17. Britannica, West Bank. Encyclopedia Britannica. 2021. Available online: https://www-britannica-com.ezproxy.library.wur.nl/place/West-Bank (accessed on 22 February 2022).
18. Yifru, B.A.; Kim, M.G.; Lee, J.W.; Kim, I.H.; Chang, S.W.; Chung, I.M. Water Storage in Dry Riverbeds of Arid and Semi-Arid Regions: Overview, Challenges, and Prospects of Sand Dam Technology. *Sustainability* **2021**, *13*, 5905. [CrossRef]
19. Melesse, A.M.; Shih, S.F. Spatially distributed storm runoff depth estimation using Landsat images and GIS. *Comput. Electron. Agric.* **2002**, *37*, 173–183. [CrossRef]
20. Alves, G.J.; de Mello, C.R.; Beskow, S.; Junqueira, J.A.; Nearing, M.A. Assessment of the Soil Conservation Service–Curve Number method performance in a tropical Oxisol watershed. *J. Soil Water Conserv.* **2019**, *74*, 500–512. [CrossRef]
21. Krois, J.; Schulte, A. AGIS-based multi-criteria evaluation to identify potential sites for soil and water conservation techniques in the Ronquillo watershed, northern Peru. *Appl. Geogr.* **2014**, *51*, 131–142. [CrossRef]
22. Adham, A.; Riksen, M.; Ouessar, M.; Ritsema, C.J. A water harvesting model for optimizing rainwater harvesting in the wadi Oum Zessar watershed, Tunisia. *Agric. Water Manag.* **2016**, *176*, 191–202. [CrossRef]
23. Mwenge Kahinda, J.; Lillie, E.S.B.; Taigbenu, A.E.; Taute, M.; Boroto, R.J. Developing suitability maps for rainwater harvesting in South Africa. *Phys. Chem. Earth* **2008**, *33*, 788–799. [CrossRef]
24. Adham, A.; Riksen, M.; Ouessar, M.; Ritsema, C.J. A Methodology to Assess and Evaluate Rainwater Harvesting Techniques in (Semi-) Arid Regions. *Water* **2016**, *8*, 198. [CrossRef]
25. Al-Adamat, R. GIS as a decision support system for siting water harvesting ponds in the Basalt Aquifer/NE Jordan. *J. Environ. Assess. Policy Manag.* **2008**, *10*, 189–206. [CrossRef]
26. Al-Adamat, R.; AlAyyash, S.; Al-Amoush, H.; Al-Meshan, O.; Rawajfih, Z.; Shdeifat, A.; Al-Harahsheh, A.; Al-Farajat, M. The Combination of Indigenous Knowledge and Geo-Informatics for Water Harvesting Siting in the Jordanian Badia. *J. Geogr. Inf. Syst.* **2012**, *4*, 366–376. [CrossRef]

27. Singh, L.K.; Jha, M.K.; Chowdary, V.M. Multi-criteria analysis and GIS modeling for identifying prospective water harvesting and artificial recharge sites for sustainable water supply. *J. Clean. Prod.* **2017**, *142*, 1436–1456. [CrossRef]
28. Saaty, T.L. Decision making with the analytic hierarchy process. *Int. J. Serv. Sci.* **2008**, *1*, 83–98. [CrossRef]
29. Al-Adamat, R.; Diabat, A.; Shatnawi, G. Combining GIS with multicriteria decision making for siting water harvesting ponds in Northern Jordan. *J. Arid. Environ.* **2010**, *74*, 1471–1477. [CrossRef]
30. Jha, M.K.; Chowdary, V.M.; Kulkarni, Y.; Mal, B.C. Rainwater harvesting planning using geospatial techniques and multicriteria decision analysis. *Resour. Conserv. Recycl.* **2014**, *83*, 96–111. [CrossRef]
31. Matomela, N.; Li, T.; Ikhumhen, H.O. Siting of rainwater harvesting potential sites in arid or semi-arid watersheds using GIS-based techniques. *Environ. Processes* **2020**, *7*, 631–652. [CrossRef]
32. Toosi, A.S.; Tousi, E.G.; Ghassemi, S.A.; Cheshomi, A.; Alaghmand, S. A multi-criteria decision analysis approach towards efficient rainwater harvesting. *J. Hydrol.* **2020**, *582*, 124501. [CrossRef]

water

Article

Economic Feasibility of Rainwater Harvesting Applications in the West Bank, Palestine

Johanna E. M. Schild [1,2,*], Luuk Fleskens [3], Michel Riksen [3] and Sameer Shadeed [4]

[1] PBL Netherlands Environmental Assessment Agency, P.O. Box 30314, 2500 GH The Hague, The Netherlands
[2] Institute of Environmental Sciences CML, Leiden University, P.O. Box 9518, 2300 RA Leiden, The Netherlands
[3] Soil Physics and Land Management Group, Wageningen University,
     P.O. Box 47, 6700 AA Wageningen, The Netherlands
[4] Water and Environmental Studies Institute, An-Najah National University,
     P.O. Box 7, Nablus 62451, Palestinian Territory
*   Correspondence: jemschild@gmail.com

Abstract: Freshwater resources are uncertain in Palestine and their uncertainty is expected to intensify due to climate change and the political situation. Yet, in this region, a stable freshwater supply is vital for domestic and agricultural uses. Rainwater harvesting could help to increase freshwater availability. This study investigates the economic feasibility of two rainwater harvesting applications in the West Bank, with eyebrow terracing in olive groves in rural areas and domestic rooftop harvesting in urban areas. Cost-effectiveness is estimated using a spatially explicit cost–benefit analysis. Three land zones varying in suitability for the implementation of eyebrow terracing in olive groves are analyzed. The potential increase in olive yield is estimated with a crop–water balance model. The potential amount of rainfall that can be harvested with domestic rooftop harvesting is calculated based on the average rooftop area for each of the 11 governorates individually. Costs and benefits are considered at the household level to calculate the economic feasibility of these two applications. Although eyebrow terracing enlarges soil moisture availability for olive trees and thereby increases olive yield by about 10–14%, construction costs are too high to make implementation cost-effective. Similarly, rooftop harvesting can harvest about 30% on average of the annual domestic water demand and is worthwhile in the northern and southern governorates. Yet, in this case, construction costs are generally too high to be cost-effective. This obstructs more widespread adoption of rainwater harvesting in the West Bank, which is urgently needed given the large impacts of climate change. Providing subsidies for rainwater harvesting could help to make adoption more attractive for households.

Keywords: rainwater harvesting; cost–benefit analysis; aridity; urban; rural; rooftop harvesting; eyebrow terracing; Palestine

Citation: Schild, J.E.M.; Fleskens, L.; Riksen, M.; Shadeed, S. Economic Feasibility of Rainwater Harvesting Applications in the West Bank, Palestine. *Water* **2023**, *15*, 1023. https://doi.org/10.3390/w15061023

Academic Editor: Enedir Ghisi

Received: 15 January 2023
Revised: 24 February 2023
Accepted: 27 February 2023
Published: 8 March 2023

1. Introduction

Freshwater availability is uncertain in Palestine and is expected to become even more limited due to climate change, rapid population growth, expansion of agricultural activities, and political implications [1]. Palestine is characterized by an arid to semi-arid climate. Access to freshwater resources from aquifers is very limited and has been decreasing over time due to Israeli control of Palestinian water resources [1,2]. As such, Palestinians are seeking to have new water alternatives such as rainwater harvesting. However, due to climate change, precipitation is expected to decrease along with an increase in dry spells and an increase in evapotranspiration due to increasing temperature [3]. Together, this is leading to increased water insecurity.

Yet, freshwater availability is vital for people's daily lives in Palestine, mainly for domestic water use and irrigation water in agriculture. Ensuring the availability and sustainable management of water and sanitation—which is described in Sustainable Development Goal 6—is not only vital in itself, but is also "essential for enhancing food security, health and wellbeing of citizens (SDG 2, 3 and 14), and the resilience of Palestinian communities in the face of water confiscation and intensifying climate change (SDG 13)" [2]. Even though the majority of homes are connected to the water grid in Palestine (i.e., 91%) [4], the water supply is irregular and intermittent, and about one-third of distributed water is lost due to leakages in the water grid [2]. As a result, the average water consumption is low and households facing intermittent supply or who are not connected to the water grid are forced to buy tanked water at highly inflated tariffs [2]. As such, coping with increasingly uncertain water supply is critical for Palestine.

To cope with increasing uncertainty in water availability, the government of Palestine is intending to promote more widespread adoption of rainwater harvesting [2,5]. Rainwater harvesting is already practiced in Palestine to some extent. It forms an additional source of water for domestic consumption and agriculture use [6]. Implementation is promoted on a small scale by local communities and non-governmental organizations to alleviate temporal and spatial water shortages. People typically collect rainwater from roofs or rock catchments and store it in cisterns in order to meet part of their water needs [6]. More widespread adoption of rainwater harvesting could help to increase water security in the region.

Rainwater harvesting is applied widely in arid regions around the world as a means to provide water for agricultural and domestic uses [7,8] and is an important method to adapt to climate change [9–11]. It is a relatively low-key solution to increase water security for households [12], and is seen as a low-regret adaptation measure to climate change [1]. It is also promising to upscale rainwater harvesting in rural areas in order to increase crop production [13]. Other benefits include—amongst others—reduced soil erosion, reduced runoff peak flow, flood mitigation, and increased groundwater recharge [14]. Many different rainwater harvesting systems exist that can be adapted to local climatic, biophysical and socio-economic conditions, and are suitable for either the urban or rural context. Rainwater harvesting is also relatively cheap to implement and a lot of experience and technical knowledge is available for successful implementation given its long history [15].

Previous studies show that the success of the implementation of rainwater harvesting systems depends heavily on their economic feasibility [14,16,17] next to their technical design and identification of suitable sites [11,18,19]. Technical design and site suitability have been studied previously for Palestine and show that a range of techniques and suitable locations exist to apply rainwater harvesting [5,20,21].

Yet, the economic feasibility of implementing rainwater harvesting for households in Palestine is unknown. The economic feasibility of rainwater harvesting can be calculated using a cost–benefit analysis [22]. Previous cost–benefit studies about rainwater harvesting in residential areas in other countries show that rainwater harvesting may be an efficient strategy, but cost-efficiency depends largely on local water prices, besides cistern size and type ([23] for Jordan; [24,25] for the USA). Often, water tariffs are heavily subsidized and very low (e.g., [23]), making it hard for alternative water sources that require initial investment and maintenance costs to be cost-efficient. Furthermore, often investment costs are the main economic barrier for households to install rainwater harvesting, especially in the global South [14,17,26], even though installation may supplement household income when harvested rainwater can be used as irrigation water for crop production [26].

In the context of climate change and increasing uncertainty in water availability, and in a region that is primarily dependent on rainwater for its freshwater input, it is essential to know whether implementation of rainwater harvesting can increase water availability and whether adoption can be cost-effective on a larger scale. Especially, given that the Palestinian government is considering promoting rainwater harvesting as a strategic option to overcome water shortages in Palestine on a larger scale [2,5]. When implementation can be economically feasible, the Palestinian government could consider to stimulate the adoption of rainwater harvesting, for example, by the means of financial stimuli, such as subsidies.

This study aims to investigate the economic feasibility of the implementation of two rainwater harvesting applications in a rural and urban setting in the West Bank of Palestine. Two types of rainwater harvesting that are commonly practiced in the West Bank will be investigated: (1) eyebrow terraces in olive cultivation in rural areas, and (2) domestic rooftop harvesting in residential areas [5]. A crop–water balance model is set up to estimate the impact of rainwater harvesting by the means of eyebrow terracing on olive yield. A rainwater harvesting calculation tool is used to estimate the amount of rainwater that can be harvested on domestic rooftops. The resulting estimates are used to calculate the costs and benefits of the two selected rainwater harvesting applications. These costs and benefits are weighed in spatially explicit cost–benefit analyses to determine whether the implementation of these two techniques is economically feasible for Palestinian households.

This study approach has two novel aspects. In contrast to previous studies that mainly focused on analyzing one technique applied in either an urban or rural setting, the current study will investigate two techniques in two distinct landscapes of the West Bank. It is expected that along with differences in the amount of rainwater that can be harvested, the costs and benefits of rainwater harvesting implementation will be different as well for rural and urban rainwater harvesting. Another novel aspect is that the cost–benefit analyses that are undertaken in this study are spatially explicit. This allows for us to account for spatial variability in local climatic, biophysical and socio-economic conditions that determine the suitability and effectiveness of rainfall capture, and ultimately cost-effectiveness of harvesting in specific regions of the West Bank. As such, it can offer insight into which parts of the country can be promising to implement rainwater harvesting.

2. Materials and Methods

2.1. Study Area

The study area is the West Bank of Palestine, which is subdivided into 11 governorates (Figure 1). The surface area of the West Bank is about 5660 km^2 [27]. The West Bank has a Mediterranean climate with climate zones ranging from dry sub-humid, semi-arid, arid to hyper-arid zones. The long-term annual average varies between 133 mm in the proximity of the Jordan River (in Jericho) and 658 mm in the central mountains (in Salfit), with an annual average value of about 420 mm for the entire West Bank [28].

In 2018, the West Bank had 2.9 million inhabitants and a population density of 522 persons/km^2 [26]. Gross Domestic Product accounted for USD 10,715.9 million and Gross Domestic Product per capita for USD 4154.2 in 2017 [26]. Agriculture (combined with fisheries) accounted for 7.1% of the national GDP in Palestine in 2018 [26].

Land use in the West Bank consists of arable land (supporting grains), irrigated farming (supporting vegetables), permanent crops (including olives, grapes, citrus, and other fruit trees), rangeland (including rough grazing and subsistence farming), woodland and forest, built-up areas and Israeli settlements [29]. The latter category is not considered in this study, since it is not controlled by the Palestinian Authority.

2.2. Implementation of Eyebrow Terraces

2.2.1. Spatial Characterization Using Archetype Analysis

A spatial characterization was made of the West Bank using archetype analysis for the implementation of eyebrow terraces in olive groves. Archetype analysis uses socio-ecological indicators to systematically identify regions with similar conditions (so-called archetypes) for the implementation of rainwater harvesting [13]. To this end, a baseline map was prepared based on land use in the West Bank [29]. Suitable land use classes were selected in which eyebrow terraces can be implemented, either in existing or new olive groves. Suitable land use classes were arable land supporting grains and permanent crops, including olives, oranges and grapes. The land use map shows that olive groves cover the largest area of the permanent crops (about 80%). It also shows that the selected area suitable for olive terracing covers 1617 km^2, which constitutes about 29% of the West Bank (Figure 2).

Subsequently, three biophysical variables were used to define suitable areas for olive cultivation on eyebrow terraces: precipitation, slope and available soil water capacity (AWC) [5].

Figure 1. Map of the West Bank, showing the 11 governorates.

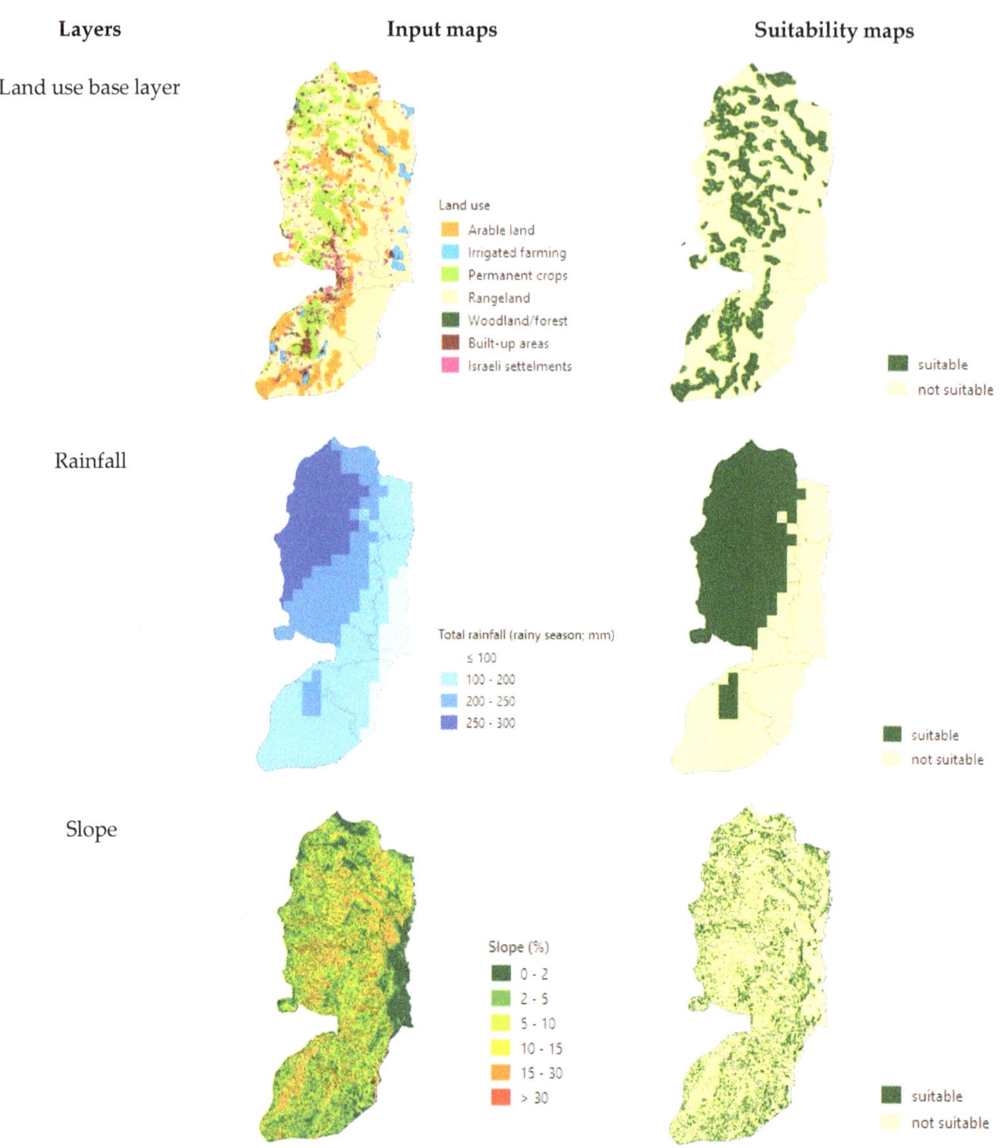

Figure 2. *Cont.*

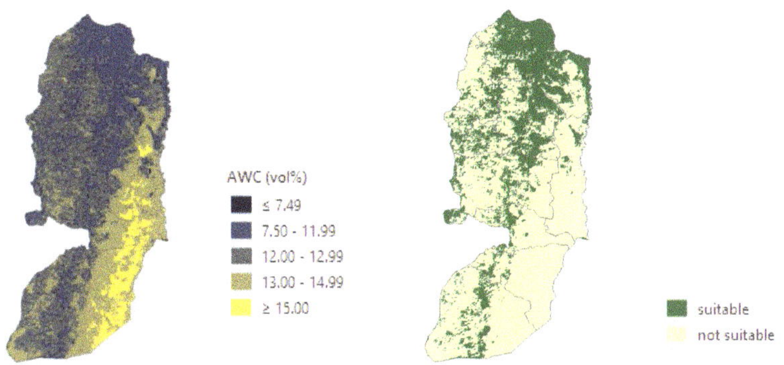

Figure 2. Archetype analysis using land use, rainfall, slope and available soil water capacity (AWC) to identify similar regions in the West Bank that are suitable for eyebrow terracing.

Precipitation

In order to define which areas are suitable for rainwater harvesting with eyebrow terracing, the precipitation (i.e., rainfall) in the wet season was analyzed. The wet season takes place from Nov–Dec–Jan–Feb–Mar in the West Bank and contributes to about 75% of the annual precipitation [30]. Monthly precipitation maps were summed up for each annual wet season to have the total precipitation of the wet season. Historical total monthly precipitation data (in mm) for the time period 2000–2018 were collected at a spatial resolution of 2.5 arc-minutes (i.e., about 5 km) from CRU-TS 4.03 [31] and downscaled with WorldClim 2.1 [32].

From the total wet season maps, the wet season with, on average, the lowest amount of precipitation was defined as the driest rainy season. The wet season with, on average, the highest amount of precipitation was defined as the wettest rainy season. Within the 2000–2018 time period, the rainy season in 2016/2017 was the driest with a total average of 194 mm (±61 S.D.), while the rainy season in 2002/2003 was the wettest with a total average of 503 mm (±175 S.D.). Based on the amount of precipitation in the driest rainy season, the West Bank was classified into two classes: (1) dry region (≤200 mm) and (2) wet region (>200 mm). The wet region was defined as suitable for eyebrow terracing with olive groves (Figure 2).

Slope

The slope was analyzed to define which areas in the West Bank can be suitable for the implementation of eyebrow terraces. The slope was calculated in percentage (%) based on the Digital Elevation Map (DEM) [29]. In the West Bank, slopes range between 0 and 82% and are on average 7.7%. A slope between 2 and 5% was defined as suitable for the construction of eyebrow terraces for olive trees (Figure 2) [5].

Available Soil Water Capacity

Available soil water capacity (AWC) was analyzed to determine which areas in the West Bank are suitable for the implementation of olive terracing. Maximum available soil moisture (in mm/m) was collected from the SoilGrids 2017 dataset having a 250 m spatial resolution [33]. Data for available soil water capacity ('AWCh1') with a field capacity at pF 2.0 were selected in order to include the widest possible range for water uptake at field capacity [34]. Data were extracted at a maximum of up to 1 m depth, which included 6 soil layers for AWC. From this data, the bulk estimate for available soil water capacity within 1 m of soil depth was calculated by taking the average of the upper and lower boundary of the depth interval (i.e., soil layer 1 at 0 cm and soil layer 6 at 100 cm; after [33]). A boundary value of ≤12 vol% was taken as suitable for the implementation of eyebrow terraces for olive cultivation (Figure 2) [5].

Suitability Zones

Based on the generated suitability maps in Figure 2, three maps with suitability zones for the implementation of eyebrow terraces were created. The land use suitability map was used as a base layer. For the first suitability map, the rainfall and slope maps were overlaid with the base layer and the areas that were overlapping in all three input maps were identified as suitable. This resulted in the rainfall and slope suitability map (Figure 3). In a similar fashion, a rainfall and AWC suitability map (using the rainfall, AWC and base-layer maps) and an AWC and slope suitability map (using the AWC, slope and base-layer maps) were generated (Figure 3; Table S1). This resulted in three different maps indicating land areas that are suitable for the implementation of eyebrow terraces for olive trees. Characteristics of the three suitability zones are listed in Figure 3.

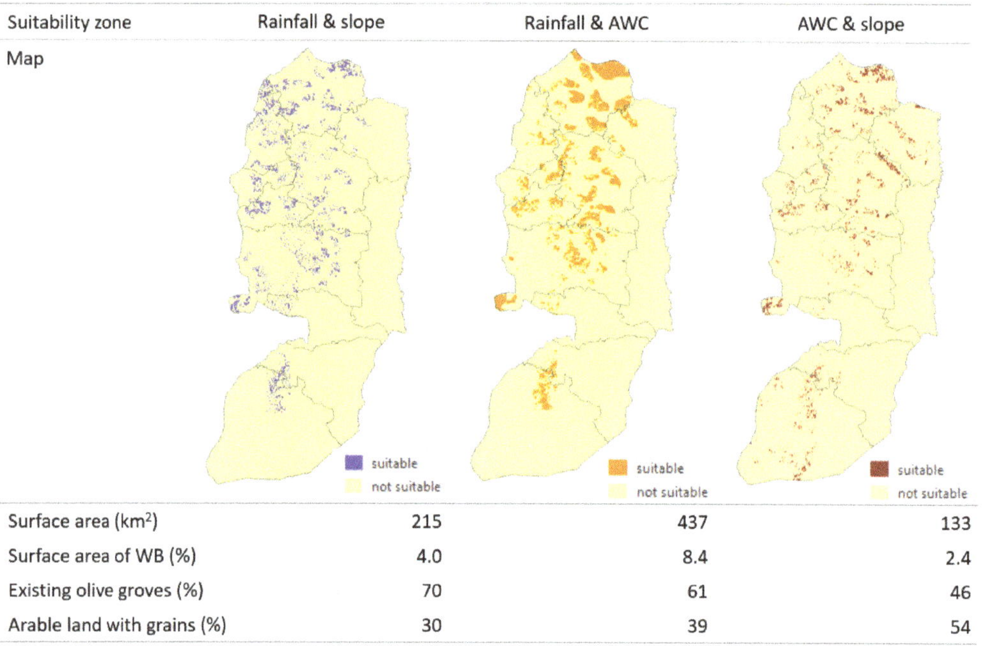

Suitability zone	Rainfall & slope	Rainfall & AWC	AWC & slope
Surface area (km²)	215	437	133
Surface area of WB (%)	4.0	8.4	2.4
Existing olive groves (%)	70	61	46
Arable land with grains (%)	30	39	54

Figure 3. Maps showing the three suitability zones for implementation of eyebrow terraces in olive groves based on (1) rainfall and slope, (2) rainfall and available soil water capacity (AWC), and (3) AWC and slope. Characteristics of the three suitability zones are listed, including surface area and percentage of coverage by olives and grains. Note that some spatial overlap occurs among the three suitability zones since they are partly selected based on the occurrence of the same suitability factor.

2.2.2. Crop–Water Balance Model

A crop–water balance model was created to estimate the impact of (limited) water availability on potential olive yield. This is a spreadsheet model that is based on a model developed by de Graaff [35], which allows for us to estimate the impact of soil water conservation measures on potential crop yield at the field-scale. For this study, the model was set up to estimate olive yield reduction (in %) at a monthly time step.

Models were created to calculate the potential yield of olive trees for the implementation of eyebrow terraces in the three defined suitability zones (see Figure 3). For each zone, a business-as-usual scenario (without terracing) and a scenario with terracing were defined to calculate the differences in olive yield.

The crop–water balance model has both static and dynamic data input. Static data input is the same for all scenarios, while dynamic data input is calculated specifically for each of the three suitability zones. Dynamic data inputs are precipitation, potential evapotranspiration, available soil water capacity and run-on.

Dynamic Data Input

The total monthly precipitation maps that were generated in Section 2.2.1 were used to calculate the mean monthly precipitation maps over the time period 2000–2018 (in mm/month). Based on these maps, the mean monthly precipitation was calculated for the three suitability zones.

Potential evapotranspiration was collected from the Global Aridity Index and Potential Evapotranspiration (ETo) Climate Database [36]. This dataset provides the monthly mean potential evapotranspiration over the 1970–2000 time period with a spatial resolution of 30 arc-seconds (i.e., about 1 km). These maps were downscaled for the West Bank. For each month, the average potential evapotranspiration (in mm/month) was calculated over the 1970–2000 time period. Based on these maps, the monthly average potential evapotranspiration was calculated for the three suitability zones.

Available soil water capacity was obtained from the SoilGrids 2017 dataset as previously described in Section 2.2.1. The mean bulk estimate (in mm/m) was calculated for the three suitability zones.

The amount of rainfall run-on on eyebrow terraces in the three suitability zones was calculated using the following formula:

$$RnR = R * \frac{RnA}{RfA} * \frac{CN}{100}$$

in which RnR is rainfall run-on (in mm/month), R is rainfall (in mm/month), RnA is a run-on area (in m^2), RfR is runoff area (in m^2) and CN is curve number (in %; Figure S1). For CN, a curve number map of the runoff area for the West Bank was used, which was obtained from Shadeed and Almasri [37] and calculated using a GIS-based Soil Conservation Service (SCS)-CN method. The curve number represents the runoff response to rainfall (in %) in which high curve numbers indicate that a large amount of the rainfall will be runoff and vice versa. Average curve numbers for the three suitability zones were calculated based on the curve number map.

The above formula assumes that runoff from the surrounding runoff area will be harvested as run-on on eyebrow terraces. The run-on area was taken as the mean surface area of eyebrow terraces for olive trees reported in the West Bank, being 17.5 m^2. The runoff area was calculated by taking an 8 m by 7 m planting distance between olive trees, resulting in a 56 m^2 surface area from which the run-on area was extracted. This resulted in a runoff area of 38.5 m^2.

Static Data Input

Several other biophysical variables needed as input for the crop–water balance model were defined. The rooting depth for olive trees without terracing was set at 1.2 m and with terracing at 1.7 m [38], since it is assumed that deeper rooting depths are created when terraces are built, based on experiences from a study in Tunisia [39] and given that terracing has been generally found to be favorable for increasing tree rooting depth [40]. The soil depth was set at 1 m for the West Bank, since soil depth should be set higher than the rooting depth in the run-on area.

Input data for crop water requirements (in mm/month) and yield response to water stress (in mm/month) were growth-stage-specific. Input for these two variables was based on mean values for olive trees in terraced areas in Tunisia [39]. The following growth stages have been distinguished (in months): initial (Mar), development (Apr–May–Jun), mid (Jul–Aug), and late (Sep–Oct–Nov).

For business-as-usual scenarios, runoff was taken as 2% during light rainfall events (<30 mm) and 20% during heavy rainfall events (≥30 mm); this was based on studies by Hammad et al. [41,42] who measured the runoff coefficient for two winter seasons in the

Ramallah Governorate. In the scenarios with terracing, runoff was taken as negligible since terraces are assumed to be very effective in capturing rainfall.

Model Validation

In order to validate whether the estimated actual soil moisture content (in mm/m) in the model was realistic, reference values for actual soil moisture content were collected from the literature. In a study at the Wadi Natuf catchment in the Ramallah Governorate, the lowest recurring soil water contents in the field at different soil depths were measured based on continuous soil moisture measurements over several years for various locations [43]. Based on location data, we calculated an average minimum soil moisture content of 44 mm/m for a business-as-usual situation, and an average minimum soil moisture content of 28 mm/m was found for terraces. The values found in our models for actual soil water content vary slightly around these reported literature values.

2.3. Implementation of Rooftop Harvesting

Calculations for the implementation of domestic rooftop harvesting are made for the 11 governorates of the West Bank individually (Figure 1) because the governorates vary greatly in the amount of built-up land (e.g., from 1.2% in Tubas to 10.5% in Jerusalem), the number of inhabitants (e.g., from about 50,000 in Jericho up to about 715,000 in Hebron) and the amount and pattern of rainfall. As such, the surface area of built-up land and rooftops was calculated for each governorate specifically.

First, the mean surface area of the rooftops of households in each governorate was calculated. The surface area of urban areas in each governorate was calculated based on the land use map.

The surface areas of urban areas were divided by the number of households in each governorate, which was estimated based on the population in each governorate and the average of six persons per household in the West Bank. As built-up land also included other buildings than domestic rooftops (such as infrastructure, office buildings and industry), the surface area was corrected by a division by four to approach the overall average rooftop surface area of 150 m^2 for the West Bank [28]. In practice, this resulted in an overall average rooftop area of 154 m^2 for the West Bank (Table 1).

Table 1. Estimated average rooftop surface area per household for each governorate in the West Bank.

Governorate	Average Rooftop Surface Area (m^2/Household)
Jenin	110.7
Tubas	123.9
Tulkarm	171.9
Nablus	106.8
Qalqiliya	90.4
Salfit	132.0
Ramallah and Al-Bireh	176.2
Jericho	307.5
Jerusalem	128.4
Bethlehem	185.7
Hebron	162.2
Total average	154.2

The potential amount of rainfall that can be harvested on average and the needed storage capacity of the water tank were calculated for each governorate using the Sam-SamWater Rainwater Harvesting Tool [44]. This tool uses rooftop size, rooftop type (with associated runoff coefficient), household size and its water demand, and spatial location (for the average amount of rainfall) to estimate the amount of rainfall that can be harvested on a monthly basis for a year with average rainfall. To point down the spatial location, centroids of the built-up land in each governorate were calculated using ArcGIS. The flat rooftop type was selected—having a runoff coefficient of 0.7—since there are typically flat roofs in the West Bank. The average household size was set at six people per household. The average water demand was set at 87 L/capita/day based on the daily consumption rate per capita in the West Bank [45], resulting in a 522 L/household/day water demand per household.

2.4. Spatially Explicit Cost–Benefit Analysis

By the means of spatially explicit cost–benefit analysis, it was calculated whether the implementation of eyebrow terraces and domestic rooftop harvesting can be economically feasible. The cost–benefit analysis for the implementation of eyebrow terraces is spatially explicit in such a way that it calculates the costs and benefits explicitly for the three land zones that are suitable for eyebrow terrace implementation (Figure 3). The cost–benefit analyses for the implementation of rooftop harvesting are calculated for each of the 11 governorates of the West Bank, individually.

Different scenarios are considered in the cost–benefit analyses as well. Cost–benefit analyses for eyebrow terraces were conducted for the business-as-usual and terracing scenarios in the three suitability zones. Cost–benefit analyses for rooftop harvesting were conducted for a scenario with 100% construction costs for installation on 100% of the rooftops and a scenario with 50% construction costs for installation on 50% of the rooftops, because households in the West Bank may already have some form of water reservoir installed at their homes. All cost–benefit analyses were calculated for a time period of 20 years and reported in EUR values for the year 2018. Data that were used as input were taken for the year 2018 (if not reported otherwise).

2.4.1. Eyebrow Terraces

In the cost–benefit analysis for eyebrow terraces, investment, maintenance and production costs, and benefits were accounted for. For the suitability zone, a business-as-usual and terracing scenario were estimated. The largest differences between these two scenarios were expected in the yield of olives and grains and labor costs. As such, these two factors were explicitly considered. The costs for fertilizers, pesticides, machinery, irrigation and seeds were assumed to be largely similar between the two scenarios. The cost–benefit analyses were calculated on a per-hectare basis.

The estimation of costs included investment, maintenance and production costs. The initial investment costs were the construction costs of terraces estimated at 3495 EUR/ha, including the manual construction of terraces estimated at 3410 EUR/ha and the incidental use of machinery estimated at 85 EUR/ha. Yearly maintenance costs of the terraces were estimated at 43 EUR/yr. These estimations were obtained from fieldwork for different projects by the Palestinian Hydrology Group (personal communication, S. Hamdan, 2020). Investment and maintenance costs for building terraces were estimated for the terracing scenarios only since no investment or maintenance costs were involved in the business-as-usual scenarios as no terraces were built. Here, the conventional cultivation of grains and olives is practiced.

Production costs were calculated for both scenarios. Costs were based on the quantity and price of labor. The price of a laborer in the West Bank was estimated at 341 EUR/month. For olive cultivation, one month of work per year was assumed, while half the amount of work was assumed to be needed for grain cultivation.

The estimation of benefits depended on crop price, crop yield and crop cover fraction. The crop price for olives, wheat and barley was calculated based on the producer price (i.e., farm gate price) which was collected from FAOSTAT [46] for the time period of 1997–2019. The annual averages for olives, wheat and barley were calculated as 1130, 386 and 320 EUR/tonne, respectively.

Crop yields for olives were calculated based on yield reduction outcomes from the crop–water balance model and the maximum recorded olive yield in the West Bank. A maximum yield of 2.4 tonne/ha was observed (for 2010) from olive yield data for the West

Bank over the time period 1994–2018 [46]. Absolute effects of yield reduction for olive production in the West Bank were calculated by assuming that the highest recorded olive yield in the West Bank was not limited by water. Crop yields for wheat and barley were collected from FAOSTAT [46] for the West Bank. The 5-year average of 2014–2018 was calculated in tonne/ha based on these data.

Crop yield was expected to be affected by the implementation of terracing in existing olive groves and on arable land producing grains in the terracing scenarios. In existing olive groves, terraces were built in year 1, due to which there was no olive production this year (e.g., due to potential damage to the olive trees). In year 2, olive trees were assumed to produce olives at full production capacity. On arable land, olive trees were planted in year 1. Since these olive trees were expected to only start bearing fruit from year 5 onwards, up to year 4, grains were still cultivated in-between the newly planted olive trees. In this setup, the production of grains was assumed to be at 80% of the normal production. In year 5, terraces were built and grains were no longer cultivated. The newly planted olive trees were assumed to start bearing fruit from year 5 onwards with a 14% annual increase in olive production up to maximum production in year 12.

To determine crop cover fraction, the ratio between olives and grains was calculated based on the three suitability zones and the land use map (Figure 3). In the West Bank, grains constitute a mix of wheat and barley. The proportion between these grains was set at 64% wheat and 36% barley, based on calculating the 5-year average of the harvested area [46].

2.4.2. Rooftop Harvesting

Cost–benefit analyses were calculated for installing rooftop harvesting in the 11 governorates of the West Bank. By accounting at the level of governorates, governorate-specific differences in rainfall, rooftop surface areas and connection to the water grid can be explicitly considered. Investment, maintenance and production costs were accounted for in the cost–benefit analyses. No direct benefits were considered since the amount of rainwater harvested was accounted for by a reduction in the costs of purchasing water.

In the cost–benefit analyses, the difference between installing rooftop harvesting as compared to conventional water use (without rooftop harvesting) was calculated. In conventional water use, no investment and maintenance costs were involved. For these scenarios, the costs of using tap water and purchasing water from other sources were solely accounted for. The cost–benefit analyses were calculated at the household level. Two scenarios were considered: one scenario with 100% construction costs and one scenario with 50% construction costs, since part of the residential buildings may have already had water tanks or cisterns installed in the West Bank.

Production costs for water use were calculated based on the percentage of households connected to the water grid and the quantity of water used (Table 2). Data were collected regarding the percentage of households connected to the water grid [4], allowing for us to calculate the percentage of water demand from the water grid and from other sources, such as water trucks (Table 2). Tap water was mostly provided by the Israelian water company Mekorot, and to a smaller extent, derived from spring discharge and water pumped from Palestinian wells. The average water tariff for tap water is 1.2 EUR/m^3. In less well-connected areas—mostly situated in rural areas—households purchase water from other sources, such as water trucks, at high water tariffs of 4 EUR/m3 on average [47]. The reduced amount of water that needed to be purchased was calculated based on governorate-specific numbers about the amount of rainwater that can be harvested. It was assumed that the remaining part of the total water demand was purchased from the water source with the lowest tariff (i.e., tap water). It should be noted that this is a conservative assumption as there is evidence that households do purchase water from trucks and rural areas rely on truck water to a large extent.

Table 2. Household connection to the water grid and water use in the West Bank.

Region	Governorates	Water Use from Water Grid (%) [1]	Water Obtained from Other Sources (%) [1]	Tap Water Use (m³/hh/yr)	Other Water Source Use (m³/hh/yr)
North	Jenin, Tubas, Tulkarm, Nablus, Qalqiliya, Salfit	87.5	12.5	166.7	23.8
Middle	Ramallah and Al-Bireh, Jericho, Jerusalem	97.8	2.2	186.3	4.2
South	Bethlehem, Hebron	83.1	16.9	158.3	32.2

Note(s): [1] Based on data for 2011 [4].

Investment and maintenance costs of the installation of rooftop harvesting were calculated. Investment costs included the costs for the construction of a cistern, purchasing and installing a pump (to pump water from the rooftop to the water reservoir), excavation costs (for installing the pump) and PVC piping (to lead the water from the rooftop to the reservoir). Maintenance costs included annual rooftop and reservoir cleaning.

Since the size of the water reservoir that needs to be constructed greatly influences construction costs, the optimal reservoir size was calculated for each governorate using the SamSamWater Rainwater Harvesting tool [44]. This tool calculates the optimal water reservoir size based on the average water use by households and the monthly amount of rainwater that can be harvested based on average monthly rainfall. Optimal storage capacities for the 11 governorates were classified into typical reservoir size classes (Table 3). Construction costs for these reservoir size classes were calculated and used to calculate governorate-specific construction costs.

Table 3. Reservoir size classes for rainwater storage capacity for the 11 governorates in the West Bank.

Reservoir Size Classes (m³)	Governorates
20	Qalqiliya, Hebron
25	Jenin, Nablus
30	Jerusalem, Tubas, Salfit
35	Bethlehem, Jericho
40	Tulkarm, Ramallah and Al-Bireh

3. Results

3.1. Eyebrow Terraces

Olive yield can be increased by 0.24 up to 0.35 tonne/ha in the West Bank with the implementation of eyebrow terraces in existing olive groves and planting new groves on suitable arable land as compared to the business-as-usual scenario (Figure 4 and Table S2). This comes down to an increase in yield of between 10 and 14% over a time period of 20 years for the three different land suitability zones (Figure 4). The maximum yield that can be reached by implementing eyebrow terraces is 2.1 tonne/ha. Although olive yield varies among the three suitability zones for the business-as-usual scenario, yield increases that can be reached with the implementation of eyebrow terracing are overall similar across the three suitability zones (Figure 4).

The crop–water balance models show that the climate in the West Bank is so dry that even after the wet winter season, the increased water harvesting capacity of the soil (due to implementing terracing) only helps to a small extent to improve olive yield. The scenarios in which terracing is implemented are very effective in capturing rainfall, as the amount of effective rainfall is doubled. However, due to the limited storage capacity of the soil, most soil moisture that is additionally harvested with terracing is rapidly lost to drainage to deeper groundwater during the wet months. This leaves little soil moisture during the dry season. Thus, yield differences between the business-as-usual and terracing scenarios are overall not that large (i.e., between 10–14% higher). As such, eyebrow terraces can capture a lot of rainfall, but soil capacity to store it is limited. In contrast, in the scenarios where no terracing is implemented (business-as-usual), there is less effective rainfall, since part of the rainfall is lost to direct runoff and there is no additional run-on. Due to this, there is no surplus of soil moisture and no drainage to deeper groundwater.

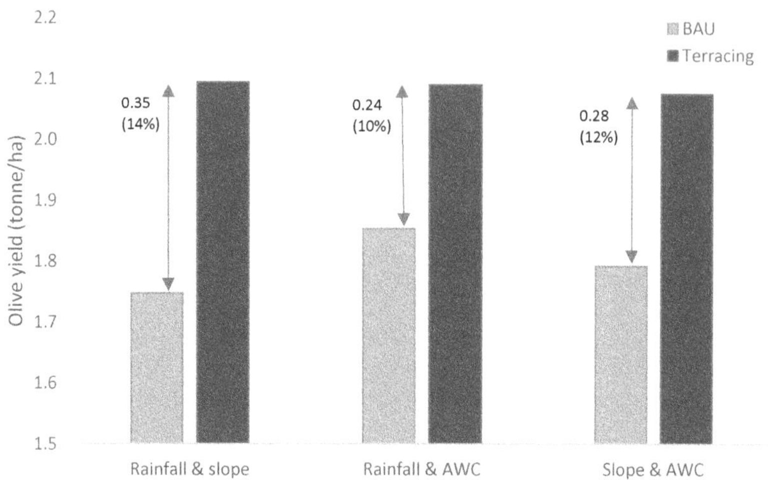

Figure 4. Estimated olive yield (tonne/ha) for the three suitability zones in the West Bank for the business-as-usual (BAU) and terracing scenarios based on results from the crop–water balance models. The arrows indicate the difference in yield between the two scenarios expressed in tonne/ha and as a percentage in-between brackets. AWC stands for available soil water capacity.

3.2. Rooftop Harvesting

On average, 56.9 m^3/hh/yr of rainwater can be harvested from domestic rooftops in the West Bank. The lowest amount of 37.5 m^3/hh/yr can be harvested in Qalqiliya, while the highest amount of 75 m^3/hh/yr can be harvested in Ramallah and Al-Bireh (Figure 5a). In the latter governorate, this is mainly due to the occurrence of relatively large rooftops and a higher amount of rainfall. On average, 30% of household demand can be met with rooftop harvesting, when taking the average household consumption in the West Bank of 191 m^3/hh/yr [45]. As such, rainfall is by far too low to meet household demand with rooftop harvesting alone in the West Bank.

The optimal water reservoir size was calculated based on the monthly amount of rainwater that can be harvested and the average water use by households (Figure 5b). The optimal size ranges from 18 m^3 in Qalqiliya up to 37 m^3 in Ramallah and Al-Bireh. This variation is due to the different amounts of rainfall that can be harvested in each governorate. Variation in optimal reservoir size has implications for the sizes of reservoirs that are ideally constructed for rooftop harvesting in the governorates and the costs incurred in the cost–benefit analyses.

3.3. Cost–Benefit Analysis

3.3.1. Eyebrow Terraces

The cost–benefit analyses of implementing eyebrow terraces in the three suitability zones show that the net results for all three scenarios are positive (Figure 6). This means that the benefits outweigh the costs. Additionally, the internal rates of return show that the annual growth rate of investing in eyebrow terraces is positive.

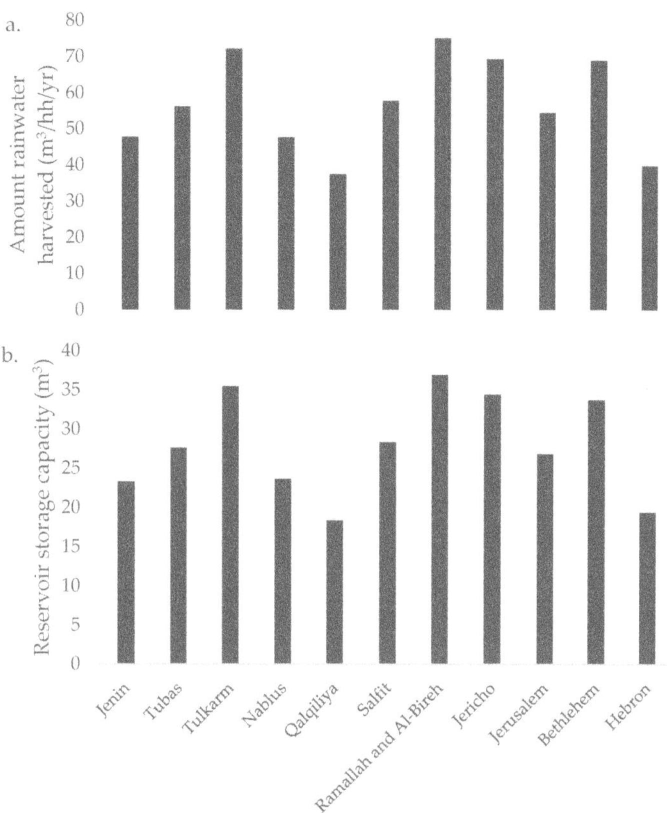

Figure 5. Figure showing (**a**) the amount of rainwater that can be harvested (in m³/hh/yr) with the installation of domestic rooftop harvesting, and (**b**) the optimal storage capacity of water reservoirs (in m³) for an average rainfall year in the 11 governorates of the West Bank.

However, when values are discounted (using a discount factor of 10%), the net present values turn out to be negative in all three scenarios. The suitability zone in which slope and AWC determine the suitability for implementing terraces turns out to be the least negative. When a lower discount rate would be used (i.e., ≤8.6%, as indicated by the internal rate of return), the implementation of eyebrow terracing would pay off in this scenario. Similarly, the other two scenarios would have a positive NPV, when the discount factor would be set at the level of the internal rates of return.

3.3.2. Rooftop Harvesting

The cost–benefit analyses in the 11 governorates of the West Bank show that investing in installing domestic rooftop harvesting does not pay off in any of the governorates when assuming that for all residential buildings new reservoirs would need to be constructed (i.e., scenario with 100% construction costs; Table 4). However, when assuming that part of the households already has water reservoirs or cisterns installed at their homes and that only in about 50% of the cases new reservoirs need to be constructed (i.e., scenario with 50% construction costs), the net present value becomes less negative and the internal rate of return is positive for the governorates located in the northern and southern governorates, making investing in installing rooftop harvesting more attractive in those regions.

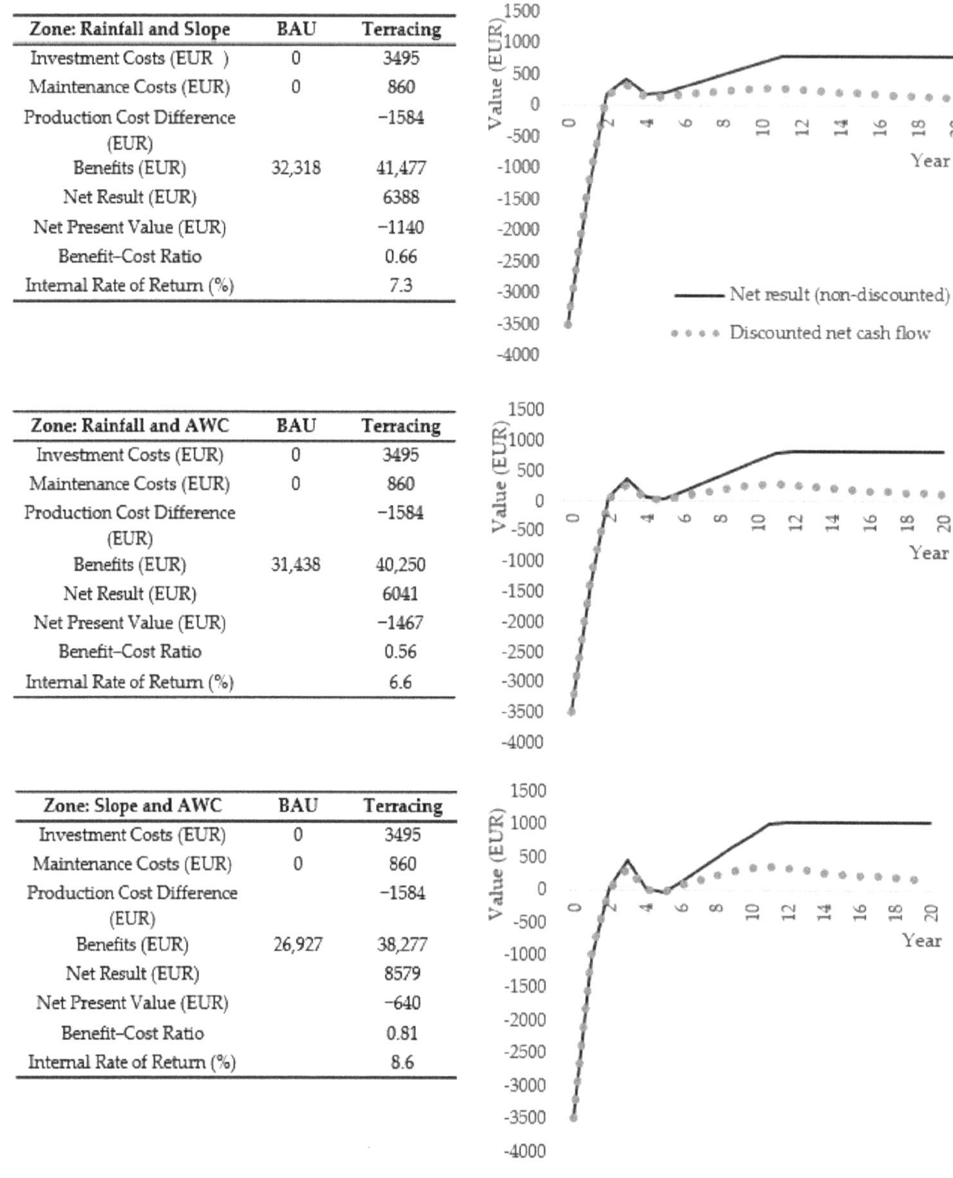

Zone: Rainfall and Slope	BAU	Terracing
Investment Costs (EUR)	0	3495
Maintenance Costs (EUR)	0	860
Production Cost Difference (EUR)		−1584
Benefits (EUR)	32,318	41,477
Net Result (EUR)		6388
Net Present Value (EUR)		−1140
Benefit–Cost Ratio		0.66
Internal Rate of Return (%)		7.3

Zone: Rainfall and AWC	BAU	Terracing
Investment Costs (EUR)	0	3495
Maintenance Costs (EUR)	0	860
Production Cost Difference (EUR)		−1584
Benefits (EUR)	31,438	40,250
Net Result (EUR)		6041
Net Present Value (EUR)		−1467
Benefit–Cost Ratio		0.56
Internal Rate of Return (%)		6.6

Zone: Slope and AWC	BAU	Terracing
Investment Costs (EUR)	0	3495
Maintenance Costs (EUR)	0	860
Production Cost Difference (EUR)		−1584
Benefits (EUR)	26,927	38,277
Net Result (EUR)		8579
Net Present Value (EUR)		−640
Benefit–Cost Ratio		0.81
Internal Rate of Return (%)		8.6

Figure 6. Results of the cost–benefit analysis for implementing eyebrow terraces in the three suitability zones in the West Bank for the business-as-usual (BAU) and terracing scenarios. Net results are not discounted, while net present values are discounted using a 10% discount rate. AWC stands for available soil water capacity and EUR values are reported for the year 2018.

Table 4. Cost–benefit analysis results for installing domestic rooftop harvesting in the 11 governorates of the West Bank, showing scenarios with 100% and 50% construction costs.

Region	Governorate	100% Construction Costs			50% Construction Costs		
		Net Result (EUR) [1]	NPV (EUR) [2]	IRR (%) [2]	Net Result (EUR) [1]	NPV (EUR) [2]	IRR (%) [2]
North	Jenin	−421	−983	−3.2	204	−358	2.3
	Tubas	−472	−1148	−3.0	278	−398	2.7
	Tulkarm	−588	−1485	−2.9	412	−485	3.1
	Nablus	−426	−985	−3.2	199	−360	2.3
	Qalqiliya	−418	−838	−4.0	82	−338	1.2
	Salfit	−433	−1132	−2.7	317	−382	3.1
Middle	Ramallah and Al-Bireh	−1615	−1922	−10.7	−615	−922	−6.4
	Jericho	−1502	−1731	−11.8	−627	−856	−7.7
	Jerusalem	−1610	−1632	−23.1	−860	−883	−20.2
South	Bethlehem	57	−1067	0.3	932	−192	7.1
	Hebron	108	−614	0.9	608	−114	7.3

Note(s): [1] Net result is the cost minus benefits, which is not discounted over time. [2] NPV is net present value, IRR is internal rate of return; both are estimated using a 10% discount rate.

In the southern region of the West Bank in particular, the net result of investing in rooftop harvesting is the most positive, demonstrating that the benefits outweigh the costs in both scenarios of 100% and 50% reservoir construction costs. Even though this region has the lowest rainfall, domestic rooftop harvesting can be of interest since households are the least well-connected to the water grid (i.e., 83.1%; see Table 2). However, when discounting values at a 10% rate over a 20-year period, the net present values become slightly negative for this region.

In contrast, in the middle region of West Bank, the net results and net present values are deeply negative, showing that investing in installing rooftop harvesting would not pay off, even when assuming that water reservoirs would need to be built only in 50% of the cases. In this region, the connection to the water grid is the best (i.e., 97.8%). Because of this, there is a low reliance on having to purchase expensive water from other sources with high water tariffs (e.g., water trucks). As such, rooftop harvesting does not pay off in this region (Table 4).

Together, the results of the cost–benefit analyses show that when using a discount rate of 10%, it does not pay off to install domestic rooftop harvesting in all governorates of the West Bank. Additionally, when assuming that new water reservoirs would have to be constructed only in about 50% of the cases (instead of in 100% of the cases), investing in rooftop harvesting does not pay off in the middle region, but does become more attractive for the northern and southern regions of the West Bank.

4. Discussion

This study aimed to investigate the economic feasibility of the implementation of the two most commonly practiced rainwater harvesting applications in a rural and urban setting in the West Bank of Palestine. Eyebrow terracing in olive cultivation and domestic rooftop harvesting in urban areas were investigated. We found that—although eyebrow terracing enlarges soil moisture availability for olive trees and thereby increases olive yield by 10 up to 14%—construction costs are too high to make implementation cost-effective. Similarly, we found that rooftop harvesting can harvest, on average, about 30% of the annual domestic water demand and is worthwhile in the northern and southern governorates in particular. Yet, also in this case, construction costs are generally too high to be cost-effective for households to implement. As such, construction costs may obstruct the implementation and upscaling of rainwater harvesting in rural and urban areas of the West Bank.

The height of the initial investment costs to install rainwater harvesting structures forms a considerable barrier to the adoption of rainwater harvesting. Similar findings for high construction costs have been found for both urban and rural applications in other low-income countries in previous studies. They found that adoption by farmers may be limited due to high initial investment costs in Kenya and Tanzania [8,48]. Similarly, initial investment costs were the main limiting factor for upscaling urban rooftop harvesting in a case study in Namibia [49].

Such high investment costs could be reduced, for example, by the means of mechanization, collective investment or including the building of rooftop harvesting infrastructure and cisterns during the construction or renovation phase of residential buildings. This would allow for the benefit/cost ratio to increase and would make the implementation of rainwater harvesting more attractive. This is demonstrated by the scenario with 50% construction costs for domestic rooftop harvesting: the internal rates of return become positive for the northern and southern governorates. These governorates are characterized by being the most arid or the least well-connected to the water grid. This also shows that rooftop harvesting is mostly of interest to those regions in the West Bank that are very arid or not (well-)connected to the water grid.

The cost-effectiveness of the rainwater harvesting techniques investigated in this study may turn out more positive in practice due to several reasons. Firstly, the construction costs for terracing estimated in this study seem relatively high when compared to other international case studies [50]. This may indicate that they might have been estimated too high. Secondly, we used a conservative assumption for calculating the costs of purchasing water for estimating the production costs of the business-as-usual scenario of rooftop harvesting. We assumed that the cheapest source of water would be bought (i.e., tap water), while people may be forced in practice to purchase water from other, more expensive water sources. There is evidence that people buy water from trucks and that especially people in rural areas depend largely on truck water [27]. Thirdly, our analysis is spatially explicit, but analyses were made for average conditions. When cost-effectiveness can be analyzed at a finer spatial scale or for a specific area, specific locations may turn out (more) positive. Future research could further investigate this. If investment costs turn out lower in practice, the implementation of these rainwater harvesting techniques can be of interest for other arid regions in the Mediterranean and elsewhere, since these regions will have to cope with increasing water security due to climate change similar to Palestine.

Despite the high investment costs, we found that considerable benefits can be obtained with both types of rainwater harvesting. By using domestic rooftop harvesting, we found that 30% of the annual domestic water demand could be met. This seems quite substantial, given that a similar study about rooftop harvesting on residential buildings in Jordan found that only 8% of the annual domestic water demand could be met on average [23].

We also found that olive production can be increased to some extent when eyebrow terracing is used, leading to increased water security and an improved livelihood. However, also a lot of harvested rainwater is lost to deep drainage due to the limited storage capacity of the soil. Because of that, terracing had only a limited impact on increasing olive yield. Therefore, it may be interesting to investigate whether harvested rainwater can be stored elsewhere, for example, by diversion to a water storage reservoir or cistern. Stored water could be used for the irrigation of olive trees in the dry season or maybe even throughout the dry season, which can lead to a higher olive yield. Previous studies also show that farm revenue can be considerably increased using rainwater harvesting techniques (e.g., [7,51]).

Overall, however, these benefits could not outweigh the high initial investment costs. Given that initial investment costs are too high for households to afford, construction costs are identified as the most important barrier to the widespread adoption of rainwater harvesting in the West Bank. Here, a role for the government may be to lower these types of costs by providing financial aid.

At the same time, societal benefits of the implementation of rainwater harvesting have not been accounted for in this study, even though these may have been substantial.

Implementation of rainwater harvesting can offer multiple societal benefits—i.e., ecosystem services—such as improved water regulation, reduced soil erosion and improved flood mitigation. Cost–benefit analyses of the implementation of rainwater harvesting in farming systems in arid regions of Jordan and Tunisia found that including environmental benefits increased the internal rates of return significantly and made rainwater harvesting a compelling case for public investment [52,53].

Such societal benefits have not been accounted for, as this study only included costs and benefits at the household level and not costs and benefits at the level of the community or society as a whole. Yet, it is expected that the implementation of rainwater harvesting will offer multiple societal benefits in the West Bank, particularly when it is applied at a large scale.

An important societal benefit is that water security would be increased in the West Bank, making it less dependent on Israel for its water supply. Another important benefit is that the resilience of farmers and households locally would be increased, which is of relevance in view of increasing water uncertainty due to climate change and the political situation. Other positive offsite benefits can include increasing the recharge of shallow groundwater and prolonged soil moisture availability downhill. This may have positive impacts on agricultural production and natural vegetation growth downslope and in the valleys, prolonged streamflow into the dry period and reduction in flood risk downstream. When these societal benefits would be considered as well, the cost-effectiveness of the two studied rainwater harvesting techniques would most likely become positive.

Given that investment costs are too high for widespread adoption, but the extensive implementation could offer multiple societal benefits, there is a role for the government of Palestine. Additionally, given the importance to take adaptation measures for the large, expected impact of climate change in the region, the government could consider developing a strategic program to promote the widespread adoption of rainwater harvesting systems. A simple mechanism for public investment, such as subsidies, may already provide the incentive for households to consider rainwater harvesting.

5. Conclusions

This study has aimed to investigate whether the implementation of two rainwater harvesting techniques—one in a rural and one in an urban context—can be economically feasible for households in the West Bank of Palestine. Eyebrow terracing in olive cultivation and domestic rooftop harvesting in urban areas has been investigated. Although eyebrow terracing enlarges soil moisture availability for olive trees (and thereby increases olive yield by about 10–14%), the costs of constructing terracing are too high for them to be cost-effective. Similarly, domestic rooftop harvesting increases water availability for domestic use, but construction costs of installing water reservoirs are too high to be economically feasible. This obstructs a more widespread adoption of rainwater harvesting, which is urgently needed given the large impacts of climate change in the region. Providing subsidies for rainwater harvesting could help to make adoption more attractive for households. This will help to increase water security in the West Bank alongside providing multiple other societal benefits, such as increasing the resilience of households and farmers.

Supplementary Materials: The following supporting information can be downloaded at: https://www.mdpi.com/article/10.3390/w15061023/s1, Table S1: three suitability types for implementation of eyebrow terraces with olive trees analyzed in this study; Figure S1. Curve number map of the West Bank; Table S2: estimated olive yield for the three suitability zones in the West Bank (see Figure 3) for the business-as-usual (BAU) and terracing scenarios based on results from the crop–water balance models.

Author Contributions: Conceptualization, J.E.M.S., L.F. and M.R.; methodology, J.E.M.S. and L.F.; formal analysis, J.E.M.S. and L.F.; writing—original draft preparation, J.E.M.S.; writing—review and editing, J.E.M.S., S.S., M.R. and L.F.; funding acquisition, M.R. and S.S. All authors have read and agreed to the published version of the manuscript.

Funding: This research was funded by the Ministry of Foreign Affairs of the Netherlands within the framework of the PADUCO 2: MORwater project (Mobile App for Optimizing, Promoting Rain WATER Harvesting for a Self-sustaining and Self-reliant Water Supply).

Data Availability Statement: All data sources are described in the article.

Acknowledgments: We would like to thank Rasha Abed for sharing spatial data of the West Bank. We would also like to thank Sami Dawoud for providing information about the local hydrological conditions of the West Bank. Three anonymous reviewers are thanked for their valuable comments that helped to improve this paper.

Conflicts of Interest: The authors declare no conflict of interest. The funders had no role in the design of the study; in the collection, analyses, or interpretation of data; in the writing of the manuscript; or in the decision to publish the results.

References

1. UNDP. *Climate Change Adaptation Strategy and Programme of Action for the Palestinian Authority*; United Nations Development Programme (UNDP) Programme of Assistance to the Palestinian People: Jerusalem, Palestine, 2010.
2. State of Palestine. *Sustainable Development Goals: Palestinian National Voluntary Review on the Implementation of the 2030 Agenda*; State of Palestine: Jerusalem, Palestine, 2018.
3. GERICS. *Climate Fact Sheet of Israel—Jordan—Lebanon—Palestine—Syria*; Climate Service Center: Hamburg, Germany, 2013.
4. PCBS. *Household Environmental Survey, 2011: Main Results (English Section)*; Palestinian Central Bureau of Statistics, Palestinian National Authority: Ramallah, Palestine, 2011.
5. Adham, A.; Riksen, M.; Abed, R.; Shadeed, S.; Ritsema, C. Assessing Suitable Techniques for Rainwater Harvesting Using Analytical Hierarchy Process (AHP) Methods and GIS Techniques. *Water* **2022**, *14*, 2110. [CrossRef]
6. Al-Salaymeh, A.; Al-Khatib, I.A.; Arafat, H.A. Towards sustainable water quality: Management of rainwater harvesting cisterns in southern Palestine. *Water Resour Manag* **2011**, *25*, 1721–1736. [CrossRef]
7. Dile, Y.T.; Karlberg, L.; Temesgen, M.; Rockström, J. The role of water harvesting to achieve sustainable agricultural intensification and resilience against water related shocks in sub-Saharan Africa. *Agric. Ecosyst. Environ.* **2013**, *181*, 69–79. [CrossRef]
8. Ngigi, S.N.; Savenije, H.H.G.; Rockström, J.; Gachene, C.K. Hydro-economic evaluation of rainwater harvesting and management technologies: Farmers' investment options and risks in semi-arid Laikipia district of Kenya. *Phys. Chem. Earth Parts A/B/C* **2005**, *30*, 772–782. [CrossRef]
9. Lebel, S.; Fleskens, L.; Forster, P.M.; Jackson, L.S.; Lorenz, S. Evaluation of in situ rainwater harvesting as an adaptation strategy to climate change for maize production in rainfed Africa. *Water Resour. Manag.* **2015**, *29*, 4803–4816. [CrossRef]
10. Pandey, D.P.; Gupta, A.K.; Anderson, D.M. Rainwater harvesting as an adaptation to climate change. *Curr. Sci.* **2003**, *85*, 46–59.
11. Velasco-Muñoz, J.F.; Aznar-Sánchez, J.A.; Batlles-delaFuente, A.; Dolores Fidelibus, M. Rainwater harvesting for agricultural irrigation: An analysis of global research. *Water* **2019**, *11*, 1320. [CrossRef]
12. Assayed, A.; Hatokay, Z.; Al-Zoubi, R.; Azzam, S.; Qbailat, M.; Al-Ulayyan, A.; Saleem, M.A.; Bushnaq, S.; Maroni, R. On-site rainwater harvesting to achieve household water security among rural and peri-urban communities in Jordan. *Resour. Conserv. Recycl.* **2013**, *73*, 72–77. [CrossRef]
13. Piemontese, L.; Castelli, G.; Fetzer, I.; Barron, J.; Liniger, H.; Harari, N.; Bresci, E.; Jaramillo, F. Estimating the global potential of water harvesting from successful case studies. *Glob. Environ. Change* **2020**, *63*, 102121. [CrossRef]
14. Rodrigues de Sa Silva, A.C.; Mendonça Bimbato, A.; Perrella Balestieri, J.A.; Nogueira Vilanova, M.R. Exploring environmental, economic and social aspects of rainwater harvesting systems: A review. *Sustain. Cities Soc.* **2022**, *76*, 103475. [CrossRef]
15. Adham, A.; Riksen, M.; Ouessar, M.; Ritsema, C. Identification of suitable sites for rainwater harvesting structures in arid and semi-arid regions: A review. *Int. Soil Water Conserv. Res.* **2016**, *4*, 108–120. [CrossRef]
16. Jan, I. Socio-economic determinants of farmers' adoption of rainwater harvesting systems in semi-arid regions of Pakistan. *J. Agr Sci Tech.* **2020**, *22*, 377–387.
17. Sheikh, V. Perception of domestic rainwater harvesting by Iranian citizens. *Sustain. Cities Soc.* **2020**, *60*, 102278. [CrossRef]
18. Al-Adamat, R.; AlAyyash, S.; Al-Amoush, H.; Al-Meshan, O.; Rawajfih, Z.; Shdeifat, A.; Al-Harahsheh, A.; Al-Farajat, M. The combination of indigenous knowledge and geo-informatics for water harvesting siting in the Jordanian Badia. *J. Geogr. Inf. Syst.* **2012**, *4*, 366–376. [CrossRef]
19. Adham, A.; Riksen, M.; Ouessar, M.; Abed, R.; Ritsema, C. Development of methodology for existing rainwater harvesting assessment in (semi-)arid regions. In *Water and Land Security in Drylands*; Ouessar, M., Gabriels, D., Tsunekawa, A., Evett, S., Eds.; Springer: Cham, Switzerland, 2016; pp. 171–184. [CrossRef]
20. Shadeed, S.M.; Judeh, T.G.; Almasri, M.N. Developing GIS-based water poverty and rainwater harvesting suitability maps for domestic use in the Dead Sea region (West Bank, Palestine). *Hydrol. Earth Syst. Sci.* **2019**, *23*, 1581–1592. [CrossRef]
21. Shadeed, S.M.; Judeh, T.G.; Riksen, M. Rainwater harvesting for sustainable agriculture in high water-poor areas in the West Bank, Palestine. *Water* **2020**, *12*, 380. [CrossRef]

22. Kim, H.W.; Li, M.-H.; Kim, H.; Lee, H.K. Cost-benefit analysis and equitable cost allocation for a residential rainwater harvesting system in the city of Austin, Texas. *Int. J. Water Resour. Dev.* **2016**, *32*, 749–764. [CrossRef]
23. Abdulla, F. Rainwater harvesting in Jordan: Potential water saving, optimal tank sizing and economic analysis. *J. Urban. Water* **2020**, *17*, 446–456. [CrossRef]
24. Dallman, S.; Chaudhry, A.M.; Muleta, M.K.; Lee, J. The value of rain: Benefit-cost analysis of rainwater harvesting systems. *Water Resourc. Manag.* **2016**, *30*, 4415–4428. [CrossRef]
25. Dallman, S.; Chaudhry, A.M.; Muleta, M.K.; Lee, J. Is rainwater harvesting worthwhile? A benefit-cost analysis. *J. Water Resour. Plan. Manag.* **2021**, *147*, 04021011. [CrossRef]
26. Mourad, K.A.; Yimer, S.M. Socio-economic potential of rainwater harvesting in Ethiopia. *Sustain. Agric. Res.* **2017**, *6*, 73–79. [CrossRef]
27. PCBS. Palestine in figures 2018. In *Palestinian Central Bureau of Statistics*; Palestinian National Authority: Ramallah, Palestine, 2019.
28. Shadeed, S.M.; Alawna, S. Optimal sizing of rooftop rainwater harvesting tanks for sustainable domestic water use in the West Bank, Palestine. *Water* **2021**, *13*, 573. [CrossRef]
29. GeoMOLG. Ministry of Local Governance. GeoMOLG. Available online: https://geomolg-geomolgarconline.hub.arcgis.com/search?collection=Dataset (accessed on 4 February 2019).
30. Shadeed, S.M. Developing a GIS-based Suitability Map for Rainwater Harvesting in the West Bank, Palestine. In *Water and Environmental Studies Institute*; An-Najah National University: Nablus, Palestine, 2011.
31. Harris, I.; Jones, P.D.; Osborn, T.J.; Lister, D.H. Updated high-resolution grids of monthly climatic observations—The CRU TS3.10 Dataset. *Int. J. Climatol.* **2014**, *34*, 623–642. [CrossRef]
32. Fick, S.E.; Hijmans, R.J. WorldClim 2: New 1km spatial resolution climate surfaces for global land areas. *Int. J. Clim.* **2017**, *37*, 4302–4315. [CrossRef]
33. Hengl, T.; Mendes de Jesus, J.; Heuvelink, G.B.M.; Ruiperez Gonzalez, M.; Kilibarda, M.; Blagotić, A.; Shangguan, W.; Wright, M.N.; Geng, X.; Bauer-Marschallinger, B.; et al. SoilGrids250m: Global gridded soil information based on machine learning. *PLoS ONE* **2017**, *12*, e0169748. [CrossRef]
34. de Jong van Lier, Q. Field capacity, a valid upper limit of crop available water? *Agric. Water Manag.* **2017**, *193*, 214–220. [CrossRef]
35. de Graaff, J. The Price of Soil Erosion: An Economic Evaluation of Soil Conservation and Watershed Development. Ph.D. Thesis, Wageningen University, Wageningen, The Netherlands, 1996.
36. Trabucco, A.; Zomer, R.J. Global Aridity Index and Potential Evapo-Transpiration (ET0) Climate Database v2 Figshare. *CGIAR Consortium for Spatial Information (CGIAR-CSI), 2018.* Available online: https://cgiarcsi.community (accessed on 14 July 2020). [CrossRef]
37. Shadeed, S.; Almasri, M. Application of GIS-based SCS-CN method in West Bank catchments, Palestine. *Water Sci. Eng.* **2010**, *3*, 1–13. [CrossRef]
38. FAO. *Crop Information Database: Olive*; Food and Agriculture Organization of the United Nations (FAO): Rome, Italy; Available online: http://www.fao.org/land-water/databases-and-software/crop-information/olive/en/#c236114 (accessed on 13 June 2020).
39. Fleskens, L.; Stroosnijder, L.; Ouessar, M.; de Graaff, J. Evaluation of the on-site impact of water harvesting in southern Tunisia. *J. Arid Environ.* **2005**, *62*, 613–630. [CrossRef]
40. Querejeta, J.I.; Roldan, A.; Albaladejo, J.; Castillo, V. Soil water availability improved by site preparation in a Pinus halepensis afforestation under semiarid climate. *For. Ecol. Manag.* **2001**, *149*, 115–128. [CrossRef]
41. Hammad, A.A.; Børresen, T.; Haugen, L.E. Effects of rain characteristics and terracing on runoff and erosion under the Mediterranean. *Soil Tillage Res.* **2006**, *87*, 39–47. [CrossRef]
42. Hammad, A.A.; Haugen, L.E.; Børresen, T. Effects of stonewalled terracing techniques on soil-water conservation and wheat production under Mediterranean conditions. *Environ. Manag.* **2004**, *34*, 701–710. [CrossRef]
43. Messerschmid, C.; Sauter, M.; Lange, J. Field-based estimation and modelling of distributed groundwater recharge in a Mediterranean karst catchment, Wadi Natuf, West Bank. *Hydrol. Earth Syst. Sci.* **2020**, *24*, 887–917. [CrossRef]
44. de Haas, S. SamSamWater Rainwater Harvesting Tool, 2013. Available online: https://www.samsamwater.com/rain/ (accessed on 23 July 2020).
45. PWA. *Water Information System*; Palestinian Water Authorities: Ramallah, Palestine, 2018.
46. FAOSTAT. *Food and Agriculture Database*; Food and Agriculture Organization of the United Nations (FAO): Rome, Italy, 2020; Available online: http://www.fao.org/faostat/en/#data (accessed on 7 December 2020).
47. PCBS. *Consumer Price Index Survey, 2018*; Palestinian Central Bureau of Statistics (PCBS): Ramallah, Palestine, 2019.
48. Senkondo, E.M.M.; Msangi, A.S.K.; Xavery, P.; Lazaro, E.A.; Hatibu, N. Profitability of rainwater harvesting for agricultural production in selected semi-arid areas of Tanzania. *J. Appl. Irrig. Sci.* **2004**, *39*, 65–81.
49. Woltersdorf, L.; Jokisch, A.; Kluge, T. Benefits of rainwater harvesting for gardening and implications for future policy in Namibia. *Water Policy* **2014**, *16*, 124–143. [CrossRef]
50. WOCAT. Global SLM database. Global Database on Sustainable Land Management (SLM) of WOCAT (the World Overview of Conservation Approaches and Technologies), 2020. Available online: https://qcat.wocat.net/en/wocat/ (accessed on 7 December 2020).

51. Biazin, B.; Sterk, G.; Temesgen, M.; Abdulkedir, A.; Stroosnijder, L. Rainwater harvesting and management in rainfed agricultural systems in sub-Saharan Africa—A review. *Phys. Chem. Earth* **2012**, *47–48*, 139–151. [CrossRef]
52. Akroush, S.; Shideed, K.; Bruggeman, A. Economic analysis and environmental impacts of water harvesting techniques in the low rainfall areas of Jordan. *Int. J. Agric. Resour. Gov. Ecol.* **2014**, *10*, 34–49. [CrossRef]
53. Ouessar, M.; Sghaier, M.; Mahdhi, N.; Abdelli, F.; de Graaff, J.; Chaieb, H.; Yahyaoui, H.; Gabriels, D. An integrated approach for impact assessment of water harvesting techniques in dry areas: The case of Oued Oum Zessar Watershed (Tunisia). *Environ. Monit. Assess.* **2004**, *99*, 127–140. [CrossRef]

Article

Rainwater Catchment System Reliability Analysis for Al Abila Dam in Iraq's Western Desert

Ammar Adham [1,2,*], Rasha Abed [2,3], Karrar Mahdi [2], Waqed H. Hassan [4,5], Michel Riksen [2] and Coen Ritsema [2]

1   Dams and Water Resources Engineering Department, College of Engineering, University of Anbar, Baghdad 55431, Iraq
2   Soil Physics and Land Management Group, Wageningen University, 6700 AA Wageningen, The Netherlands
3   Anbar Technical Institute, Middle Technical University, Baghdad 10074, Iraq
4   College of Engineering, University of Warith Al-Anbiyaa, Kerbala 56001, Iraq
5   College of Engineering, University of Kerbala, Kerbala 56001, Iraq
*   Correspondence: engammar2000@uoanbar.edu.iq

Abstract: Rainwater Catchment System Reliability (RCSR) is the chance that a system will deliver the required water for an interval of time. Rainwater Harvesting (RWH) is gaining popularity as a potential alternative water source for household or agricultural use. The reliability of the Al Abila dam in the western desert of Iraq was analyzed using a water budget simulation model and two explanations of reliability, time-based reliability, and volumetric reliability. To evaluate rainwater harvesting system performance, comprehensive software utilizing a method for everyday water balance using data from 20 years of daily rainfall. According to the findings, volumetric reliability, and for the three climate scenarios (wet, average, and dry year), increased as the storage volume increased until a threshold accrued on the storage capacity of 11.7×10^5 m^3. While time-based reliability shows an increase up to a storage volume of 10.2×10^5 m^3. Volumetric reliability of roughly 34–75% may be achieved, while only 14–28% time-based reliability may be achieved. Water saving efficiency decreases with increasing demand fraction, while the runoff coefficient has no significant influence on water effectiveness. While growing storage fraction value increases the effectiveness of water conservation and the value of the runoff coefficient influences the water saving efficiency. For both cases, water saving efficiency for the dam does not reach 50%. Using daily rainfall data, the technique given in this paper might be applied to predict water savings and the RWH systems' reliability in different arid and semi-arid areas.

Keywords: reliability; rainwater catchment; Al-Abila dam; Iraq

Citation: Adham, A.; Abed, R.; Mahdi, K.; Hassan, W.H.; Riksen, M.; Ritsema, C. Rainwater Catchment System Reliability Analysis for Al Abila Dam in Iraq's Western Desert. *Water* **2023**, *15*, 944. https://doi.org/10.3390/w15050944

Academic Editor: Luís Filipe Sanches Fernandes

Received: 23 January 2023
Revised: 24 February 2023
Accepted: 27 February 2023
Published: 1 March 2023

1. Introduction

Iraq was seen as having abundant water supplies until the 1970s. However, water shortages in Iraq have been brought on by the building of dams in the Tigris and Euphrates Rivers and their tributaries outside the Iraqi border, as well as by increasing water consumption, population growth, and urban and industrial expansion. Therefore, water scarcity is among the most serious issues in arid and semi-arid regions, especially in developing areas such as the western desert of Iraq [1,2]. As a result, both developing policies and technology to locate alternate water supplies, and enhancing water resource management and planning, will be crucial. Rainwater Harvesting (RWH) systems are gaining popularity as an alternative water supply, and are seen to be viable approaches for storing water for home or farming purposes [3–5]. However, little is known about their effect on the chance that these systems provide the required water for a period of time, here defined as the Rainwater Catchment System Reliability (RCSR)A water balance model based on input and output flow may be used to evaluate available water supply [6–8]. The water balance modeling is also useful for calculating RWH reliability. Determining the reliability of an

RWH system is a key aspect to assess demand reliability, the probability that the system will satisfy the water demand for a certain timeframe is described as reliability [9–12].

Baek and Cole [9] evaluated the impact of the modeling time period for the water balance models and the concept of reliability to determine the variation in the watershed systems' reliability. Five concepts of reliability and weekly and daily modeling time periods are used to assess the reliability of catchment dam systems at ten sites in Western Australia's dryland rural regions.

According to evaluation findings, the possibility of underestimation makes utilizing yearly period-based prediction for reliability unsuitable for dry and semi-arid regions of Western Australia. Volume-based predictions as well as weekly and daily period-based predictions have the possibility of being overestimated when the pattern of growing crops and water requirements in Western Australia is taken into account. For the design of water harvesting schemes in the dryland agricultural lands of the southwest of Western Australia, it is therefore advised to employ monthly period-based prediction.

Jafarzadeh et al. [10] evaluated the future reliability of RWH. Using the outcomes from General Circulation Models (GCMs) for both historical and future eras, monthly rainfall was predicted in the first stage. Data was then spatially downscaled. For each month rainfall was interpolated for future periods using the standard kriging approach. Finally, the reliability of RWH was evaluated and investigated for various roof areas and storage tank capacities. Findings demonstrate that a reliability band of 0.05–0.45 RWHS was calculated for the historical period and that this reliability range will increase for the future period based on the best GCMs. Additionally, a variance in RWH reliability revealed that, in general, RWH reliability under Representative Concentration Pathway (RCP), 2.6 rcp will be greater than 8.6 rcp in the future.

Imteaz et al. [13] created a tank tool using daily water balance simulation. In several Australian cities, including Melbourne, the advanced approach was regularly employed to assess RWH tank's reliability. Researchers looked at how reliable a specific magnitude of rain water tank is in relation to annual volume and meeting daily estimated requirements in Bangladesh's megacity [14]. In conjunction with the town water delivery systems in Dhaka City, this article examines the economic viability, adaptability, and reliability of rainwater harvesting (RWH) systems to partially balance the daily water requirements in multistory buildings. To evaluate the reliability and viability of the RWH systems in an urban setting, extensive computer program was created. By examining daily rainfall data for the previous 20 years, three distinct climate scenarios—rainy, normal, and dry years—were chosen. Results showed that within the wet climatic condition, roughly 15–25% reliability may be reached [14].

Male and Kennedy [15] investigated the probable role of rainwater usage for home uses in Portland, Oregon, with a focus on rainwater collection reliability. Applying the water balance, they detailed the technique using the amount of rainfall collected, domestic demand, and capacity of storage tanks. The capacity of the storage tank, in addition to the catchment area's size, was shown to be essential in determining the system's reliability.

The reliability of RWH is critical to residents' desire to know that it might be necessary for their source of water [15,16]. Nevertheless, no comprehensive research has been undertaken as yet on the feasibility and RWH reliability collection technology in the sub-catchment [17,18].

Liuzzo et al. [19] looked at the efficiency of a potential RWH tank for a model single-family home in a neighborhood. Information from more than 100 locations in Sicily was used to test performance for various yearly precipitation amounts. The performance was evaluated for three uses of the rainwater collected and three storage sizes (10, 15, and 20 m^3). The system's reliability was examined as a function of average annual rainfall after the system's performance for the full research area had been assessed. This analysis allowed for the development of mathematical equations with regional applicability and implementation. To determine the degree of uncertainty surrounding the regional model provisions, a data resampling approach was used. To determine the payback period for the capital cost associated with the installation of the RWH system, a cost-benefit analysis was lastly carried out.

In contrast, in a developing country such as Iraq, specifically in the western desert, where the technique for supplying water is below enormous stress and can be vanished, there are few comprehensive studies on the possibility of harvesting rainwater. In addition, there is no in-depth research about the sustainability, reliability, or efficiency of any planned RWH systems in the area, the studies are limited to investigating the suitability of the sites and the techniques to implement RWH.

The major goal of this study is to evaluate RWH systems reliability in a sub-catchment applying a water balance technique, by examen how much the reliability of RWH in the sub-catchment of the Al Abila dam is influenced by the storage capacity. This will be done by estimating the volumetric reliability and time-based reliability for wet, average, and dry-year climate scenarios.

2. Methodology

2.1. Area and Data Utilised

Wadi Horan is placed in Al-Anbar region in western Iraq, around 450 km west of Baghdad (Figure 1). The watershed is about 370 km^2 in size and has a dry environment by dry summers and mild winters. The average yearly rainfall is just 115 mm, where the winter months get around 49% of the rain, the spring months 36%, the fall 15%, and the summer months see no rain. The average annual temperature is 21 degrees Celsius, with July being the warmest month and January being the coldest [3,20]. The yearly evaporation possible is 3200 mm on average.

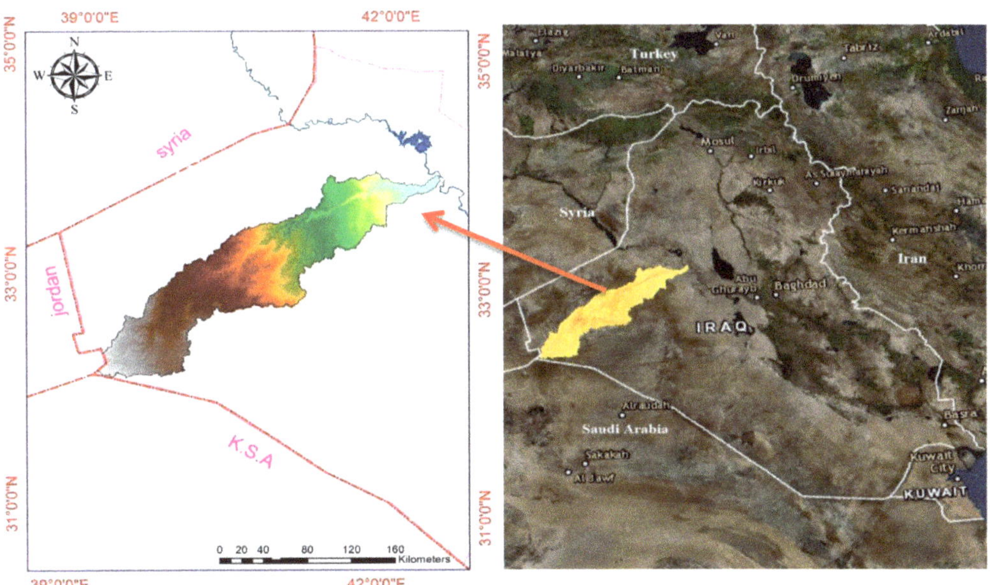

Figure 1. Study area, showing the Al-Abila dam location after [3].

Hard limestone makes up the majority of the wadi Horan's exposed rocks [20]. They can be utilized to protect the front edge of barriers and make a nice foundation for dams or other barriers. In order to minimize the number of building materials required for the dams, reduce evaporation losses, and guarantee the necessary storage, the catchment area and potential for a hard, narrow cross-section of the wadi with vertical shoulders were taken into con-sideration while deciding where to place the dams.

To test the RCSR of the Al Abila dam, a 30-year rainfall dataset (1990–2020) was used to determine the three distinct meteorological years: wet (max. rainfall), average, and dry (min. rainfall). An average year was defined as one equivalent to a regular yearly rainfall over a period of approximately 20 years.

The physical characteristics of the Al Abila catchment were assessed. Using a tape measure and the Global Positioning System (GPS), the height of the dike and spillways for the Abila dam were determined. These data were used to determine the entire volume of water that might be gathered behind the dam. In the Abila catchment, the texture of soil was evaluated by aggregating samples, and the slope and area were observed in the field. DEMs and a Geographic Information System (GIS) were used.

A rainfall simulator was used to quantify runoff coefficients at many places in the Al Abila catchment. A Kamphorst, [21], rainfall simulator was used to simulate rainfall in the Abila catchments' area. A device called a "rainfall simulator" attempt to replicate the physical traits of natural rainfall as accurately as possible [22]. The instrument was calibrated in line with Kamphorst's instructions (1987). Each test took three minutes to monitor the water level, taking readings every 30 s. A tube was used to catch any runoff, and the amount was measured. At the completion of each simulation, the runoff coefficient (C) value was computed.

The infiltration rates of the Al Abila dam were evaluated. The infiltration rate was determined using a double-ring infiltrometer [23]. We made use of infiltrometers with 18/30 cm internal and external rings. In order to assure a trustworthy result, tests were often conducted twice for each site. Preliminary filling of internal and external rings was to a depth of 15 cm. More water was given to maintain equal levels once the water level in the external ring dipped just below the level in the internal ring. A scale mounted on the internal ring was used to measure the water level as a function of time during the test. We kept doing this until the level of water fell to less than 5 cm, the water was then supplied for the following iteration. In most cases, one to four repeats were carried out to guarantee that a steady infiltration rate was attained. These results were used to assess the Al Abila catchment's average infiltration rate over a specific time period. The Water Harvesting model (WHCatch) [24] uses these data types as input.

2.2. Water Harvesting Model (WHCatch)

To analys the effectiveness of the RWH approaches based on current climatic circumstances, we used the WHCatch basic model [24] for the Al Abila dam watershed. Based on the water demands, the water supply, and the structures design of the RWH, the water balance of the Al Abila dam was examined. The variance between the total input and output was used to compute the variation in water storage volume. A runoff area and a reservoir area are the two basic components of a catchment. To increase the RWH system's reliability, we evaluated the performance of RWH throughout the entire system and examined the water balance of these two components. The water storage change in the Al Abila dam was calculated by subtracting the entire inflow from the total outflow [25]:

[Water balance calculation of any location in m^3:]

$$\Delta S = I - Q \tag{1}$$

where ΔS is the storage change during a certain time period, I is the input flow, and Q is the output flow, all together in m^3. The details of this model (WH Catch) and its application with the manual were explained and published in two articles [24]. In MS Excel, the Al-Abila dam's monthly water balance study was carried out to assess the dam's effectiveness in satisfying local water demands. We used the WHcatch program, a straightforward Visual Basic for Applications (VBA) macro in Excel, since all input data were logged and available there. The computations were carried out by this macro, which then recorded the results in the appropriate cells. A WHCatch module and a Sub-catchment Class module made up the code. The last one included a routine to carry out certain fundamental calculations as well as all the attributes of a sub-catchment. Three general and a few private subroutines made up the WHCatch module. The VBA macro won't be visible to the Excel workbook's regular users. Only when the further capability is needed, entering the code section will be crucial. This tool allows for the reading of data into GIS applications, as well as all outcome is kept and shown in the same Excel workbook. Nearly all circumstances, the shape file containing the location's layout and the IDs of its sub-catchments are accessible. Thus, the ID in the Excel workbook (column name) and the ID in the sub-catchment ID in the shape file can be collective.

2.3. The SCS–CN Method

The Soil Conservation Service (SCS) technique, created in the USA by Soil Conservation Service (SCS) in 1969, is a straightforward, dependable, and consistent intellectual approach for the determination of runoff based on rainfall. It simply depends on the variable CN. We calculated the runoff depth using the Curve Number (CN). Following that, the depth of runoff is applied to calculate the probable water supply following runoff. The influence of soil and land use on precipitation and runoff makes CN dependable. An expression for runoff depth is:

$$Q = \frac{(P - I_a)^2}{(P - I_a) + S} \tag{2}$$

where Q is the depth of runoff (in mm), P is the amount of rainfall (in mm), S represents the possible maximum retention (in mm), and I_a is the initial abstraction (in mm), which accounts for all losses prior to the start of runoff, infiltration, evaporation, and water interception. The rainfall data for numerous small rural regions were analyzed to arrive at $I_a = 0.2S$.

2.4. Reliability Analysis

This research revealed two reliability categories. The following Imteaz equation is used to calculate time-based reliability [13]:

$$R_t = \frac{T_d - U_d}{T_d} \times 100 \tag{3}$$

where R_t refers to time-based reliability (percentage), U_d represents the number of days where RWH was inadequate for daily demand, and T_d is how many days there are in a year (365).

The second type of reliability is volumetric reliability (R_v) which is given by:

$$R_v = \frac{\sum (VW_d - VD_d)}{\sum VW_d} \tag{4}$$

where VD_d is the annual water deficit and VW_d is the annual water demand.

2.5. Sensitivity Analysis

Sensitivity assessments were carried out to examine the effect of the coefficient of runoff on the efficiency of RWH storage and demand fraction. Four values of runoff coefficient (from 0.3 to 0.6) were considered (C1 = 0.3, C2 = 0.4, C3 = 0.5, C4 = 0.6) taking into account the infiltration and the spilling losses from the rainfall.

The relation of water saving efficiency with demand fraction and storage fraction, respectively, were shown in sensitivity figures. Demand fraction (D/Q) is calculated using the formula below:

$$\frac{D}{Q} = \frac{VW_d}{VW_s} \tag{5}$$

where VW_d refers to annual water demand and VW_s refers to the annual volume of rainwater supply.

The storage fraction (S/Q) is calculated using the following formula:

$$\frac{S}{Q} = \frac{SW_d}{VW_s}$$

(6)

where SW_d is the annual storage capacity of rainwater supply.

3. Results and Discussion

The initial evaluation was carried out to see how storage capacity affects daily reliability and to determine the amount of storage that offers the Abila Dam's highest average reliability value. The model of water balance was run on a daily basis, and throughout the investigation, the Abila dam's daily average reliability was calculated. The associated percentile values were then calculated.

3.1. The Al Abila Dam Assessment

Based on the greatest depth of daily rainfall measured from the Al-Rutba station for the years (1990–2020), the data was examined using the water balancing approach in the Abila watershed utilizing the Water Harvesting model (WHCatch) under Microsoft Excel to determine the variation in water storage within the volume.

The Abila Dam has a 4×10^6 m^3 design capacity. Only once, in 1994, did the dam's reservoir fill to its intended level, as illustrated in Figure 2. Additionally, there was essentially little runoff from 2000 to 2009, which left the dam's reservoir dry and the dam inoperable.

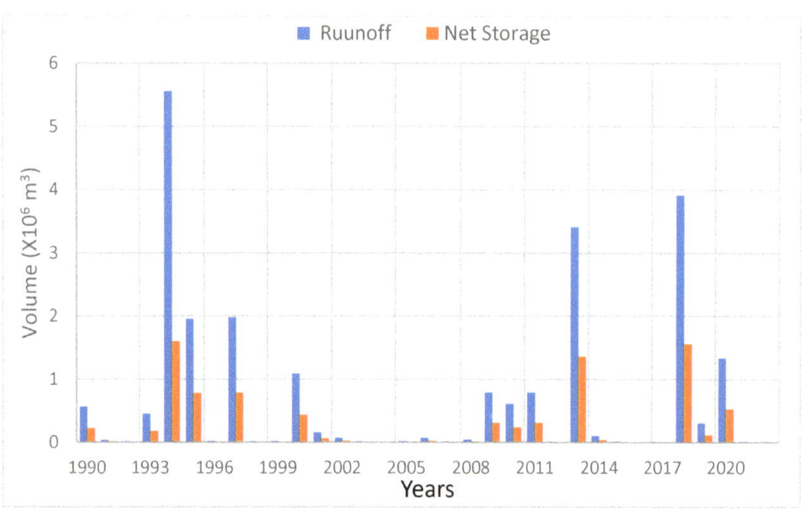

Figure 2. The total annual runoff volume and net storage volume.

The analysis showed that out of the total years (1990–2020) where surface runoff occurred, the number of years where it entered the dam reservoir and caused the dam to store all of its intended 4×10^6 capacity was 5.2%. According to the findings, the dam's reservoir is greater than the available storage. Since the base of the trench has not been deepened to reach the hard rocky layers, the Abila dam has a problem with regular seepage along its body. A back trench (Toe drain) downstream of the dam was also missing.

3.2. Reliability Analysis

The relation between the reliability and the storage volume was illustrated in Figure 3, which shows where the reliability increased and then stabilized over the increasing of the storage volume, for (a) wet, (b) average, and (c) dry year climate scenarios. Each scenario shows similar profiles over the course of the increasing storage volume, while wet scenarios show more reliability than average and dry scenarios due to the additional rainwater in the wet year, which keeps the reservoir full. With the dry-year scenario, reliability was within an acceptance rate as well. With no rainfall period with around 5% to 18% R_t with a storage volume that varies from approximately 4×10^5 to 10×10^5 m^3 the dam can still provide water.

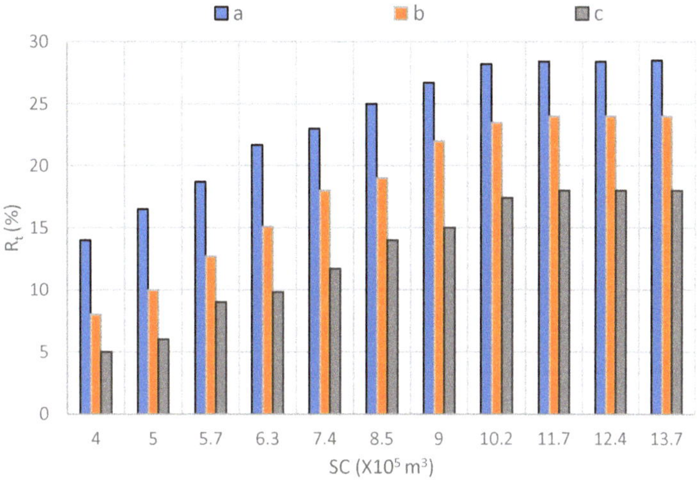

Figure 3. The percentage of time-based reliability (R_t)vs storage capacity (SC) for (a) wet, (b) average, and (c) dry-year climate scenarios.

The figure also shows that in all three scenarios, a threshold point accrued around a storage volume of 10.2×10^5 m^3. The threshold point specifies the stage at which storage capacity does not affect reliability. In a wet year, if the dam received all the harvested water and its reservoir become full, then increasing the storage beyond this volume is not needed and does not influence the dam's reliability. This also implies the dry-year scenario.

Figure 4 represents the volumetric reliability or the proportion of water saved for a catchment area with varied storage volume capacity. It can be seen that the influence of increasing the storage capacity on volumetric reliability follows a similar pattern as time-based reliability. For wet, average, and dry-year climate scenarios, the volumetric reliability tends to a considerable increase as the storage volume increase until it remains steady around 11.7×10^5 m^3, where increasing the storage volume has no more influence on the reliability of the dam.

For wet and average scenarios, the reliability results were more convergent for the same storage size, for instance: at a storage capacity of 9×10^5 m^3, the percentage of volumetric reliability was 69% and 64% for the wet and average scenarios, respectively. This may mainly be because, according to the calculation, the annual average water demand for the catchment is considered to be constant, and it has the main influence on the value of the volumetric reliability. After then, each distinct scenario's variances in the amount of captured rainfall accumulated.

According to Figures 3 and 4, the volumetric reliability value for the catchment is discovered to be greater than the time-based reliability. Where the time-based reliability varies from 14% to 28%, volumetric reliability ranges from around 34% to 75% for the wet year scenario. This is mostly due to the time-based reliability was computed using the number of days overall during which the rainwater harvest is sufficient to provide the re-quired water demand, and this is relatively not much, due to the local's climatic conditions compared to the amount of water demand required to meet the daily needs.

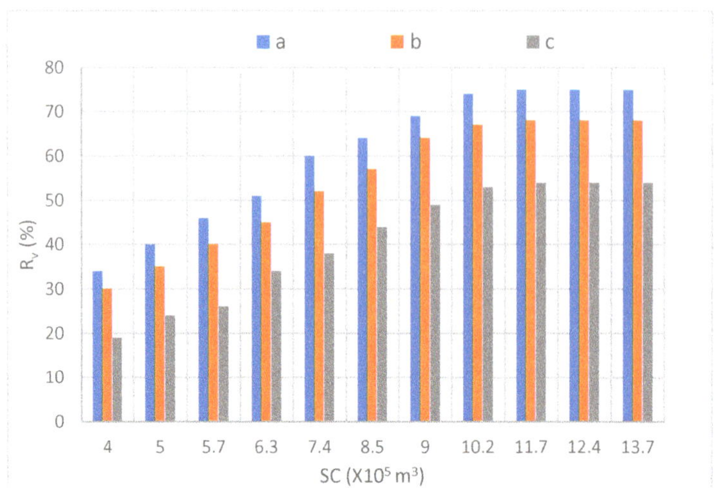

Figure 4. The percentage of volumetric reliability (R_v) vs. storage capacity (SC) for (a) wet, (b) average, and (c) dry-year climate scenarios.

3.3. Sensitivity Analysis

Figure 5 shows how the runoff coefficient (C) affects water-saving efficiency in wet year conditions. Sensitivity analysis was implemented on arrange of runoff coefficient values starting from 0.3 to 0.6. According to the result shown in Figure 5, the coefficient of runoff has no significant influence on water effectiveness in wet climatic conditions, and the variety of C does not imply a significant change in the influence pattern of efficiency as well.

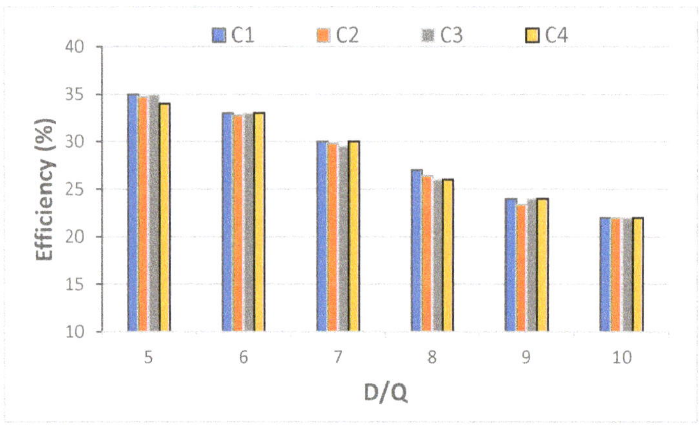

Figure 5. Water saving efficiency dealings with runoff coefficients vs. demand fraction (D/Q).

The influence of the coefficient of runoff on water-saving efficiency is seen in Figure 6. The results show that increasing the storage fraction leads to increased efficiency. So, when the *S/D* value increases from 0.07 to 0.09, the efficiency increases by about 2%. In addition, the results show that for the same storage fraction, the value of C has an obvious influence on efficiency. For C1 and C4 there is an average increase of 2% in efficiency.

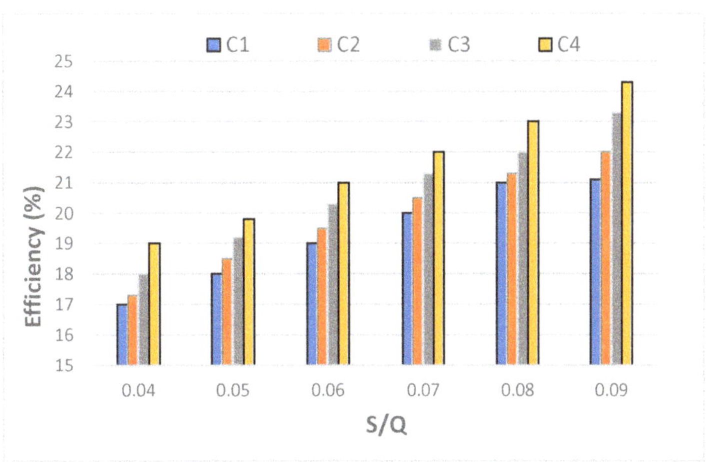

Figure 6. Water saving efficiency dealings with runoff coefficients vs. storage fraction (D/Q).

The sensitivity analysis shows that the water-saving efficiency for the dam does not reach 50% for both cases (demand fraction and storage fraction) and for the different C values.

The efficiency has been influenced differently by changing the runoff coefficient as shown in Figures 5 and 6. In the case of demand fraction, the change in the C value has almost no effect on the efficiency within the same demand fraction value. While the efficiency increases slightly with increasing the runoff coefficient for each storage fraction value.

4. Conclusions

Recently, awareness of RWH systems as a substitute water supply has grown. These systems can be linked with existing traditional water sources to provide supplemental water supplies in various locations, or they can act as the primary water source in arid and semi-arid regions in which water availability is a major concern. Additionally, using RWH is a successful adaptation technique to combat the decline in water availability caused by climate change. This study evaluates RWH systems reliability in a sub-catchment applying a water balance technique by examen how much the reliability of RWH in the sub-catchment of the Al Abila dam is influenced by the storage capacity. The reliability was investigated using a water balance model, time-based reliability, and volumetric reliability followed by sensitivity analysis. According to reliability correlations with varying storage volumes, time-based reliability shows an increase throughout the wet year, average year, and dry year, equal to a storage volume of 10.2×10^5 m^3, and beyond that, the threshold point accrued when reliability doesn't rise with riding storage volume. The volumetric reliability, and for the three climate scenarios (wet year, average year, and dry year), increased as the storage volume increased until 11.7×10^5 m^3, then increasing the storage volume has no more in-fluence on the reliability of the dam. For the three scenarios, volumetric reliability value for the catchment is detected to be greater than the time-based reliability. The time-based reliability varies from 14% to 28%, while volumetric reliability varies from about 34% to 75% for the wet year scenario. Despite the large storage capacity of the dam, the reliability of the dam does not show significant value.

Water saving efficiency decreases with increasing demand fraction, while the runoff coefficient has no significant influence on water effectiveness in wet climatic conditions. Growing storage fraction value increases the water-saving efficiency, and the value of the runoff coefficient influences the water-saving efficiency. For instance, there is an average increase of 2% in water efficiency for the same value of storage fraction. For both cases, the water-saving efficiency of the dam does not reach 50%.

For a reliable rainwater harvesting system, this kind of study should be considered as a design guideline to scientifically highlight the system elements that play a significant role in increasing the reliability of any system. For instance, thresholds enable landowners to get the most out of their systems, while saving additional costs of increasing the storage of the rainwater harvesting system (the dam in this case). While the storage capacity is attempting to grow, reliability does not improve. In all situations, volumetric reliability was shown to be greater than time-based reliability, and thus could not be neglected in the design of any new systems. Using daily rainfall data, the technique given in this paper might be applied to forecast water conservation and the RWH systems' reliability in different arid and semi-arid areas.

It is important to include the impact of climate change on rainfall in future analyses. The equations described here hold for both historical and contemporary environmental circumstances. The performance of an RWH system may be considerably impacted by trends. In particular, a significant decline in system efficiency may be caused by a drop in rainfall volume and a variation in the timing of rainfall throughout the year. Deriving future climatic scenarios from regional climate models should therefore be considered while designing the RWH systems.

Finally, understanding the reliability of a rainwater harvesting system is critical, which informs decision-makers and households about the design criteria of the system and how much they expect from the system to meet water demands throughout the year.

Author Contributions: Conceptualization A.A. and R.A.; methodology, A.A, R.A. and K.M.; software, A.A. and R.A.; formal analysis, K.M. and M.R.; investigation, A.A., W.H.H. and R.A.; resources, A.A. and R.A.; writing—original draft preparation, A.A., C.R., and R.A.; writing—review and editing, C.R., K.M., W.H.H. and M.R.; visualization, A.A., K.M., W.H.H. and M.R.; supervision, C.R.; project administration, C.R.; funding acquisition, C.R. All authors have read and agreed to the published version of the manuscript.

Funding: This research received no external funding.

Data Availability Statement: Some data in this manuscript was obtained from the Ministry of Agriculture and the Ministry of Water Resources, Iraq. The other data from the fieldwork and previous studies.

Acknowledgments: This study was funded by the NUFFIC Orange Knowledge Program (OKPIRA-104278) and coordinated by Wageningen University & Research, The Netherlands.

Conflicts of Interest: The authors declare no conflict of interest.

References

1. Adham, A.; Wesseling, J.G.; Abed, R.; Riksen, M.; Ouessar, M.; Ritsema, C.J. Assessing the impact of climate change on rainwater harvesting in the Oum Zessar watershed in Southeastern Tunisia. *Agric. Water Manag.* **2019**, *221*, 131–140. [CrossRef]
2. Abdullah, M.; Al-Ansari, N.; Laue, J. Water harvesting in Iraq: Status and opportunities. *J. Earth Sci. Geotech. Eng.* **2020**, *10*, 199–217.
3. Adham, A.; Sayl, K.N.; Abed, R.; Abdeladhim, M.A.; Wesseling, J.G.; Riksen, M.; Ritsema, C.J. A GIS-based approach for identifying potential sites for harvesting rainwater in the Western Desert of Iraq. *Int. Soil Water Conserv. Res.* **2018**, *6*, 297–304. [CrossRef]
4. Fowler, H.J.; Kilsby, C.G.; O'Connell, P.E. Modeling the impacts of climatic change and variability on the reliability, resilience, and vulnerability of a water resource system. *Water Resour. Res.* **2003**, *39*, 1222–1233. [CrossRef]
5. Alhadithi, A.A.; Alaraji, A.A. Rainwater harvesting of Hauran valley, west of Iraq. *Iraqi J. Sci.* **2016**, *57*, 456–468.
6. Adham, A.; Riksen, M.J.P.M.; Ouessar, M.; Ritsema, C.J. A methodology to assess and evaluate rainwater harvesting techniques in (semi-) arid regions. *Water* **2016**, *8*, 198. [CrossRef]
7. Pachpute, J.S.; Tumbo, S.D.; Sally, H.; Mul, M.L. Sustainability of rainwater harvesting systems in rural catchment of Sub-Saharan Africa. *Water Resour. Manag.* **2009**, *23*, 2815–2839. [CrossRef]
8. Hashim, H.Q.; Sayl, K.N. Detection of suitable sites for rainwater harvesting planning in an arid region using geographic information system. *App. Appl. Geomat.* **2021**, *13*, 235–248. [CrossRef]

9.   Baek, C.W.; Coles, N.A. Defining reliability for rainwater harvesting systems. In Proceedings of the International Congress on Modelling and Simulation MODSIM, Perth, Australia, 12–16 December 2011; pp. 12–16.

10. Jafarzadeh, A.; Bilondi, M.P.; Afshar, A.A.; Yaghoobzadeh, M. Reliability estimation of rainwater catchment system using future GCM output data (case study: Birjand city). *Eur. Water* **2017**, *59*, 169–175.

11. Dongol, R.; Bohora, R.C.; Chalise, S.R. Sustainability of rainwater harvesting system for the domestic needs: A case of Daugha Village Development Committee, Gulmi, Nepal. *Nepal J. Environ. Sci.* **2017**, *5*, 19–25. [CrossRef]

12. Chartzoulakis, K.; Bertaki, M. Sustainable Water Management in Agriculture under Climate Change. *Agric. Agric. Sci. Procedia* **2015**, *4*, 88–98. [CrossRef]

13. Imteaz, M.A.; Ahsan, A.; Naser, J.; Rahman, A. Reliability analysis of rainwater tanks in Melbourne using daily water balance model. *Resour. Conserv. Recycl.* **2011**, *56*, 80–86. [CrossRef]

14. Karim, M.R.; Bashar, M.Z.I.; Imteaz, M.A. Reliability and economic analysis of urban rainwater harvesting in a megacity in Bangladesh. *Resour. Conserv. Recycl.* **2015**, *104*, 61–67. [CrossRef]

15. Male, J.W.; Kennedy, M.S. Reliability of rainwater harvesting. *WIT Trans. Built Environ.* **2006**, *86*, 20–29.

16. Rahman, A.; Aurib, K.; Datta, D.; Yunus, A. Reliability Analysis and Water Modeling of Optimum Tank Size for Rainwater Harvesting in Two Salinity Affected Areas of Bangladesh. *J. Environ. Eng. Stud.* **2017**, *2*, 1–12.

17. Zhang, S.; Zhang, J.; Jing, X.; Wang, Y.; Wang, Y.; Yue, T. Water saving efficiency and reliability of rainwater harvesting systems in the context of climate change. *J. Clean. Prod.* **2018**, *196*, 1341–1355. [CrossRef]

18. Lawrence, D.; Lopes, V.L. Reliability analysis of urban rainwater harvesting for three Texas cities. *J. Urban Environ. Eng.* **2016**, *10*, 124–134. [CrossRef]

19. Liuzzo, L.; Notaro, V.; Freni, G. A reliability analysis of a rainfall harvesting system in southern Italy. *Water* **2016**, *8*, 18. [CrossRef]

20. Sayl, K.; AL-Ani, A.; Oliwi, S. Hydrologic study for Iraqi Western Desert to Assessment of Water Harvesting Projects. *Iraqi J. Civ. Eng.* **2006**, *7*, 16–27.

21. Kamphorst, A. A small rainfall simulator for the determination of soil erodibility. *Neth. J. Agric. Sci.* **1987**, *35*, 407–415. [CrossRef]

22. Aksoy, H.; Unal, N.; Cokgor, S.; Gedikli, A.; Yoon, J.; Koca, K.; Inci, S.; Eris, E. A rainfall simulator for laboratory-scale assessment of rainfall-runoff-sediment transport processes over a two-dimensional flume. *Catena* **2012**, *98*, 63–72. [CrossRef]

23. Al-Qinna, M.; Abu-Awwad, A. Infiltration rate measurements in arid soils with surface crust. *Irrig. Sci.* **1998**, *18*, 83–89. [CrossRef]

24. Wesseling, J.G.; Adham, A.; Riksen, M.J.; Ritsema, C.J.; Oostindie, K.; Heidema, N. A Microsoft Excel Application to Simulate Water Harvesting in a Catchment. *Water Harvest. Res.* **2021**, *4*, 1–18.

25. Boers, T.; Zondervan, K.; Ben-Asher, J. Micro-catchment-water-harvesting (MCWH) for arid zone development. *Agric. Water Manag.* **1986**, *12*, 21–39. [CrossRef]

Article

Sustainability of the Al-Abila Dam in the Western Desert of Iraq

Ammar Adham [1,*], Shwan Seeyan [2], Rasha Abed [3], Karrar Mahdi [3], Michel Riksen [3] and Coen Ritsema [3]

1   Dams and Water Resources Engineering Department, College of Engineering, University of Anbar, Baghdad 55431, Iraq
2   Soil and Water Department, Agricultural Engineering Sciences College, University of Salahaddin, Erbil 44002, Iraq; shwan.seeyan@su.edu.krd
3   Soil Physics and Land Management Group, Wageningen University, 6700 AA Wageningen, The Netherlands; rha.abed@gmail.com (R.A.); karrar.mahdi@wur.nl (K.M.); michel.riksen@wur.nl (M.R.); coen.ritsema@wur.nl (C.R.)
*   Correspondence: engammar2000@uoanbar.edu.iq; Tel.: +96-479-0320-6401

Abstract: Water scarcity is a major problem in the arid climate of Iraq's Western Desert and people struggle to manage the precarious water supply. Harvesting rainwater is one sustainable method that can be used to increase the supply of water. Rainwater harvesting systems (RWH) are considered to be sustainable "if they can continue collecting, utilising, and consuming natural water resources for maximum livelihood development". This study assessed the sustainably of the Al- Abila dam in Iraq's Western Desert by determining its level of functionality in harvesting water and using it effectively. The reliability of the water supply and its potential productivity and water use efficiency were investigated as well. The balancing storage at the end of each runoff shows that dam storage of this magnitude is insufficient to fulfil the water demand. This research highlighted constraints that have affected system functioning or sustainability and provided suggestions and recommendations for risk-managed rainwater harvesting system installation methods and designs. The water conveyance factor and adequacy of the system were low, with 60% conveyance losses. This research helps policymakers to conduct large-scale, high-level assessments and answer basic problems about small earth dam development and management in Anbar's Western Desert.

Keywords: Al-Abila dam; rainwater harvesting systems; sustainability of reservoir; Western Iraqi Desert

Citation: Adham, A.; Seeyan, S.; Abed, R.; Mahdi, K.; Riksen, M.; Ritsema, C. Sustainability of the Al-Abila Dam in the Western Desert of Iraq. *Water* **2022**, *14*, 586. https:// doi.org/10.3390/w14040586

Academic Editor: Enedir Ghisi

Received: 3 January 2022
Accepted: 13 February 2022
Published: 15 February 2022

Publisher's Note: MDPI stays neutral with regard to jurisdictional claims in published maps and institutional affiliations.

1. Introduction

Until the 1970s, Iraq was considered to have rich water resources due to the proximity of the Tigris and Euphrates Rivers. Unfortunately, dam construction on the tributaries of the rivers in Turkey and Syria has resulted in water scarcity in Iraq [1]. Additionally, Iraq's growing population has raised the demand for water, while climate change and declining rainfall rates have further restricted the water supply since 2007 [2]. Iraq's Western Desert is the area most affected by water scarcity. This arid area has had major challenges in providing and managing water. Rainfall distribution is erratic, and the country suffers from high evaporation rates, high temperatures, and a shortage of groundwater and surface water [3]. Iraq's desert accounts for around 55% of Iraq's total land area, most of it uninhabitable because of the lack of water. Many larger valleys, such as the Wadi Horan, Wadi Amije, and Wadi Al Awaje, receive comparatively high amounts of floodwater [4]. The need to develop new water sources or to optimize the use of existing sources has become critical. Many studies show that sustainable rainwater collection is essential for better water management, from a socio-economic as well as a biophysical development perspective. Sustainability is a new term currently used when talking about development methods, and it may be interpreted in a variety of ways depending on the context. It has become an all-encompassing word that is used for nearly every system on the planet. Sustainability is defined as "enhancing human well-being while respecting ecological limits", according to the International Union for Conservation of Nature (IUCN), the United Nations Environment Programme (UNEP), and the World Wildlife Fund (WWF) [5]. Sustainability assessment (SA) is becoming a more widely recognized technique for assisting in the transition to sustainability [6]. A sustainability assessment evaluates the influence of a proposed or existing policy, plan, program, project, law, or current practice or activity on sustainability [6]. How to measure sustainability is a recurring topic in sustainability evaluations [7].

Rainwater harvesting systems are sustainable if they can be used and the water consumed in the future to improve livelihood [8]. According to [9], rainwater harvesting systems should have the following qualities to be sustainable: reliable water supply and production potential; efficient water use; and minimal environmental consequences. Rainfall amount and quality of runoff, social and economic variables such as expertise and investment capability, labor accessibility, and institutional backing are all elements that influence the long-term viability of rainwater harvesting systems [8].

Rainwater harvesting systems (RWH) have been evaluated using a variety of sustainability criteria. Hashimoto [10] proposed the assessment of a reservoir's sustainability using reliability, resilience, and vulnerability. Kjeldsen [11] evaluated several techniques for estimating reliability, resilience, and vulnerability, concluding that the maximum values of deficiency duration and volume provide more consistent findings than average values. To assess sustainability, Sandoval [12] presented a geometric average of reliability, resilience, and vulnerability.

Park [13] investigated the long-term sustainability of RWH systems in six major US cities with variable rainfall data. There are several differences between the characteristics of a reservoir and the characteristics of an RWH, and thus, the sustainability index (SI) assessment procedure for the RWH was adjusted. This study introduces a new RWH performance model as well as a method for assessing sustainability indicators.

The management and sustainability of water resources in Iraq's Western Desert were investigated by [14]. The system was simulated using remote sensing and numerical analysis tools. The yearly water harvesting rate for each basin in the research region was established, and new water collecting places were discovered for diverse agricultural uses and community development.

Abdulhameeda [15] investigated the idea of building a series of small dams to retain rainfall water in order to sustain and enhance the ecological system. The construction of 13 optimal-height dams in the Horan valley would increase the water surface area of reservoirs in this valley from 15 to 90 km^2, replenishing groundwater and sustaining rainwater collection systems.

Various disciplines, from engineering to business to policymaking, have suggested and created a large number of methods and conceptual frameworks. Most of these frameworks were formulated in the last decade or so but were never tested. The conceptual frameworks have two primary features: creating objectives and evaluation criteria using sustainable principles, and providing quantitative indicators for each assessment criterion.

This study's objective was to evaluate the sustainability of a rainwater harvesting system, the Al-Abila dam in the Western Desert of Iraq. In this research, we consider a system to be sustainable if it can provide a reliable supply of water that can be used efficiently and has productivity potential. Moreover, special attention was paid to identifying defects or constraints that have reduced the functionality or sustainability of the system.

2. Materials and Methods

In this study, the Abila Dam in the Wadi Horan (Figure 1) was taken as a case study. Catchment characteristics and physical characteristics were collected and measured to evaluate the study area. To evaluate the sustainability of the dam, we assessed the reliability of the water supply and its production potential (the water balance model was used to estimate water supply and demand) and the effectiveness of the water use.

2.1. Study Area

The Wadi Horan is Iraq's biggest Wadi, extending from the Saudi border to the Jordanian and Syrian borders. As illustrated in Figure 1, the Wadi Horan is located south of the Euphrates River, at 32°10′44″ to 34°11′00″ N, 39°20′00″ to 42°30′00″ E. The watershed area of the Wadi Horan is 13,370 km^2, with a length of 362 km [16]. The average annual rainfall is about 120 mm, of which around half falls in the winter, 35% in spring, and 15% in autumn [17].

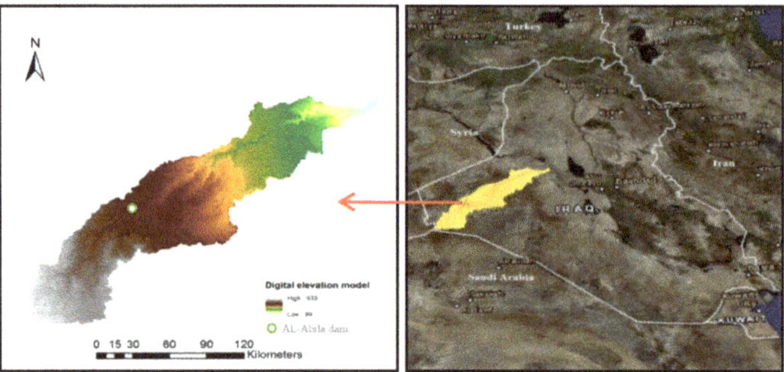

Figure 1. Wadi Horan and the location of the Al-Abila dam (study area) [17].

The temperature of the research area is similar to that of a continental hot desert [18]. July has the highest monthly mean temperature of 31 °C, whereas December has the lowest monthly mean temperature of 8 °C. The average yearly temperature is around 20 °C. Extreme temperatures and dry conditions result in a high rate of evaporation, estimated to be approximately 3000 mm year. Evaporation averages vary month to month from April through October. The highest monthly average evaporation recorded was 433 mm in July, while the lowest was 79 mm in January.

Designing and implementing RWH in these regions is crucial to enhancing the water supply as well as the quality of life in these communities. Small dams are major structures that have been built in the Western Desert of Iraq to capture and store rainfall for use during the dry seasons. The Wadi Horan's exposed rocks are mostly solid limestone [17]. The limestone provides a great foundation for dams or barriers, and it can also be used to cover the front side of the barrier, as seen in Figure 2. For building purposes, the dams were placed in hard, narrow valley cross-sections with high shoulders to decrease the need for construction material, minimize evaporation losses and ensure efficient storage.

Figure 2. Al-Abila Dam.

2.2. Catchment Characteristics

The structural dimensions of the Al-Abila dam were gathered from the dam design report, the small dam management in the Western Desert, and the global positioning system (GPS). Table 1 shows the total amount of water that could potentially be collected behind the dam.

Table 1. The characteristics of Al-Abila dam [19].

Name of Dam	Catchment Area (km^2)	Storage Area (km^2)	Max. Spillway Height (m)	Storage Capacity (m^3)
Al-Abila dam	580	1.5	11	310,142

2.3. Physical Characteristics

Soil texture influences both infiltration and runoff. The textural class is determined by sand, silt, and clay content. The actual soils in the research region range from sandy loam to silty sand [20], with sandy and sandy loam soils being two of the most common.

The Horan valley watershed was mapped using remote sensing satellite data. The Landsat 8 image (23 June 2019) with 30-metre spatial resolution was used for land use/cover mapping.

2.4. Water Balance Model

The Al-Abila dam's water balance was evaluated to estimate runoff and change in water storage volume. A catchment has two main components: a drainage area and a retention area [21]. An area's water balance equation may be stated as [22]:

$$\Delta v = I - O \tag{1}$$

where:

Δv = a storage change over time, in m^3.
I = input volume, in m^3.
O = output volume, in m^3.

A more sophisticated water balance equation is possible with different inflow and outflow variables:

$$\Delta v = Vrunoff + Vrainfall - Inf - Evp \tag{2}$$

where:

Vrunoff represents the amount of upstream runoff collecting in the storage basin per unit of time. The Soil Conservation Service Curve Number (SCS CN) model was applied to compute *Vrunoff*.

$$Vrainfall = P \times As \tag{3}$$

P = is the max. daily precipitation (mm).
As = is the storage basin's area (m^2).
Inf = the storage basin's infiltration loss (mm/day).
Evp = the maximum evaporation (mm/day).

$$Evp = Ev \times As \tag{4}$$

where:

Ev = the average yearly potential evaporation (mm/year).
In the storage area, this volume (Δv) is added to the existing volume (*Si*).

$$Si = Si + \Delta v \tag{5}$$

If the maximum storage height is *hs*, then the maximum storage capacity *Smax* is:

$$Smax = hs \times As \tag{6}$$

If $Si > Smax$, an outflow to the next sub-catchment of $Vout$ occurs.

$$Vout = Si - Smax \tag{7}$$

Using the aforementioned method, the overall change in storage over time (Δvt) is:

$$\Delta vt = \Delta v - Vout \tag{8}$$

Evaporation and infiltration are two types of water losses in the reservoir. Evaporation losses account for a considerable portion of total storage capacity in arid and semi-arid areas. Water depths equal to a reservoir's yearly evaporation range from 1.50 to 3.00 m, making it an essential parameter in water balance calculations [23]. Evaporation might cause up to half of the water held in dams to evaporate, resulting in a massive loss of resources. Calculating lake and reservoir evaporation is a complicated process, since many variables impact the rate of evaporation, including the water body's climate and physiography, and its surroundings [24–26]. As direct measurements of lake evaporation are difficult to take and usually restricted to extremely short time periods, the percentage of evaporation losses (40%) each year was calculated based on prior research and conversations with local experts during this study. Infiltration differs depending on the soil texture, which includes sand, silt, and clay. The infiltration rate of fine-textured soil is lower than that of coarse-textured soil. The infiltration mechanism and rate are also influenced by soil structure [27]. The design reports of the dams in the Western Desert state that the percentage of infiltration losses in the dam reservoirs is more than 20% [28], and this is the percentage that was used in this study.

The Al-Abila dam's monthly water balance analysis in MS Excel was used to assess the dam's ability to satisfy local water needs.

2.5. Assessing the Sustainability of the Al-Abila Dam

2.5.1. Reliability of Water Supply and Potential

For the Al-Abila dam, a supply and demand analysis was conducted. Using an ARCINFO GIS 30 m resolution Digital Elevation Model (DEM), the catchment area, mean slope, type of soil, land use, and land cover of the Al-Abila dam's source streams were estimated. The surface runoff produced by the dam's catchment regions was calculated using the SCS CN model. The Curve Number values were selected based on land use, hydrologic soil group (HSG), and antecedent soil moisture (ASM) status [29].

To determine runoff potential, soils were categorized into four HSGs. A-group (>90% sand and <10% clay) had the lowest runoff potential, B-group had a moderately low runoff potential (10–20% clay and 50–90% sand), C-group had a moderately high runoff potential (20–40% clay and <50% sand), and D-group had a high runoff potential (>40% clay and <50% sand). The hydrological soil group map from the data in previous studies [30] was used in this study.

The Al-Abila dam's water demand was calculated. The Penman–Monteith equation was used to calculate the potential evapotranspiration of widely grown crops [21]. The water needs of the most important crops were calculated. Based on the observed water conveyance factor values and a field application factor of 0.6, the gross irrigation requirement of cultivated crops was estimated [8]. The domestic water requirements were calculated for a home with an average-sized animal herd. To verify the feasibility of fulfilling local water demand, a monthly water balance study of the Al-Abila dam system was performed in MS Excel.

2.5.2. Effectiveness of Water Use

Field observations and conversations with local residents and consumers provided information about water consumption for a specified RWH system. Interviews and a GPS survey were used to acquire data on dam operations such as irrigation size, user count, cropping patterns, farm distance, and allocations. The water conveyance factor and adequacy have both been used to evaluate the effectiveness of dam irrigation systems. The following relationships were used to estimate the values [8].

$$Water\ conveyance = \frac{Vr}{Vd} \tag{9}$$

where: Vr is the volume of water received at field in m^3, and Vd is the volume of water diverted from the dam in m^3.

$$Adequacy = \frac{ET \times A}{Vr \times 0.6} \tag{10}$$

Data on the use of water were gathered. Irrigation timing and crop yields were documented. The selection and production of principle crops grown by local farmers were studied to measure possible water consumption.

3. Results and Discussions

3.1. Land Use/Cover (LU/LC)

A land use/cover map of the Horan valley watershed was created using remote sensing (RS) satellite data. The Landsat 8 image (23-June-2019) with 30-meter spatial resolution was chosen. After downloading the US Geological Survey (USGS) picture, we obtained the composite band using the ArcGIS program, selecting Arc Toolbox > Data Management Tools > Raster > Raster Processing and, finally, Composite Band. A land use/cover map was produced using supervised classification. Bare soil, built-up land in Al-Rutba city, and water bodies such as dam reservoirs, agricultural land, and grass land in the highlands and upstream and downstream of the Horan valley were among the land use/land cover categories in the research area. Bare soil occupyied approximately 70% of the area, and only a small portion of the land was taken up by Al-Rutba city and water bodies, as shown in Figure 3. These findings are consistent with our past research [17].

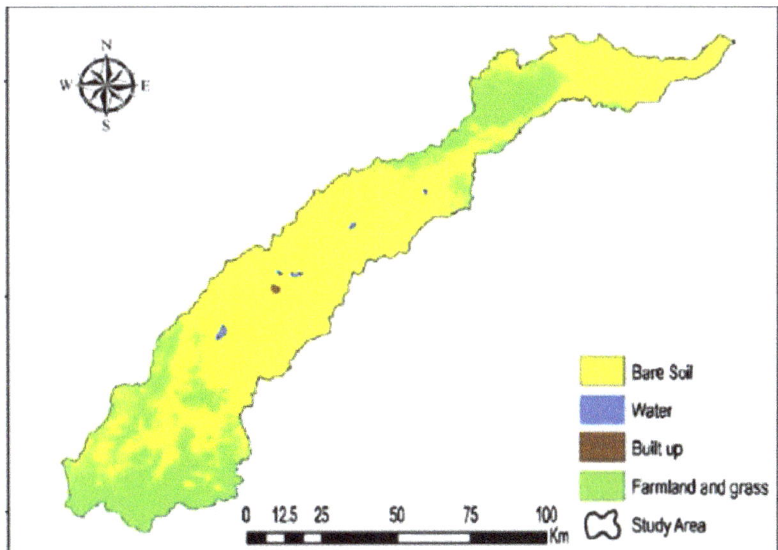

Figure 3. Land use/land cover map.

3.2. Rainwater Harvesting Characteristics

In this study, the hydrological soil group map from earlier studies [30] was employed. For the hydrological soil groups, soil properties were identified, and a GIS map was created.

Groups A, B, C, and D are the four types of hydrological soil groups found in the research area, as illustrated in Figure 4.

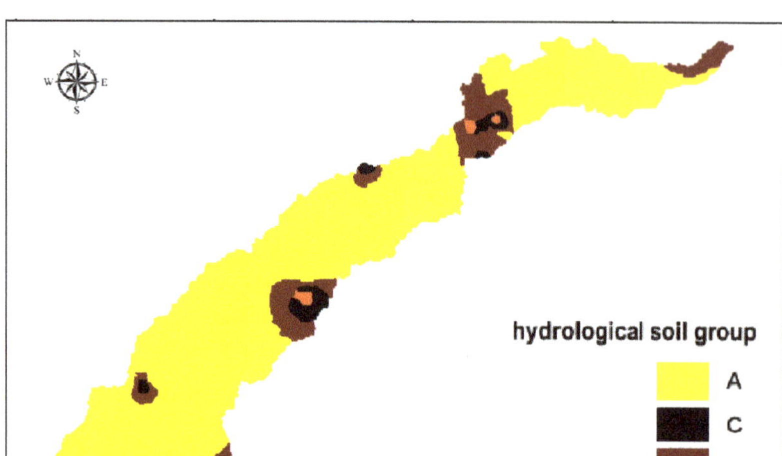

Figure 4. Hydrological soil group map of the Wadi Horan.

To obtain the Curve Number (CN) map of the Wadi Horan, the HSGs map and the land use map were overlayed, as shown in Figure 5.

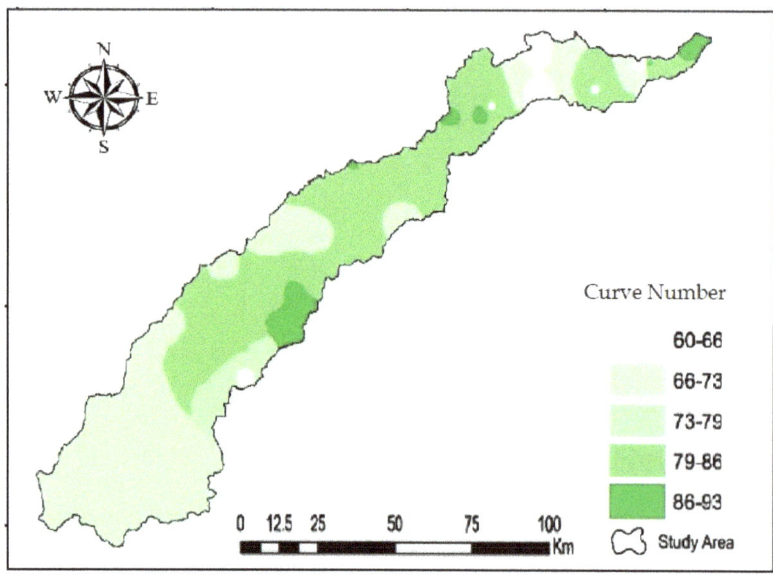

Figure 5. CN map for the Wadi Horan.

The USGS categorization method was used to produce the Curve Number values for each polygon in the land–soil map, as shown in Table 2. The CN value in Wadi Horan varied from 60 to 92 throughout the whole Horan valley.

Table 2. Runoff Curve Number [30].

	Hydrologic Soil Group (*HSG*)			
Land Cover	**A**	**B**	**C**	**T**
Bare Soil	77	86	91	94
Built up	61	75	83	87
Water	100	100	100	100
Farmland	72	81	88	91
Grass	43	65	76	82

Table 3 shows the average characteristics and other key elements such as water availability, catchment area, and average slope of the Al-Abila dam.

Table 3. Source streams of the Al-Abila dam catchment.

Average Slope (m/km)	Main Soil Type	Significant *LU/LC*	Water Availability	Catchment Area (km²)	CN	Max. Retention (S)
1.46	Sandy loam	Bare land	Ephemeral	580	76.5	78.2

According to Table 3, the average CN was 76.5, and based on land use analysis, bare soil accounted for more than 70% of the total area with a CN range from 77 to 94 (as indicated in Table 2). As a consequence, our findings are compatible with the standard CN (Table 2) and other research such as [14,30].

3.3. Assessing the Sustainability of the Al-Abila Dam

3.3.1. Reliability of the Water Supply and Potential

The SCS CN model was used to determine the surface runoff generated by the dam's catchment areas. We estimated the water demand for the RWH system. The Penman–Monteith equation was used to compute the potential evapotranspiration of regularly planted crops. The water requirements of the most important crops were calculated. The gross irrigation demand of cultivated crops was calculated using observed water conveyance factor values and a field application factor of 0.6 [8]. Then, the domestic water needs for a residence with an average-sized animal herd were determined.

In MS Excel, a monthly water budget calculation of the RWH system was performed to see if it was enough to fulfil local water demand. Figure 6 depicts the water demand and supply for the runoff years.

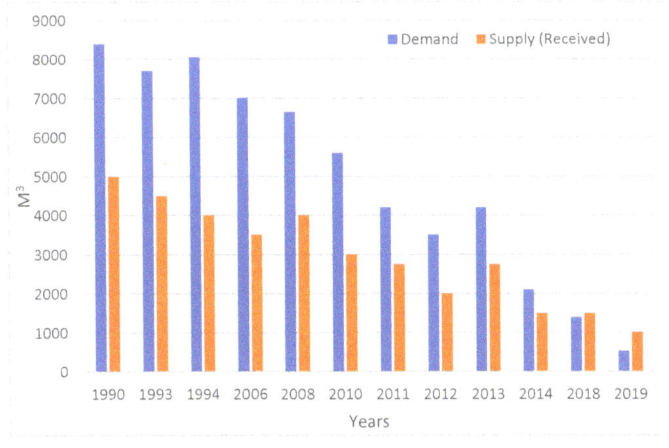

Figure 6. A water balance (supply and demand) of the Al-Abila dam.

The end-of-runoff balance storage indicates that dam storage does not meet the demand. Furthermore, Figure 6 shows a decrease in both water supply and demand from 1990 to 2013, followed by a steep drop in both from 2014 to 2019, as expected. Water supply reduced and fluctuated as a result of unpredictable rainfall, increasing evaporation losses and runoff losses from overland flow, whereas demand declined as a result of the emigration of people from this area due to the ISIS war and agricultural abandonment. This might affect the dam's long-term sustainability. Furthermore, surface runoff from catchments contains a significant quantity of sediments, and the unlined irrigation channels from the dam have a negative influence on the dam's sustainability.

Based on statistics, surveys, and interviews with local residents conducted during the inquiry year of 2019, when rainfall was below normal and distribution was poor, it was revealed that each household in this area has around 10 members and owns approximately 2500 sheep. According to previous studies [4,15], the average daily human consumption is 120 L, while the average daily animal consumption is 13 L, so human consumption accounted for around 16% of the total water stored in the reservoir, while animal watering accounted for about 84%, and the storage capacity was depleted in late June, as shown in Table 4.

Table 4. Monthly water use (m^3) due to human use and animal watering in the year 2019.

Month	Human Uses (m^3) $\times 10^3$	Animal Watering (m^3) $\times 10^3$	Residual Water (m^3) $\times 10^3$
May	4	23	0.3
June	12	64	0.1
July	8	41	0
August	0	0	0
September	0	0	0
Total	24	128	

3.3.2. Effectiveness of Water Use

Throughout dry seasons, the Al-Abila dam allows for the supplemental irrigation of wheat and barley crops, while perennial streams are abstracted and utilized to irrigate vegetable crops during the dry season. In the current scenario, water is transported through unlined earthen channels. As indicated in Table 5, the water conveyance factor and adequacy are poor in region 0.3, which experienced a loss of 60% from the conveyance. Due to erratic rainfall and a lack of assistance and resources for farmers, water from the dam is rarely used. In the investigation year of 2019, the overall volume of water supply was around 498×10^3 m^3, 33% of the total supply being used. Although losses accounted for 67%, evaporation was the major contributor, accounting for 56% of the total volume provided.

Table 5. The water conveyance factor for the runoff years 1990–2019.

Year	Volume of Water Diverting (m^3)	Volume of Water Receiving (m^3)	Water Conveyance
1990	15,000	5000	0.33
1993	13,000	4500	0.35
1994	14,500	4000	0.28
2006	12,000	3500	0.29
2008	11,000	4000	0.36
2010	10,000	3000	0.3
2011	8500	2750	0.32
2012	7800	2000	0.26
2013	9000	2750	0.31
2014	6000	1500	0.25
2018	5500	1500	0.27
2019	4500	1000	0.22
		Average	0.3

Constraints

The main constraints of the study were:

1. After examining the design structure maps and conducting field observations, we discovered that there is no bottom outlet utilized to operate the dam in order to deliver water for irrigation.
2. We detected seepage along the dam's body based on prior studies and field interviews with engineering supervisors.
3. Unlined irrigation canals cause a lot of water loss, erosion, and sedimentation.
4. The assessment of water projects in Iraq's Western Desert confirmed that the Western Desert's water resources are poorly managed. A huge volume of water flows into the river without being used in the Wadies upstream.
5. The Iraqi Western Desert, like all arid and semi-arid regions, suffers from a lack of data, particularly for surface runoff in valleys.
6. Bedouins and those who live near water harvesting systems face a lack of assistance. People are still using simple ways to convey and use water.

Suggestions and Recommendations

1. Maintain and control losses by lining the canals and grouting the concrete along the body of the dam.
2. We propose installing hydrometric stations at the valley outlets in order to obtain accurate surface runoff data.
3. We urge additional researchers to conduct field studies and field observations at small dam sites to develop a better understanding of all the data needed to construct and maintain dams in the Western Desert.
4. Organize seminars and meetings with Bedouins and local farmers to apply modern technology and smart agriculture to this region, resulting in the best possible use of dam irrigation systems to support local livelihoods.

4. Conclusions

The level of functionality and sustainability of the Al-Abila dam was assessed. Rainwater harvesting techniques used in the watershed would help smallholders improve their livelihoods. Dams for stream water abstraction would enable year-round agriculture, which would boost household incomes while also enhancing food supply in the region.

The balancing storage at the end of each runoff shows that dam storage of this magnitude is insufficient to fulfil the water demand. This study indicates that, in the investigation year 2019, when rainfall was below average and distribution was poor, the storage capacity was depleted in late June and negatively affected dam sustainability. Moreover, the results showed that surface runoff from catchments contained a significant quantity of sediments, and the unlined irrigation channels from the dam produced a poor water conveyance factor and adequacy, thus restricting the efficacy of water distribution and usage. This research highlighted faults or constraints that have affected system functioning or sustainability and provided suggestions and recommendations for risk-managed rainwater harvesting system installation methods and/or designs.

This data is critical for managing small reservoirs and storage capacity. It was possible to determine how much water a reservoir could retain at any given time. Furthermore, knowing the storage capabilities will allow planners and water managers to quickly determine how to use and manage the available water in light of alternative applications.

Author Contributions: Conceptualization A.A. and R.A.; methodology, A.A.; software, A.A. and R.A.; validation, A.A. and K.M.; formal analysis, A.A. and M.R.; investigation, A.A. and S.S.; resources, A.A. and C.R.; data curation, S.S.; writing—original draft preparation, A.A. and R.A.; writing—review and editing, C.R.; visualization, A.A., S.S. and K.M.; supervision, C.R.; project administration, C.R. and M.R.; funding acquisition, C.R. All authors have read and agreed to the published version of the manuscript.

Funding: This Research is fully funded by Nuffic Orange Knowledge program OKP-IRA-104278.

Institutional Review Board Statement: Not applicable.

Informed Consent Statement: Not applicable.

Data Availability Statement: Some data in this manuscript were obtained from Ministry of Agriculture and Ministry of Water Resources, small dam management in the Western Desert. The other data are from the field work and previous studies.

Acknowledgments: This study has been conducted in cooperation between Wageningen University & Research (The Netherlands) and the University of Anbar, which was done under the part of Orange Knowledge Programme (OKP-IRA-104278) which is fully funded by Nuffic.

Conflicts of Interest: The authors declare no conflict of interest.

References

1. Wesseling, J.; Adham, A.; Riksen, M.; Ritsema, C.; Oostindie, K.; Heidema, N. A Microsoft Excel Application to Simulate Water Harvesting in a Catchment. *Water Harvest. Res.* **2021**, *4*, 1–18.
2. Sayl, K.N.; Mohammed, A.S.; Ahmed, A.D. GIS-based approach for rainwater harvesting site selection. *IOP Conf. Ser. Mater. Sci. Eng.* **2020**, *737*, 012246. [CrossRef]
3. Hussien, B.M. Hydrogeologic condition within al-Anbar governorate. *J. Anbar Univ. Pure Sci.* **2010**, *4*, 97–111.
4. Kamel, A.; Sulaiman, S.O.; Sayl, K.N. Hydrologic study for the Iraqi Western Desert to assess water harvesting projects. *Iraqi J. Civ. Eng.* **2011**, *7*, 16–27.
5. Dongol, R.; Bohora, R.C.; Chalise, S.R. Sustainability of rainwater harvesting system for the domestic needs: A case of Daugha Village Development Committee, Gulmi, Nepal. *Nepal J. Environ. Sci.* **2017**, *5*, 19–25. [CrossRef]
6. Pope, J.; Annandale, D.; A Morrison-Saunders, A. Conceptualising sustainability assessment. *Environ. Impact Assess Rev.* **2004**, *24*, 595–616. [CrossRef]
7. Waheed, B.; Khan, F.; Veitch, B. Linkage-Based Frameworks for Sustainability Assessment: Making a Case for Driving Force-Pressure-State-Exposure- Effect-Action (DPSEEA) Frameworks. *Sustainability* **2009**, *1*, 441–463. [CrossRef]
8. Pachpute, J.S.; Tumbo, S.D.; Sally, H.; Mul, M.L. Sustainability of rainwater harvesting systems in rural catchment of Sub-Saharan Africa. *Water Resour. Manag.* **2009**, *23*, 2815–2839. [CrossRef]
9. Ngigi, S.; Savenije, H.H.G.; Glehuki, F.N. Hydrological impacts of flood storage and management on irrigation water abstraction in Upper Ewaso Ng'iro river basin, Kenya. *J. Water Resour. Manag.* **2008**, *22*, 1859–1879. [CrossRef]
10. Hashimoto, T.; Stedinger, J.R.; Loucks, D.P. Reliability, resiliency, and vulnerability criteria for water resource system performance evaluation. *Water Resour. Res.* **1982**, *18*, 14–20. [CrossRef]
11. Kjeldsen, T.R.; Rosbjerg, D. Choice of reliability, resilience and vulnerability estimators for risk assessments of water resources systems. *Hydrol. Sci. J.* **2004**, *49*, 755–767. [CrossRef]
12. Sandoval-Solis, S.; McKinney, D.C.; Loucks, D.P. Sustainability index for water resources planning and management. *J. Water Resour. Plan. Manag.* **2011**, *137*, 381–390. [CrossRef]
13. Park, D.; Um, M.-J. Sustainability Index Evaluation of the Rainwater Harvesting System in Six US Urban Cities Daeryong. *Sustainability* **2018**, *10*, 280. [CrossRef]
14. Sulaiman, S.O.; Kamel, A.H.; Sayl, K.N.; Alfadhel, M.Y. Water resources management and sustainability over the Western desert of Iraq. *Environ. Earth Sci.* **2019**, *78*, 495. [CrossRef]
15. Abdulhameeda, I.M.; Mohammeda, A.S.; Kamela, A.H.; ALmafrchib, K.Y.; Alhadeethia, I.K.; Fayyadha, A.S.; Al-Esawia, J.S.E. Sustainable Development of Wadi Houran-Western Iraqi Desert. *Iraqi J. Civ. Eng.* **2020**, *15*, 45–53.
16. Sayl, K.; Adham, A.; Ritsema, C.J. A GIS-based multicriteria analysis in modeling optimum sites for rainwater harvesting. *Hydrology* **2020**, *7*, 51. [CrossRef]
17. Adham, A.; Sayl, K.N.; Abed, R.; Abdeladhim, M.A.; Wesseling, J.G.; Riksen, M.; Ritsema, C.J. A GIS-based approach for identifying potential sites for harvesting rainwater in the Western Desert of Iraq. *Int. Soil Water Conserv. Res.* **2018**, *6*, 297–304. [CrossRef]
18. Consortium-Yugoslavia. *Hydrogeological Explorations and Hydrotechnical Work-Western Desert of Iraq*; Blook7; Directorate of Western Desert Development Projects: Baghdad, Iraq, 1977; *Unpublished*.
19. Ministry of Water Resources. *General Authority for Dams and Reservoirs*; Water Resources Department in Al-Anbar Governorate: Baghdad, Iraq, 2015; *Unpublished data*.
20. Ministry of Agriculture. *Hydrological Explorations and Hydro Technical Works. (Vol 1 Climatology and Hydrology Works); Vol 5: Hydrology Iraq Western Desert—Block 7*; Directorate of Western Desert Development Projects: Baghdad, Iraq, 1977.
21. Adham, A.; Wesseling, J.G.; Abed, R.; Riksen, M.; Ouessar, M.; Ritsema, C.J. Assessing the impact of climate change on rainwater harvesting in the Oum Zessar watershed in Southeastern Tunisia. *Agric. Water Manag.* **2019**, *221*, 131–140. [CrossRef]
22. Boers, T.M.; Zondervan, K.; Ben-Asher, J. Micro-catchment-water-harvesting (MCWH) for arid zone development. *Agric. Water Manag.* **1986**, *12*, 21–39. [CrossRef]

23. Charman, J.; Kostov, L.; Minetti, J.; Stoutesdijk, D.; Tricoli, D.; Olivier, A. *Small Dams and Weirs in Earth and Gabion Materials*; Land and Water Development Division, Food and Agriculture Organization of the United Nations: Rome, Italy, 2001.

24. Maestre-Valero, J.F.; Martínez-Granados, D.; Martínez-Alvarez, V.; Calatrava, J. Socio-economic impact of evaporation losses from reservoirs under past, current and future water availability scenarios in the semi-arid Segura Basin. *Water Resour. Manag.* **2013**, *27*, 1411–1426. [CrossRef]

25. Martinez-Granados, D.; Francisco Maestre-Valero, J.; Calatrava, J.; Martinez-Alvarez, V. The economic impact of water evaporation losses from water reservoirs in the Segura Basin, SE Spain. *Water Resour. Manag.* **2011**, *25*, 3153–3175. [CrossRef]

26. Gallego-Elvira, B.; Martínez-Alvarez, V.; Pittaway, P.; Brink, G.; Martín-Gorriz, B. Impact of micrometeorological conditions on the efficiency of artificial monolayers in reducing evaporation. *Water Resour. Manag.* **2013**, *27*, 2251–2266. [CrossRef]

27. Laxman, J. An Artificial Neural Network Approach for the Determination of Infiltration Model Parameters. Ph.D. Thesis, Dept. Faculty of Engineering, Swami Ramanand Teerth Marathwada University, Nanded, India, 2016.

28. *Western Desert Dams, Design Report of Horan Dam H/3, H/2*; Ministry of Irrigation: Colombo, Sri Lanka, 1994.

29. Yuan, Y.; Mitchell, J.; Hirschi, M.; Cooke, R. Modified SCS curve number method for predicting subsurface drainage flow. *Trans. ASBE* **2001**, *44*, 1673–1682. [CrossRef]

30. Sayl, K.N.; Afan, H.A.; Muhamad, N.S.; Elshafie, A. Development of a Spatial Hydrologic Soil Map Using Spectral Reflecatance Band Recognition and a Multiple-Output Artificial Neural Network Model. *Hydrol. Earth Syst. Sci. Discuss* **2017**, 1–16. [CrossRef]

MDPI

St. Alban-Anlage 66

4052 Basel

Switzerland

www.mdpi.com

Water Editorial Office

E-mail: water@mdpi.com

www.mdpi.com/journal/water